Vladimir G. Baidakov
**Explosive Boiling of
Superheated Cryogenic Liquids**

1807–2007 Knowledge for Generations

Each generation has its unique needs and aspirations. When Charles Wiley first opened his small printing shop in lower Manhattan in 1807, it was a generation of boundless potential searching for an identity. And we were there, helping to define a new American literary tradition. Over half a century later, in the midst of the Second Industrial Revolution, it was a generation focused on building the future. Once again, we were there, supplying the critical scientific, technical, and engineering knowledge that helped frame the world. Throughout the 20th Century, and into the new millennium, nations began to reach out beyond their own borders and a new international community was born. Wiley was there, expanding its operations around the world to enable a global exchange of ideas, opinions, and know-how.

For 200 years, Wiley has been an integral part of each generation's journey, enabling the flow of information and understanding necessary to meet their needs and fulfill their aspirations. Today, bold new technologies are changing the way we live and learn. Wiley will be there, providing you the must-have knowledge you need to imagine new worlds, new possibilities, and new opportunities.

Generations come and go, but you can always count on Wiley to provide you the knowledge you need, when and where you need it!

William J. Pesce
President and Chief Executive Officer

Peter Booth Wiley
Chairman of the Board

Vladimir G. Baidakov

Explosive Boiling of Superheated Cryogenic Liquids

WILEY-VCH Verlag GmbH & Co. KGaA

The Author

Vladimir G. Baidakov
Institute of Thermal Physics
Russian Academy of Sciences
Ekaterinburg, Russian Federation
bai@itp.uran.ru

Consultant Editor

Jürn W. P. Schmelzer
Physics Department
University of Rostock
Rostock, Germany
juern-w.schmelzer@uni-rostock.de

Cover illustration
Explosive boiling-up of gas-saturated liquid solutions

by V. S. Uskov,
Institute of Thermal Physics of the Ural Branch
of the Russian Academy of Sciences

All books published by Wiley-VCH are carefully produced. Nevertheless, authors, editors, and publisher do not warrant the information contained in these books, including this book, to be free of errors. Readers are advised to keep in mind that statements, data, illustrations, procedural details or other items may inadvertently be inaccurate.

Library of Congress Card No.:
applied for

British Library Cataloguing-in-Publication Data
A catalogue record for this book is available from the British Library.

Bibliographic information published by the Deutsche Nationalbibliothek
The Deutsche Nationalbibliothek lists this publication in the Deutsche Nationalbibliografie; detailed bibliographic data are available in the Internet at <http://dnb.d-nb.de>.

© 2007 WILEY-VCH Verlag GmbH & Co. KGaA, Weinheim

All rights reserved (including those of translation into other languages). No part of this book may be reproduced in any form – by photoprinting, microfilm, or any other means – nor transmitted or translated into a machine language without written permission from the publishers. Registered names, trademarks, etc. used in this book, even when not specifically marked as such, are not to be considered unprotected by law.

Typesetting   Uwe Krieg, Berlin
Printing   Strauss GmbH, Mörlenbach
Binding   Litges & Dopf GmbH, Heppenheim
Wiley Bicentennial Logo   Richard J. Pacifico

Printed in the Federal Republic of Germany
Printed on acid-free paper

ISBN   978-3-527-40575-6

Foreword

The present monograph is written by an outstanding specialist in the field of first-order phase transitions, Professor Vladimir G. Baidakov. He studied physics and completed his PhD thesis in 1973 at the Ural Institute of Technology in Ekaterinburg, Russia. In the same year, he started working at the Institute of Thermal Physics of the Ural Branch of the Russian Academy of Sciences in Ekaterinburg. In the recent years, he was the Vice-Director for Research and presently he is the Director of this institute and Head of the Laboratory of Cryogenics and Energetics. He is a pupil and long-standing co-worker of the late Professor Vladimir P. Skripov, who, about 45 years ago, initiated the research on the properties of substances in the metastable state and the kinetics of phase transformation processes performed at the Institute of Thermal Physics of the Russian Academy of Sciences in Ekaterinburg.

Professor Baidakov's work focuses on experimental and theoretical investigations of the kinetics of first-order phase transitions and, in particular, on the investigation of the kinetics of boiling, the analysis of bulk and surface properties of fluids in thermodynamically stable and metastable states, and the investigation of the properties of bubbles of critical sizes determining the rate of bubble formation. The results of the work of V. G. Baidakov and his co-workers are outlined in six monographs and 125 journal publications. In acknowledgment of his results, he was awarded with the Russian State Prize in Science and Technology in 1999.

The results of the long-standing, highly original investigations have been presented by the author and his co-workers and discussed in several research workshops (*Nucleation Theory and Applications*) which take place in Dubna (near Moscow) at the Bogoliubov Laboratory of Theoretical Physics of the Joint Institute for Nuclear Research and have been organized by the editor of the present book each year since 1997. The first comprehensive accounts of these results are published in English in the workshop proceedings (*Nucleation Theory and Applications*, Dubna, Russia, 1999, 2002, 2005) and in the monograph *Nucleation Theory and Applications* published by Wiley-VCH in 2005 (V. G. Baidakov: *Boiling-Up Kinetics of the Solutions of Cryogenic Liquids*, pp. 126–177).

It is a real pleasure to be able to now present the extended English translation of the monograph of the author published in Russian language in 1995. The present book gives for the first time the opportunity to the interested reader to get a comprehensive overview in English on the experimental and theoretical investigations performed by the author and his co-workers on the thermodynamic properties of fluids and the kinetics of boiling of one-component liquids and liquid solutions. Finally, I would like to express my gratitude to Mrs. Antje Schmelzer for her support in the preparation of the final version of the manuscript for publication.

Rostock (Germany) & Dubna (Russia), August 2006 *Jürn W. P. Schmelzer*

Contents

Foreword *V*

Preface *XI*

1   **Introduction** *1*

2   **Equilibrium, Stability, and Metastability** *7*
2.1   Types of Equilibria: Stability Criteria *7*
2.2   Boundary of Essential Instability *13*
2.3   Elements of Statistical Theory *16*
2.4   Phase Stability Against Finite Perturbations *19*
2.5   Critical Heterophase Fluctuations *23*
2.6   Relaxation Processes in Metastable Phases *26*
2.7   Dynamics of Heterophase Fluctuations *30*
2.8   Kinetic Nucleation Theory (Multiparameter Version) *35*
2.9   Approximations and Limitations of Classical Nucleation Theory *41*
2.10   Nucleation at a High Degree of Metastability *47*
2.11   Nucleation Bypassing the Saddle Point *51*
2.12   Some Comments on Nucleation Theory *57*

3   **Attainable Superheating of One-Component Liquids** *61*
3.1   Two Approaches to the Determination of the Work of Formation of a Critical Bubble *61*
3.2   Boiling-Up Kinetics of Superheated Liquids *65*
3.3   Elements of the Stochastic Theory of Nucleation *74*
3.4   Experimental Procedures in the Analysis of Boiling in Superheated Liquids *78*
3.5   Quasistatic Methods of Investigating Limiting Superheatings of Liquids *82*
3.6   Dynamic Methods of Investigating Explosive Boiling-Up of Liquids *88*

Explosive Boiling of Superheated Cryogenic Liquids. Vladimir G. Baidakov
Copyright © 2007 WILEY-VCH Verlag GmbH & Co. KGaA, Weinheim
ISBN: 978-3-527-40575-6

3.7 Results of Experiments on Classical Liquids *91*
3.8 Superheating of Quantum Liquids *107*
3.9 Surface Tension of Vapor Nuclei *116*
3.10 Cavitation Strength of Cryogenic Liquids *123*
3.11 Attainable Superheating of Liquid Argon at Negative Pressures *138*
3.12 Initiated Nucleation *145*
3.13 Heterogeneous Nucleation *151*

4 Nucleation in Solutions of Liquefied Gases *159*
4.1 Critical Nucleus and the Work of its Formation *159*
4.2 Theory of Nucleation in Binary Solutions *165*
4.3 Attainable Superheating of Solutions of Hydrocarbons *172*
4.4 Methods of Experimentation on Solutions of Cryogenic Liquids *174*
4.5 Solutions with Complete Solubility of the Components *177*
4.6 Solutions with Partial Solubility of the Components *182*
4.7 Equation of State and Boundaries of Thermodynamic Stability of Solutions *190*
4.8 Properties of Critical Bubbles in Binary Solutions *195*
4.9 Comparison of Theory and Experiment for Binary Solutions *204*
4.10 Kinetics and Thermodynamics of Nucleation in Three-Component Solutions *209*
4.11 Attainable Superheatings of Ternary Solutions of Cryogenic Liquids *213*

5 Nucleation in Highly Correlated Systems *219*
5.1 Introduction *219*
5.2 Critical Configuration and its Stability *222*
5.3 Steady-State Nucleation *228*
5.4 Peculiarities of New Phase Formation in the Critical Region *232*
5.5 Experimental Investigations of Nucleation in the Vicinity of Critical and Tricritical Points *240*
5.6 Comparison of Theory and Experiment *245*
5.7 Nucleation in the Vicinity of a Spinodal Curve *253*
5.8 Theory of Spinodal Decomposition *259*
5.9 Experimental Studies of Spinodal Decomposition *266*

6 Nucleation Kinetics Near the Absolute Zero of Temperature *273*
6.1 Quantum Tunneling of Nuclei *273*
6.2 Limiting Supersaturations of ^{4}He–^{3}He-Solutions *278*
6.3 Formation of Quantum Vortices in Superfluid Helium *284*

6.4	Quantum Nucleation Near the Boundary of Essential Instability *289*	
6.5	Quantum Cavitation in Helium *294*	
6.6	Some Other Problems of Phase Metastability *299*	
7	**Explosive Boiling-Up of Cryogenic Liquids** *309*	
7.1	Superheating in Outflow Processes *309*	
7.2	Vapor Explosion at the Interface of Two Different Liquids *314*	

List of Symbols *321*

References *323*

Index *337*

Preface

Dedicated to the memory of my teacher, Academician Vladimir Pavlovich Skripov

The monograph is devoted to the description of the kinetics of spontaneous boiling of superheated liquefied gases and their solutions. Experimental results are given on the temperature of accessible superheating, the limits of tensile strength of liquids connected with the existence of processes of cavitation, and the rates of nucleation of classical and quantum liquids. The kinetics of evolution of the gas phase is analyzed in detail for solutions of cryogenic liquids and gas-saturated fluids. The properties of the critical clusters of the newly evolving gas phase are analyzed for initial states near the equilibrium coexistence curves of liquid and gas, for states near the limits of accessible superheating and for initial states near the respective spinodal curves. Particular attention is devoted to the peculiarities of nucleation in highly correlated systems, e.g., for states in the vicinity of critical and tricritical points and near the spinodal curve and to some of the problems of the kinetics of spinodal decomposition in liquid solutions. Experimental results are given on the limiting values of the supersaturation of helium solutions ^{3}He–^{4}He and the tensile strength of one-component fluids of the different helium isotopes (^{3}He and ^{4}He) for temperatures near the absolute zero. The experimental results are compared with the kinetic theories of thermally activated and quantum nucleation. Initiated and heterogeneous nucleation of boiling is discussed as well as nucleation on electron bubbles and ring vortices in quantum liquids. Finally, processes of explosive boiling of cryogenic liquids are analyzed occurring as a result of outflow processes and intensive interactions with high-temperature liquid masses.

The monograph is based mainly on the results of experimental and theoretical analyses performed by the author and his co-workers at the Institute of Thermal Physics of the Ural Branch of the Russian Academy of Sciences in Ekaterinburg, Russia. These results are partly supplemented by data reported by other research groups for completeness. In particular, this procedure is employed in the book with respect to quantum nucleation, nucleation near critical points, and spinodal decomposition.

The topics discussed in the present book are partly outlined in an earlier publication (V. G. Baidakov, *Superheating of Cryogenic Liquids*, Publisher of the Ural Branch of the Russian Academy of Sciences, Ekaterinburg, Russia, 1995) written in Russian. The respective chapters have been revised, however, considerably taking into consideration the new results that have been obtained since the publication of the Russian version. They are supplemented in the present monograph by new chapters devoted to the analysis of boiling in binary and ternary solutions of cryogenic liquids, the limits of tensile strength of liquefied gases, the analysis of the surface tension of bubbles of critical sizes in multicomponent systems, and the kinetic theory of boiling in superheated liquid solutions.

Finally, I would like to express my deep gratitude for Dr. Jürn W. P. Schmelzer, who performed a large work in the preparation of the monograph for publication and for Mrs. E. V. Urakova for helping in the translation of the Russian manuscript.

Ekaterinburg (Russia), August 2006 *Vladimir G. Baidakov*

1
Introduction

When dealing with the thermodynamic properties of matter, the systems under consideration are usually investigated in thermodynamic equilibrium states. However, in a variety of cases configurations are realized which do not correspond to global but local maxima of the entropy or minima of the internal energy or other relevant thermodynamic functions. These states are metastable and have a finite lifetime. Metastability is a common property of first-order phase transformations and manifests itself in systems of quite different nature such as nuclear matter and quark–gluon plasmas, electron–hole fluids, biological systems near the self-organization threshold, as well as in more conventional evaporation, condensation, segregation, crystallization, and melting processes.

Phase transformations represent a macroscopic manifestation of the action of the intermolecular forces in systems consisting of a large number of particles. The similarity of the interaction forces results in an essentially universal picture of phases and phase transitions, at least, in simple systems. Properties of substances may behave differently in the course of phase transformations. Phase transitions can roughly be separated into first and second orders. Correlations of anomalously growing fluctuations in the vicinity of points of second-order phase transformations result in the scale invariance of the properties of systems, which differ profoundly with respect to the structure and character of the interactions between the particles of the system. In contrast, such fluctuations may be disregarded at first-order phase transformations, and the properties of the substances under consideration are characterized by a lower order universality at this point, i.e., by the thermodynamic similarity, which is a reflection of the similarity of intermolecular forces. Therefore, the microscopic theory of phase transformations of first order and phase metastability encounters the same difficulties as the physics of the condensed state.

Since the law of particle interactions dominates first-order phase transformations, for the study of principal problems of phase metastability and phase transitions, those systems are of particular interest whose molecules have a spherically symmetric or an almost spherically symmetric shape and a simple dispersion type dependence of the intermolecular forces. Such (simple)

Explosive Boiling of Superheated Cryogenic Liquids. Vladimir G. Baidakov
Copyright © 2007 WILEY-VCH Verlag GmbH & Co. KGaA, Weinheim
ISBN: 978-3-527-40575-6

substances serve as a "touchstone" in the theoretical analyses of problems directed to condensed media. Since the interparticle bonding forces are weaker than those in complex molecular compounds, simple substances exist in the liquid state at temperatures lower than normal temperatures on the Earth, i.e., they represent liquefied gases.

Liquefied gases with a normal boiling temperature of below 120 K are classified as cryogenic fluids. Local heat supply and pressure pulses in storage and transport systems of cryogenic fluids can cause considerable superheats of the fluids. The establishment of such high local degrees of superheating is facilitated by good wettability of most of the solids with cryogenic fluids and a small content of dissolved gases contained in them.

A superheated fluid represents a particular case of a system in a metastable state and a very convenient object of study. A low viscosity of superheated fluids ensures quick relaxation of the structure. This is not always the case in supercooled fluids. Unlike the liquid–crystal phase transformation, where the interfacial energy is unknown in most cases of interest, the surface tension in the liquid–vapor system can be measured directly. If the condensation of a supersaturated vapor is studied, it is much more difficult to remove the initiating effect of walls, which represent ready and/or easily activated centers of condensation, than active centers during boiling.

The observation of a fluid in the state of a metastable equilibrium was first mentioned in the second half of the 17th century when experiments, which were performed by Huygens and Boyle, revealed the existence of water and mercury at negative pressures. An interpretation of this phenomenon in physical terms became possible only after van der Waals derived his famous equation of state (1873). The terms "metastable" and "unstable" were introduced by Ostwald in order to distinguish sections of different degree of stability in the van der Waals isotherms. Pioneering reproducible quantitative data concerning the accessible temperature of superheating of some fluids were obtained by Wismer et al. (1922–1927). Still earlier, in 1878, Gibbs [1] explained phase transformations with the appearance of nuclei of a new phase. The ideas expressed by Gibbs laid the foundation of the classical theory of thermally induced fluctuation nucleation. It was formulated by Volmer and Weber [2], Farkas [3], Becker and Döring [4], Zeldovich [5], and Frenkel [6]. This theory is universal in its thermodynamic principles and is applicable to various types of phase metastability.

The study of phase metastability includes a wide range of tasks. The present monograph is dedicated mainly to the analysis of the nucleation kinetics in metastable liquefied gases and their solutions. Nuclei may be formed spontaneously in systems which are free of inclusions which may initiate a phase transformation. Such a mechanism of initiating a phase transformation is denoted as homogeneous nucleation. The knowledge of the mechanisms of homogeneous nucleation is highly significant since it determines the upper limits of stability of a metastable phase with respect to discontinuous changes of the state parameters. In the theoretical description of thermodynamic and kinetic aspects of these processes, we employ both Gibbs' thermodynamic method and the continuum's approach, which goes back to van der Waals' works on the theory of capillarity [7] (Chapter 2).

The fluctuation-induced boiling-up of a superheated fluid is preceded by the formation of nuclei having a characteristic size of 5 to 10 nm. Properties of such small objects can be expected to depend considerably on the character of interparticle interactions and should differ from the properties of the respective bulk phases. By comparing experimental results with the theory of homogeneous nucleation it is possible to develop estimates of the lower bound of the validity of the thermodynamic description of such small molecular systems.

In a variety of cases, the theoretical limits of accessible supersaturation determined by the homogeneous nucleation theory may not be reached in real systems due to the presence of foreign nucleation centers in the system under consideration, which favor the formation of nuclei of the new phase. Beyond them, specific nucleation centers, which are not found in ordinary fluids, may be present in liquefied gases at low temperatures. Such specific nucleation centers include electron bubbles in liquid He, H_2 and Ne, vortex lines, and rings in superfluid helium, thermal spikes resulting from the ortho–para conversion in normal hydrogen. These and other topics of nucleation in one-component liquefied gases are discussed in Chapter 3.

Chapter 4 deals with the fluctuation-induced boiling-up kinetics in superheated solutions. Two types of solutions are considered: solutions with a complete solubility of the different components, and gas-saturated systems with partial solubility. In contrast to one-component fluids, the problem of the theoretical description of the spontaneous boiling-up kinetics in liquid solutions is so far not completely solved. In addition to size effects, both adsorption and the time scale of establishment of the absorption equilibrium may considerably affect nucleation in solutions.

The critical point of one-component fluids or solutions is the only point where the boundaries of the regions of essential instability of a metastable phase are adjacent to the region of stable states of the substances under consideration. As this point is approached, the width of the metastable region tends to zero, and the radius of correlations and the relaxation time of hydrodynamic modes increase unlimitedly. These factors have a specific effect on the character of the phase transformations in the vicinity of the critical point. In this range of initial states, not only the metastable region becomes more or less easily accessible to experimental studies, but also the region of unstable states. The theoretical description of nucleation in intensively fluctuating systems near critical and tricritical points is discussed in Chapter 5. This chapter also deals with relaxation near the spinodal and the theoretical description of spinodal decomposition.

The occurrence of nucleation induced by thermal fluctuations is impossible near the absolute zero temperature, and the decay of a homogeneous metastable phase can proceed here only via quantum tunneling of a heterophase fluctuation through the respective thermodynamic potential barrier. In its basic premises, the theory of quantum nucleation was developed by Lifshitz and Kagan [8]. In the recent years, considerable attention was devoted to the experimental detection of processes of quantum nucleation in liquefied gases and solutions as well as in other systems. The achievements and problems encountered in this field of research constitute the contents of Chapter 6.

The design and creation of large cryogenic systems for transportation and storage of cryogenic fluids, bubble chambers, ultrahigh frequency cooling systems and molecular quantum generators have set up some fundamentally new scientific problems including the problem of fast phase transformations, which are accompanied by deviations from phase equilibrium. The cessation of superheating may in some cases lead to considerable hydraulic shocks. The prevention of any superheating of the fluids is one of the possible methods for making the cryogenic equipment more reliable. At the same time, it is desirable to maintain phase metastability of fluids for a long time in order to ensure the cavitation-free operation of cryogenic pumps. A strong metastability is required for the operation of bubble chambers, too. Superheated fluids represent an interesting object not only for theoretical studies but also for new technological applications. Problems regarding the dynamics of boiling-up of cryogenic fluids are discussed in Chapter 7.

Since 1961, systematic studies of the phenomenon of superheating of fluids and boiling have been performed by Skripov and his co-workers first at the Ural Polytechnical Institute and then at the Institute of Thermal Physics of the Ural Branch of the Russian Academy of Sciences. Results of those studies were summarized in the monographs [9–13]. After the publication of Skripov's monograph "Metastable Liquids" [9], published in its English translation by Wiley in 1974, a large number of investigators joined the study of these types of problems, resulting in an increase in the number of publications. The results of the research performed in the last few years are generalized in reviews and monographs [14–16]. They are extended and systematized in the present monograph.

The demand for such a systematic presentation of the state-of-the-art problems and prospects of experimental investigations of phase metastability is felt not only by experimental physicists but also by theorists who, in the course of their analyses, come across with a variety of problems regarding the interpretation of experimental results concerning the properties of different systems in the metastable state, and also by engineers who have to deal with processes of intensification of heat and mass exchange, and fast phase transitions. One can confidently say that the requirements concerning the level of knowledge of the properties of matter in metastable states and the processes taking place in them will increase in the course of further progress in science and technology.

This book is primarily meant for readers who are interested in the liquid state of substances, liquid–vapor phase transitions, and boiling. It is based on studies performed at the Institute of Thermal Physics of the Ural Branch of the Russian Academy of Sciences. A considerable part of the experimental data was obtained jointly with Kaverin, Rubshtein, and Sulla. I would also like to express my sincere thanks to Professor Vladimir P. Skripov, who acquainted me with the "world of phase metastability," for his advice and support.

2
Equilibrium, Stability, and Metastability

2.1
Types of Equilibria: Stability Criteria

Under certain conditions, a one-component system is capable of decaying into macroscopically coexisting phases. If external fields are absent, an equilibrium in a two-phase system is established under the conditions that temperature, T, pressure, p, and chemical potentials, μ, are the same in each of the phases under consideration, i.e., the relations

$$\mu'(T,p) = \mu''(T,p) \tag{2.1}$$

have to be fulfilled.

The condition given by Eq. (2.1) results from the general conditions for an equilibrium of thermodynamic systems formulated by Gibbs [1]. The derivation of these conditions is based on the method of thermodynamic potentials and the principle of virtual displacements. According to Gibbs, the equilibrium state corresponds to extreme values of the entropy, S, or one of the other appropriate thermodynamic potentials. The choice of the potential is hereby determined by the boundary conditions imposed on the system. For a system at constant entropy, S, and constant volume, V, the equilibrium is established when the condition

$$(\delta U)_{\text{eq}} = 0 \tag{2.2}$$

is fulfilled. Here U is the internal energy while the symbol δ refers to infinitesimally small and otherwise arbitrary changes of the function following it.

One can distinguish different types of equilibrium states. If, at $S = \text{const}$ and $V = \text{const}$, the internal energy of a system has the lowest of all possible values, then such an equilibrium is absolutely stable (i.e., the system exists in *the state of a stable equilibrium*; see Fig. 2.1a). Provided that another local minimum exists with a higher value of the internal energy, this local minimum of the internal energy corresponds to a *state of metastable equilibrium*. Such states are stable with respect to infinitesimal perturbations not changing qualitatively the initial state of aggregation of the system. Following Frenkel [6],

Explosive Boiling of Superheated Cryogenic Liquids. Vladimir G. Baidakov
Copyright © 2007 WILEY-VCH Verlag GmbH & Co. KGaA, Weinheim
ISBN: 978-3-527-40575-6

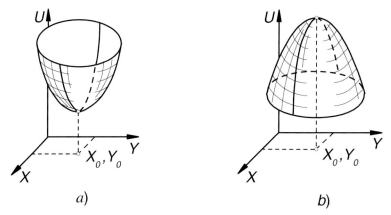

Fig. 2.1 Shape of the surface of the internal energy in the vicinity of states of stable or metastable (a) and unstable (b) equilibria (specified by X_0 and Y_0).

we shall call such perturbations *homophase fluctuations*. In the configurational space, the minima of the internal energy, corresponding to metastable and stable states of the system, are separated by a certain barrier. The height of such a barrier depends on the thermodynamic state of the ambient phase and may vary in a wide range. Under certain conditions, the system is capable to overcome the barrier and to go over into an energetically more favorable (stable) state. Such processes may proceed due to the formation of *heterophase fluctuations*, which take the system out of the initial state of aggregation and lead to the formation of a nucleus of a new phase capable to grow further. Any metastable system is unstable with respect to such (finite) changes in the state parameters of the ambient phase, and, after some more or less extended period of time, relaxes into a state corresponding to the stable phase.

The homogeneous state of a system is unstable (*in a state of unstable equilibrium*) if the internal energy at given values of entropy, S, and volume, V, has reached a maximum (Fig. 2.1b). Relaxation processes at the initial stages of phase separation proceed here without the formation of nuclei and do not require overcoming of activation barriers. It is assumed that such unstable equilibrium states may be described by the equation of state of homogeneous phases similar to the states of stable and metastable equilibrium.

Metastable states are realized in first-order phase transitions. According to Eq. (2.1), in the (μ, T, p)-space the curve of phase equilibrium (i.e., the *binodal*) is determined by the intersection of the surfaces of the chemical potentials of both the possible phases (Fig. 2.2). A homogeneous system may pass through the line of phase equilibrium without undergoing a phase transition. At the point E, the phase with the chemical potential μ' is metastable, i.e., stable against infinitesimal perturbations, but thermodynamically less stable

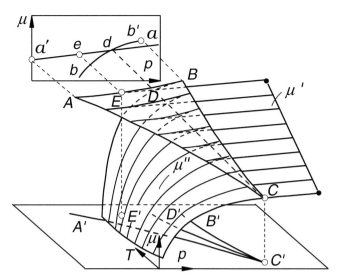

Fig. 2.2 Surface of the chemical potential of a one-component system in the region of liquid–gas phase transition: DC is the binodal curve; AC, BC are the different branches of the spinodal curve. In the (p, T)-plane, the projections of these lines are shown, and in the plane, $T = $ const, the traces of the surface $\mu(p, T)$ for liquid (a, a') and vapor (b, b') are shown.

compared to the phase with the chemical potential μ'' at the same values of T and p.

In order to distinguish between stable, metastable, and unstable states, the condition of equilibrium (2.2) has to be supplemented by the requirement of stability [1] with respect to finite perturbations,

$$(\Delta U)_{eq} > 0. \tag{2.3}$$

The symbol Δ denotes a finite change of a given quantity, in the present case, of the internal energy. Condition (2.3) implies that the state under consideration corresponds to a minimum of the internal energy. If this condition is fulfilled for any arbitrary perturbations, the considered equilibrium state is absolutely stable. For infinitesimal perturbations, inequality (2.3) takes the form

$$(\delta^2 U)_{eq} > 0. \tag{2.4}$$

In contrast, if inequality (2.4) is not fulfilled, the system is in an unstable equilibrium state.

Equations (2.2) and (2.4) represent the necessary and sufficient conditions for the existence of a stable thermodynamic equilibrium state with respect to continuous changes of the state parameters. It gives the possibility of formulating specific explicit stability criteria. These criteria are independent of the system size and nature of the perturbations. In discussing these specific stability criteria, we will therefore always consider a well-defined amount of the substance under consideration (either a mole or a unit volume, etc.). For the specification of the thermodynamic parameters referred to such a unit we will employ lowercase letters. The particular choice of the unit employed will be specified separately. In the present section, we assume that the unit of substance is equal to one mole.

Let us examine a two-component homogeneous and isotropic system and formulate explicit stability criteria for them. Stability criteria for one-component systems may be obtained as a special case when the number of particles of one of the components tends to zero. We again assume that external fields are absent and that the thermodynamic system is sufficiently large, so that its surface energy is negligible compared with the bulk terms. If, in a binary system, the molar fraction of the second component, $c = c_2 = n_2/(n_1 + n_2)$, is taken as the independent parameter, the first variation of the molar internal energy, u, is given by

$$\delta u = u_s \delta s + u_v \delta v + u_c \delta c. \tag{2.5}$$

In this case, the stability condition (2.4) reads

$$\delta^2 u = u_{ss}(\delta s)^2 + 2u_{sv}\delta s \delta v + u_{vv}(\delta v)^2 \\ + 2u_{sc}\delta s \delta c + 2u_{vc}\delta v \delta c + u_{cc}(\delta c)^2 > 0. \tag{2.6}$$

In Eqs. (2.5) and (2.6), the following notations have been employed:

$$u_s = \left(\frac{\partial u}{\partial s}\right)_{v,c} = T, \qquad u_v = \left(\frac{\partial u}{\partial v}\right)_{s,c} = -p,$$

$$u_c = \left(\frac{\partial u}{\partial c}\right)_{s,v} = \mu_1 - \mu_2 = \Delta\mu, \qquad u_{ss} = \left(\frac{\partial T}{\partial s}\right)_{v,c},$$

$$u_{vv} = -\left(\frac{\partial p}{\partial v}\right)_{s,c}, \qquad u_{s,v} = \left(\frac{\partial T}{\partial v}\right)_{s,c} = -\left(\frac{\partial p}{\partial s}\right)_{v,c}, \tag{2.7}$$

$$u_{cc} = \left(\frac{\partial \Delta\mu}{\partial c}\right)_{s,v}, \qquad u_{sc} = \left(\frac{\partial T}{\partial c}\right)_{v,s} = \left(\frac{\partial \Delta\mu}{\partial s}\right)_{v,c},$$

$$u_{vc} = -\left(\frac{\partial p}{\partial c}\right)_{s,v} = \left(\frac{\partial \Delta\mu}{\partial v}\right)_{s,c}.$$

The necessary and sufficient conditions that the real quadratic form (2.6) is positive definite consist in the requirement that the determinant, det u_{svc}, be composed of the coefficients of the quadratic form, and all minor subdeterminants be larger than zero. In order to formulate the stability conditions in a simple way, let us convert the quadratic form given by Eq. (2.6) into the canonical form. Using Eqs. (2.7), inequality (2.6) can be rewritten as

$$\delta^2 u = \delta T \delta s - \delta p \delta v + \delta \Delta \mu \delta c > 0. \tag{2.8}$$

Expressing the variation of the difference in the chemical potentials of the two components of the solution in terms of the variables T, p, and c, we get [17]

$$\delta^2 u = \left(\frac{\partial T}{\partial s}\right)_{v,s} (\delta s)^2 - \left(\frac{\partial p}{\partial v}\right)_{T,c} (\delta v)_c^2 + \left(\frac{\partial \Delta \mu}{\partial c}\right)_{p,T} (\delta c)^2 > 0, \tag{2.9}$$

where

$$(\delta v)_c = \left(\frac{\partial v}{\partial s}\right)_{p,c} \delta s + \left(\frac{\partial v}{\partial p}\right)_{s,c} \delta p. \tag{2.10}$$

Homogeneous states of thermodynamic systems are, consequently, stable if the conditions

$$\det u_{svc} = -\left(\frac{\partial T}{\partial s}\right)_{v,c} \left(\frac{\partial p}{\partial v}\right)_{T,c} \left(\frac{\partial \Delta \mu}{\partial c}\right)_{p,T}$$
$$= -\left(\frac{\partial T}{\partial s}\right)_{p,c} \left(\frac{\partial p}{\partial v}\right)_{s,c} \left(\frac{\partial \Delta \mu}{\partial c}\right)_{p,T} > 0 \tag{2.11}$$

and

$$-\left(\frac{\partial p}{\partial v}\right)_{T,c} = (v\beta_{T,c})^{-1} > 0, \quad \left(\frac{\partial T}{\partial s}\right)_{v,c} = \frac{T}{c_{v,c}} > 0 \tag{2.12}$$

or

$$-\left(\frac{\partial p}{\partial v}\right)_{s,c} = (v\beta_{s,c})^{-1} > 0, \quad \left(\frac{\partial T}{\partial s}\right)_{p,c} = \frac{T}{c_{p,c}} > 0 \tag{2.13}$$

are fulfilled. Here $c_{v,c}$ and $c_{p,c}$ are the isochoric and isobaric heat capacities at constant composition, and $\beta_{T,c} = -v^{-1}(\partial v/\partial p)_{T,c}$ and $\beta_{s,c} = -v^{-1}(\partial v/\partial p)_{s,c}$ are the isothermal and the adiabatic compressibilities at constant composition, respectively.

The stability determinant (2.11) consists of elements which may be written as second-order derivatives of the thermodynamic potentials with respect to the generalized coordinates. Some derivatives are taken at constant values of the generalized coordinates (*adiabatic stability coefficients*), others at constant values of the generalized thermodynamic force and concentration (*mixed stability coefficients*), or at constant values only of generalized forces (*isodynamic stability coefficients*) [17,18].

The first inequalities in Eqs. (2.12) and (2.13) express the conditions of mechanical stability for a homogeneous system, respectively, against isothermal and adiabatic perturbations, the second terms specify the conditions of thermal stability at constant volume or pressure. In order to ensure stability of a binary system, the fulfillment of these conditions is necessary, but not sufficient. Besides thermal and mechanical perturbations, variations of composition may occur here and, according to Eq. (2.11), inequalities (2.12) and (2.13) should be supplemented by the condition of stability against diffusion

$$\left(\frac{\partial \Delta \mu}{\partial c}\right)_{p,T} > 0. \tag{2.14}$$

In the limiting case of a one-component system ($c \to 0$ or $c \to 1$), Eqs. (2.11)–(2.13) yield

$$\det u_{sv} = -\left(\frac{\partial T}{\partial s}\right)_v \left(\frac{\partial p}{\partial v}\right)_T = -\left(\frac{\partial p}{\partial v}\right)_s \left(\frac{\partial T}{\partial s}\right)_p > 0, \tag{2.15}$$

$$-\left(\frac{\partial p}{\partial v}\right)_T > 0 \quad \text{or} \quad \left(\frac{\partial T}{\partial s}\right)_p > 0. \tag{2.16}$$

These inequalities (2.15) and (2.16) have a simple geometrical interpretation. The Gaussian curvature, K_g, of the surfaces of the internal energy in the vicinity of the equilibrium state is given by the expression [19]

$$K_g = \frac{\det u_{sv}}{(1 + u_s^2 + u_v^2)^2}. \tag{2.17}$$

Since the denominator in Eq. (2.17) is positive, Eq. (2.15) yields the inequality $K_g > 0$, and the point of stable equilibrium $u(s_0, v_0)$ is an elliptic point. The main radii of curvature R_1 and R_2 are proportional to the derivatives $(\partial T/\partial s)_v$ and $-(\partial p/\partial v)_T$.

The response of a thermodynamic system to changes of its state determines its degree of stability. Such changes in the state of the system may be of both internal and external nature. In the absence of external perturbations, the actual degree of perturbations is determined by the level of thermal fluctuations. The probability, P_{eq}, of a given fluctuation in an isolated thermodynamic system in thermodynamic equilibrium is determined by the Einstein formula [20]

$$P_{eq} \sim \exp\left(\frac{\Delta s}{k_B}\right) \sim \exp\left(-\frac{W}{k_B T}\right), \tag{2.18}$$

where Δs is the variation of the entropy of the system due to the evolution of the fluctuation considered, and W is the minimum work required to generate such variation of the thermodynamic state in a reversible process. In application to homophase fluctuations, Eq. (2.18) can be written as follows:

$$P_{eq} \sim \exp\left[-\frac{1}{2}\frac{(\delta^2 u)_{eq}}{k_B T}\right].\tag{2.19}$$

According to Eq. (2.19), the coefficients of the determinant in Eq. (2.11) determine the probability and the magnitude of the fluctuations. In a one-component system, the moments of second order for volume and entropy fluctuations are proportional to the isodynamic stability coefficients

$$\langle(v-\langle v\rangle)^2\rangle = k_B T v \beta_T, \qquad \langle(s-\langle s\rangle)^2\rangle = k_B c_p.\tag{2.20}$$

The squared averages of the fluctuations are connected, consequently, with the correlation functions [21] as

$$\rho k_B T \beta_T = 1 + \rho \int G(\vec{r})d\vec{r} = \left[1 - \rho \int C(\vec{r})d\vec{r}\right]^{-1},\tag{2.21}$$

where $G(\vec{r})$ and $C(\vec{r})$ are the pair and the direct correlation functions, and $\rho = v^{-1}$ is the density. Stability of a thermodynamic system in a homogeneous state can be realized, consequently, only if the pair-correlation function is not equal to zero in a finite volume of the system.

2.2 Boundary of Essential Instability

On a thermodynamic surface, the line of phase equilibrium divides the regions of stable and metastable states and, as a result, determines the boundary of the region in thermodynamic phase space where a given phase is absolutely stable. A metastable system retains its restoring reaction with respect to infinitesimal perturbations of arbitrary wavelength. The value of determinant (2.11) may be regarded as a measure of stability of the metastable state with respect to continuous long wavelength perturbations. A value of the determinant equal to zero corresponds to the boundary of metastability and the transition of the system to states of essential instability of the phase under consideration (*spinodal*). According to Eq. (2.11), a crossover process like this one takes place if at least one of the stability coefficients becomes zero. It can be shown [17] that in the stable phase the adiabatic stability coefficient is always not less than the isodynamic stability coefficient. Hence, penetrating deeper into the metastable region of a solution, first one will observe the manifestation of the diffusion instability. The *diffusion spinodal*, or simply the spinodal, of a binary solution is determined by the condition

$$\left(\frac{\partial \Delta \mu}{\partial c}\right)_{p,T} = 0.\tag{2.22}$$

If the stability determinant is equal to zero and a stability coefficient of some kind becomes zero, then all other stability coefficients of this kind are also equal to zero.

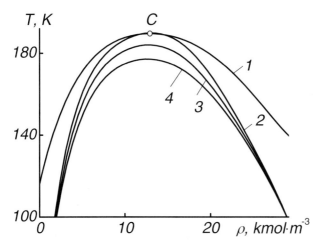

Fig. 2.3 Phase diagram of an argon–krypton solution ($c = 0.5$): 1—binodal; 2—diffusional spinodal; 3—mechanical spinodal; 4—adiabatic spinodal; C—critical point.

Besides Eq. (2.22) [17], the following relations are fulfilled at the diffusion spinodal:

$$\left(\frac{\partial \Delta\mu}{\partial s}\right)_{p,T} = \left(\frac{\partial T}{\partial c}\right)_{p,\Delta\mu} = 0, \qquad \left(\frac{\partial \Delta\mu}{\partial v}\right)_{p,T} = -\left(\frac{\partial p}{\partial c}\right)_{T,\Delta\mu} = 0,$$
$$\left(\frac{\partial T}{\partial s}\right)_{p,\Delta\mu} = 0, \quad -\left(\frac{\partial p}{\partial v}\right)_{T,\Delta\mu} = 0, \qquad \left(\frac{\partial T}{\partial v}\right)_{p,\Delta\mu} = -\left(\frac{\partial p}{\partial s}\right)_{T,\Delta\mu} = 0. \tag{2.23}$$

If, as an additional limitation, constancy of composition and homogeneity is imposed on a system, the stability of the equilibrium state will be determined only by the mixed and adiabatic stability coefficients. In this case, the boundary of stability (*mechanical spinodal*) is determined by equality to zero of the minors of the quadratic form (2.9), which does not contain isodynamic derivatives

$$\det u_{sv} = -\left(\frac{\partial T}{\partial s}\right)_{v,c}\left(\frac{\partial p}{\partial v}\right)_{T,c} = -\left(\frac{\partial T}{\partial s}\right)_{p,c}\left(\frac{\partial p}{\partial v}\right)_{s,c} = 0. \tag{2.24}$$

Condition (2.24) does not presuppose zero values of the adiabatic stability coefficients, but requires vanishing of all mixed derivatives which are determined through the minor of $\det u_{sv}$, i.e.,

$$-\left(\frac{\partial p}{\partial v}\right)_{T,c} = 0, \quad \left(\frac{\partial T}{\partial s}\right)_{p,c} = 0, \quad \left(\frac{\partial T}{\partial v}\right)_{p,c} = -\left(\frac{\partial p}{\partial s}\right)_{T,c} = 0. \tag{2.25}$$

Simultaneously with the mixed stability coefficients, the inverse isodynamic stability coefficient $(\partial \Delta \mu / \partial c)_{p,T}^{-1}$ becomes zero. This property ensures a nonzero value of $\det u_{svc}$ on the mechanical spinodal. In (T, ρ)-coordinates, the mechanical spinodal of a two-component system is "inserted" into the diffusion spinodal and in the general case has no contact points with it (Fig. 2.3). In the limit $c \to 0$ or $c \to 1$, the diffusion and the mechanical stability boundaries merge into one line. In the case of a one-component system, the mechanical spinodal, or simply spinodal, is determined by the equations

$$-\left(\frac{\partial p}{\partial v}\right)_T = (v\beta_T)^{-1} = 0, \qquad \left(\frac{\partial T}{\partial s}\right)_p = \frac{T}{c_p} = 0. \tag{2.26}$$

If perturbations develop strictly adiabatically and homogeneity of the composition is retained, the stability of a binary solution is lost at

$$-\left(\frac{\partial p}{\partial v}\right)_{s,c} = 0, \qquad \left(\frac{\partial T}{\partial s}\right)_{p,c}^{-1} = \frac{c_{p,c}}{T} = 0, \qquad \left(\frac{\partial \Delta \mu}{\partial c}\right)_{p,s}^{-1} = 0. \tag{2.27}$$

Such boundary is called *adiabatic spinodal* [22]. On this boundary, the inequalities $\det u_{svc} \neq 0$ and $\det u_{sv} \neq 0$ hold. The adiabatic spinodal is located inside the mechanical one if $(\partial s/\partial T)_{v,c} \neq 0$ holds (Fig. 2.3). At $c_{v,c} \to \infty$, all coefficients of the minor $\det u_{sv}$ are equal to zero, and the adiabatic spinodal coincides with the mechanical spinodal. Further properties of stability boundaries of a homogeneous binary solution are examined in Ref. [17].

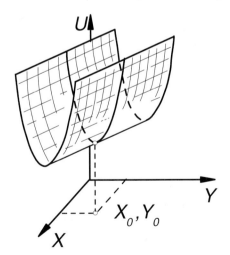

Fig. 2.4 Shape of the surface of the internal energy at the boundary of thermodynamic stability.

Penetrating the metastable region of a solution, the diffusion spinodal is reached first. In a one-component system, this statement applies to the mechanical spinodal (note, however, that not every metastable state of a system is necessarily bounded by a spinodal curve, a boundary of essential instability; such a boundary is absent, for instance, in a one-component supercooled liquid [23]). These boundaries of thermodynamic stability correspond to parabolic points of the surface of the nonequilibrium thermodynamic potential (Fig. 2.4). At these points, the Gaussian curvature, K_g, is equal to zero. One of the main radii of curvature takes on here an infinitely large value, whereas the other has a finite positive value. The spinodal has one single point of crossing from a region of metastable states into a stable region. This point corresponds to the critical point in a one-component system. In a binary system, the crossing proceeds along some line. The critical point belongs both to the spinodal and to the binodal. The line of critical points of a two-component system is determined by the equation [1]

$$\frac{\partial(T, -p, \det u_{svc})}{\partial(s, v, c)} = 0. \tag{2.28}$$

Using the technique of Jacobians and expressing the derivatives of the stability determinant in terms of the second-order derivatives of the thermodynamic variables on the line of critical points, we get the following conditions in addition to Eqs. (2.22) and (2.23):

$$\left(\frac{\partial^2 \Delta\mu}{\partial c^2}\right)_{p,T} = 0, \quad \left(\frac{\partial^2 \Delta\mu}{\partial v^2}\right)_{p,T} = 0, \quad \left(\frac{\partial^2 \Delta\mu}{\partial s^2}\right)_{p,T} = 0,$$

$$\left(\frac{\partial^2 p}{\partial c^2}\right)_{T,\Delta\mu} = 0, \quad \left(\frac{\partial^2 p}{\partial v^2}\right)_{T,\Delta\mu} = 0, \quad \left(\frac{\partial^2 p}{\partial s^2}\right)_{T,\Delta\mu} = 0, \tag{2.29}$$

$$\left(\frac{\partial^2 T}{\partial c^2}\right)_{T,\Delta\mu} = 0, \quad \left(\frac{\partial^2 T}{\partial v^2}\right)_{T,\Delta\mu} = 0, \quad \left(\frac{\partial^2 T}{\partial s^2}\right)_{T,\Delta\mu} = 0.$$

At the critical point of a one-component system, besides Eq. (2.26), the following relations hold:

$$\left(\frac{\partial^2 p}{\partial v^2}\right)_T = 0, \quad \left(\frac{\partial^2 T}{\partial s^2}\right)_p = 0. \tag{2.30}$$

According to Eqs. (2.20) and (2.21), at the critical point and on the spinodal both the intensity of density fluctuations and their correlation radius are infinite.

2.3
Elements of Statistical Theory

The theory of thermodynamic stability assumes that the thermodynamic potential, for example, the chemical potential (see Fig. 2.2), of each of the phases is uniquely determined on both sides of the line of phase equilibrium. This assumption implies that the extensions of all substance properties from the stable into the metastable region are unique and smooth. Such definiteness will not be observed in the statistical interpretation of metastable states.

In a statistical interpretation, the following quantity is identified with the density of the Helmholtz free energy:

$$f(\rho, T) = -k_B T \lim_{V \to \infty, N \to \infty, (N/V) = \rho = \text{const}} \frac{\ln Z(N, V, T)}{V}, \quad (2.31)$$

where $Z(N, V, T)$ is the partition function. The thermodynamic limit is equivalent to the requirement of an infinitely long time of observation of a system during which the phase point has to pass all regions of phase volume allowed by the boundary conditions of the system. In general, the finite lifetime of a metastable state makes it impossible to realize the thermodynamic limit and to determine the thermodynamic functions in a metastable region.

With regard to stable states, the existence of a finite value of the density of the Helmholtz free energy at requirements not too rigid with respect to the intermolecular potential has been proven by van Hove (cited by [21]). As shown by Penrose and Lebowitz [24], the function $f(\rho, T)$ may be determined in a metastable region only, if, in calculating the partition function, a limiting transition to infinitely weak forces with an unlimited radius of interaction is simultaneously performed. In this case, a subregion $\mathfrak{R}$, which corresponds to the metastable state, can be specified in the phase space of the system. Although it is not very probable to find the phase point in this subregion, the limiting probability of leaving it is equal to zero. In systems with long-range forces one can also observe genuine first-order phase transitions described by the van der Waals equation of state in combination with the Maxwell rule [25]. The transition to attractive forces with long-range macroscopic interaction radii reduces the many-particle problem to the problem of the behavior of one particle in a certain averaged force field. The mean-field theory does not take into consideration fluctuations and yields thermodynamic functions of a hypothetical system, which are necessarily homogeneous at distances of action of the attractive forces. This theory becomes rigorous in the limit of long-range attractive forces and describes not only stable and metastable but also unstable states (Fig. 2.5). S-shaped loops on precritical isotherms of statistical models of the liquid state are the result of the application (in many cases in an implicit way) of the mean-field approximation and the condition of density homogeneity.

Attempts to extend the approach of Penrose and Lebowitz [24] to systems with real interaction potentials were not successful [26–28]. It turned out that in the phase space of systems of particles with short-range attractive forces there is no localized subregion, $\mathfrak{R}$, which would correspond to metastable

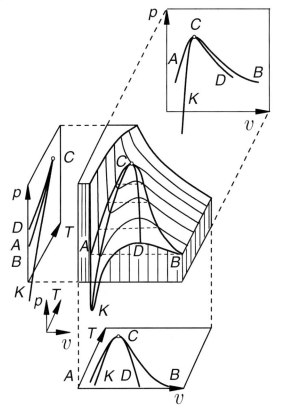

Fig. 2.5 Thermodynamic (p, v, T)-surface of a van der Waals fluid and its projections: ACB is the binodal; KCD is the spinodal; C is the critical point.

states. Nevertheless, the results of the mean-field theory may also be obtained in systems with short-range forces if the dimensionality of space, d, is formally considered to be infinite [29]. The limiting transition, $d \to \infty$, is equivalent to a transition to long-range forces.

According to the well-known results of van Hove [21], Yang and Lee [30], the point of a first-order phase transition in systems with a finite interaction radius is a peculiar point of the thermodynamic potential. The character of this peculiarity was first investigated by Andreev [31]. Similar results were later obtained in Refs. [32–35]. According to these papers, the points of the thermodynamic potential corresponding to a phase transition of first order are essentially peculiar (infinitely differentiable) points. The physical reason of such peculiarity consists in heterophase fluctuations in the metastable phase. Apparently, the infinitely differentiable peculiarity is not observed experimentally; it does not interfere with an extension of equilibrium functions from the stable into the metastable region, but leads to an uncertainty in the character of such extension. If we present the pressure in the form of a series with respect to the deviations of the parameters of the system from total equilibrium, this series is separated in a metastable region [35]. At first the terms decrease up to the nth component, and then increase infinitely. The value of the nth component determines the limiting accuracy up to which one can calculate the pressure in a metastable region. The value of n decreases with increasing depth of penetration into a metastable region. On the spinodal, the uncertainty in the thermodynamic properties becomes infinitely large. Purely phenomenological evaluations lead to the conclusion [36] that in regions of low and high metastability the value of such uncertainty is small compared to the level of thermal fluctuations. This property makes it possible to speak about the uniqueness of extensions of the properties of the substance beyond the line of phase equilibrium and the application of the same equations of state for both the stable and the metastable regions.

At present, the thermophysical properties of cryogenic liquids have been analyzed comprehensively not only in the stable, but also in the metastable (superheated) region. The following properties have been studied at high degrees of superheating: (p, ρ, T)-properties [37–42], velocity [42–47], and absorption [48, 49] of ultrasound, isochoric heat capacity [50, 51]. On the basis of existing experimental data, unified equations of state have been constructed valid for both the stable and the metastable region. The results of this work are generalized in the reviews [52, 53]. Analytical parametrizations of experimental data made it possible to approximate the boundary of essential phase instability, the spinodal, and to suggest equations determining the density of the free energy of simple substances in stable, metastable, and unstable states [54].

2.4
Phase Stability Against Finite Perturbations

Retaining the stabilizing reaction with respect to infinitesimal perturbations, a metastable system displays instability against discontinuous finite changes of the state parameters. The result of such instability is the emergence of a new phase in the system. In this case, the conditions of stability determine the minimum length, R_*, and energy, W_*, scales, at which a heterophase fluctuation becomes a viable center (nucleus) of the newly evolving phase. In order to determine these parameters, R_* and W_*, the methods of local thermodynamics in its point formulation are no longer applicable.

While examining the conditions of stability for a metastable system against finite changes in the state parameters it is more convenient to use instead of the internal energy, U, the Helmholtz free energy, F, expressed via the appropriate variables V and T. These variables are the independent parameters of the thermal equation of state. We shall restrict ourselves to the case of one-component isotropic systems and assume that the state of the system is fully determined by specifying the chosen variable, the density (number of particles per volume) $\rho(\vec{r})$, assuming the temperature to be constant ($T = $ const). The local density $\rho(\vec{r})$ is supposed to be homogeneous on the scale of the radius of action of the attractive forces, r_0. The discontinuity of changes in the state parameters implies the emergence of a finite inhomogeneity in the system. In this case, the thermodynamic potential has to contain a term specifying its linear scale. Following van der Waals [7], we shall present the Helmholtz free energy of a unit volume of an inhomogeneous system as a series in terms of derivatives of the local density as

$$f = f\left(\rho, \nabla\rho, \nabla^2\rho, \ldots\right) = f(\rho) + \kappa_1 \left(\nabla\rho\right)^2 + \kappa_2(\nabla^2\rho) + \cdots. \tag{2.32}$$

Here $f(\rho)$ is the density of the free energy of an inhomogeneous system, and κ_1 and κ_2 are the expansion coefficients which depend, in general, on temperature and density. Terms, linear in $\nabla\rho$, do not occur in Eq. (2.32) since they do not form a scalar. Considering weakly inhomogeneous systems, where the characteristic scales of the inhomogeneities are much larger than the radius of action of the intermolecular forces r_0, in expanding Eq. (2.32) one may restrict oneself only to terms that are quadratic in $\nabla\rho$ (*square-gradient approximation*). In this case, the Helmholtz free energy is written as follows:

$$F[\rho] = \int \left[f(\rho) + \kappa(\nabla\rho)^2\right] d\vec{r}, \tag{2.33}$$

where the notation $\kappa = \kappa_1 + d\kappa_2/d\rho$ is introduced. In Eq. (2.33), the surface contribution to the Helmholtz free energy from the system boundaries is excluded. According to the conditions of stability for a homogeneous state, the inequality $\kappa > 0$ must hold. The value of the expansion coefficient is determined outside the framework of thermodynamics; it can be expressed as [55]

$$\kappa = \frac{k_B T}{3} \int r^2 C(\vec{r}) d\vec{r} = r_0^2 k_B T \rho^{-1}. \tag{2.34}$$

Considering the total volume of the system, V, as being fixed and taking into account the condition of conservation of the number of particles

$$\int (\rho - \rho_0) d\vec{r} = 0, \tag{2.35}$$

where ρ_0 is the density of a homogeneous metastable phase, for the change in the Helmholtz free energy, $\Delta F = F[\rho] - F[\rho_0]$, connected with a heterophase fluctuation, $\Delta\rho = \rho - \rho_0$, we have

$$\Delta F[\rho] = \int \left[\Delta f(\rho) + \kappa (\nabla \rho)^2\right] d\vec{r}, \tag{2.36}$$

$$\Delta f = f(\rho) - f(\rho_0) - (\rho - \rho_0)\mu_0. \tag{2.37}$$

The functional (2.36) has the meaning of the excess grand thermodynamic potential; $\mu_0 = (\partial f/\partial\rho)_0$ is the chemical potential of the metastable phase.

The shape of the hypersurface of the functional, $\Delta F(\rho)$, in the configurational space of functions of local density is determined by the character of the dependence of the density of free energy of a homogeneous system on ρ and T. In the subsequent derivations it is assumed that this dependence is described by the van der Waals equation of state (Figs. 2.6a and b). The conditions of validity of introduction of such approximation are analyzed in Refs. [56, 57]. Figures 2.6c and d show the dependences $\Delta f(\rho)$ in application to the cases when the initial density of the metastable phase, ρ_0, is close to the liquid orthobaric density, ρ'_s, and the liquid density on the spinodal, ρ'_{sp}. The minima of $\Delta f(\rho)$ correspond to homogeneous states with the same values of the chemical potential. The inflection points of the curve $\Delta f(\rho)$, where $(\partial^2 f/\partial\rho^2)_T = 0$ holds, separate metastable and unstable states. If perturbations do not disturb the homogeneity of a metastable phase, the second component in the integrand in Eq. (2.35) is equal to zero. In this case, $F = fV$, and the conditions of stability are written as Eqs. (2.2) and (2.4).

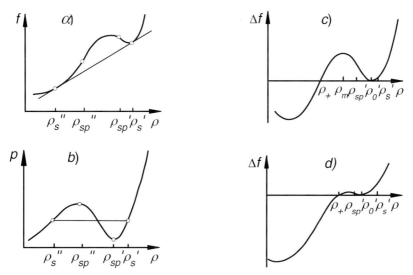

Fig. 2.6 Density of the Helmholtz free energy (a), pressure (b) and excess density of the free energy of the van der Waals fluid for states close to the binodal (c) and the spinodal (d).

According to Eq. (2.35), the emergence of a local inhomogeneity in a system results in an increase of the Helmholtz excess free energy. The more diffusive the heterophase fluctuation, the smaller its contribution to $\Delta F[\rho]$. For a stable phase, the first component in the integrand in Eq. (2.35) is always positive, and homogeneities of any amplitude and spatial extent will disperse. In a metastable region, heterophase fluctuations with an amplitude smaller than $|\rho_0 - \rho_m|$ result in an increasing excess density of free energy (see Fig. 2.6c). The emergence of a stronger density inhomogeneity results in a decrease in $\Delta f(\rho)$. In this case, at a certain value of $\rho = \rho_+$, the excess density of free energy becomes negative. A detailed analysis proves the existence of a certain critical heterophase fluctuation. Precritical heterophase fluctuations are thermodynamically disadvantageous, therefore they decay and reappear at other parts of the system. Supercritical heterophase fluctuations (nuclei) are viable centers of a new phase; their growth is thermodynamically irreversible. For a critical heterophase fluctuation, the following conditions should be fulfilled:

$$(\delta F[\rho])_* = 0, \qquad (\delta^2 F[\rho])_* = 0. \tag{2.38}$$

At small supersaturations, the function $\Delta f[\rho]$ has a large maximum and becomes negative only at densities close to the equilibrium density of the competing phase (see Fig. 2.6c). The characteristic linear size of a transition layer in a heterophase fluctuation is proportional to the correlation radius $\xi = [\kappa/(\partial^2 f/\partial \rho^2)_T]^{1/2}$. Far from the critical point, the correlation radius, ξ, coincides in the order of magnitude with the radius of action of intermolecular forces. To compensate the abrupt increase of energy in a transition layer, a viable heterophase fluctuation has to contain a large number of particles, i.e., it needs to have linear dimensions which considerably exceed the thickness of the transition layer.

When the metastable phase is close to the boundary of essential instability, the correlation radius is large. The presence of a gradient term in Eq. (2.30) is disadvantageous with respect to the formation of a nucleus of a stable phase of small size. Here, a critical heterophase fluctuation should have larger spatial dimensions but a small density amplitude, which proves to be sufficient to compensate for the small hump with regard to the dependence $\Delta f(\rho)$ (see Fig. 2.6d).

There is an infinite number of ways how a heterophase fluctuation may evolve into a viable nucleus. With respect to energetic aspects, those are the most advantageous which are passing the lowest point of the Helmholtz free energy barrier separating the metastable phase from the stable one. This point is the saddle point of the functional (2.36). Thus, the conditions given by Eqs. (2.38) are necessary, but not sufficient for answering the question whether the heterophase fluctuation they determine is a critical nucleus.

In an isotropic system a critical heterophase fluctuation will be a spherically symmetric formation (the validity of such an assumption is discussed in Section 5.2). In order to describe such nuclei, let us introduce spherical coordinates with the origin of the system of coordinates at the center of a heterophase fluctuation. The first condition in Eq. (2.38) (the condition of extremum of the functional (2.36)) yields [58]

$$\frac{d^2\rho}{dr^2} + \frac{2}{r}\frac{d\rho}{dr} + \frac{1}{2\kappa}\frac{\partial \kappa}{\partial \rho}\left(\frac{d\rho}{dr}\right)^2 = \frac{1}{2\kappa}\frac{\partial \Delta f}{\partial \rho} = \frac{1}{2\kappa}(\mu - \mu_0). \tag{2.39}$$

Equation (2.39) describes the density distribution in a heterophase fluctuation corresponding to an extremum of the thermodynamic potential. This equation has to be solved accounting for the boundary conditions

$$\begin{aligned} \rho &\to \rho_0 \quad \text{for} \quad r \to \infty, \\ \frac{d\rho}{dr} &= 0 \quad \text{for} \quad r \to 0 \quad \text{and} \quad r \to \infty. \end{aligned} \tag{2.40}$$

These conditions imply as essential requirements inhomogeneity, locality, and smoothness of the function $\rho(r)$ at all points, including the point $r = 0$. Besides, it is assumed that the emergence of a heterophase fluctuation corresponding to an extremum of the thermodynamic potential in a metastable system does not change its thermodynamic state parameters.

2.5
Critical Heterophase Fluctuations

All density distributions, $\rho_*(\vec{r})$, satisfying Eq. (2.39), correspond to minima, maxima, or saddle points of the functional $\Delta F[\rho]$. To solve the question as to whether the function $\rho_*(\vec{r})$ describes a critical nucleus, it is necessary to investigate the sign of the second variation of $\Delta F[\rho]$ at extremum. Let $\delta\rho(\vec{r}) = \rho(\vec{r}) - \rho_*(\vec{r})$ be a small perturbation of the extremal distribution, $\rho_*(\vec{r})$. Expanding $\Delta F[\rho]$ into a series and restricting ourselves to terms quadratic in $\delta\rho(\vec{r})$, we have

$$\begin{aligned} \delta^2 \Delta F &= \frac{1}{2} \iint \frac{\delta^2 F}{\delta\rho(\vec{r})\delta\rho(\vec{r}')}\bigg|_{\rho=\rho_*(\vec{r})} \delta\rho(\vec{r})\delta\rho(\vec{r}')d\vec{r}d\vec{r}' \\ &= \frac{1}{2} \int \delta\rho(\vec{r})\widehat{L}\delta\rho(\vec{r})d\vec{r}, \end{aligned} \tag{2.41}$$

where

$$\widehat{L} = \left[\frac{d^2\Delta f}{d\rho^2} - \kappa\nabla^2\right]_{\rho=\rho_*(\vec{r})} \tag{2.42}$$

is the differential operator (operator of stability).

At the saddle point of the functional (2.33), the sign of $\delta^2 \Delta F$ depends on the choice of the variation of $\delta\rho(\vec{r})$, i.e., on the direction of deviation from the extremum. Such kind of behavior is possible only if the spectrum of the operator, $\hat{L}$, contains both positive and negative eigenvalues. We shall expand an arbitrary variation of $\delta\rho(\vec{r})$ into a series in terms of the eigenfunctions, $\psi_{\vec{s}}(\vec{r})$, of the Hermitian operator, $\hat{L}$, as

$$\delta\rho(\vec{r}) = \sum_{\vec{s}=0}^{\infty} x_{\vec{s}} \psi_{\vec{s}}(\vec{r}). \tag{2.43}$$

Here, $\vec{s}$ is the number of the eigenfunction, and $x_{\vec{s}}$ is the expansion coefficient.

We shall restrict ourselves to the spherically symmetric case and present $\psi_{\vec{s}}(\vec{r})$ as the product of the radial and the angular parts as

$$\psi_{\vec{s}}(r) = \psi_{n,l}(r) Y_{l,m}(\theta, \varphi), \tag{2.44}$$

where (r, θ, φ) are the spherical coordinates, and $Y_{l,m}(\theta, \varphi)$ is the spherical Laplace function. The subscript n is a series of positive numbers, including zero. At a given l, the subscript m takes the values $m = 0, \pm 1, \pm 2, \ldots, \pm l$. The functions $\psi_{n,l}(r)$ and $Y_{l,m}(\theta, \varphi)$ are solutions of the eigenvalue problem of the operator, $\hat{L}$,

$$\left[-\frac{1}{r^2} \frac{d}{dr} \left(r^2 \frac{d}{dr} \right) + \frac{l(l+1)}{r^2} + V_{\text{eff}}(r) \right]_{\rho=\rho_*(r)} \psi_{n,l} = e_{n,l} \psi_{n,l}, \tag{2.45}$$

$$\hat{l}^2 Y_{l,m} = l(l+1) Y_{l,m}. \tag{2.46}$$

Here,

$$\hat{l}^2 = -\left[\frac{1}{\sin^2\theta} \frac{\partial^2}{\partial\varphi^2} + \frac{1}{\sin\theta} \frac{\partial}{\partial\theta} \left(\sin\theta \frac{\partial}{\partial\theta} \right) \right] \tag{2.47}$$

is the operator of the momentum squared, and the effective potential is given by $V_{\text{eff}} = (\partial^2 \Delta f / \partial \rho^2)/2\kappa$.

Substitution of Eq. (2.43) into Eq. (2.41) taking into account Eq. (2.45) and the orthonormalization of the eigenfunctions of the operator $\hat{L}$ transforms the quadratic form into the canonical expression

$$\delta^2 \Delta F = \frac{1}{2} \sum_{\vec{s}}^{\infty} e_{\vec{s}} |x_{\vec{s}}|^2. \tag{2.48}$$

The sign of the eigenvalue of $e_{\vec{s}}$ determines the sign of the curvature of the hypersurface $\Delta F[\rho]$ at the extremum point in the direction specified by the vector $\vec{s}$.

Equation (2.45) gets the form of the Schrödinger equation for a particle moving in a spherically symmetric field with the potential $V_{\text{eff}}(r)$ [59]. The character of the dependence $V_{\text{eff}}(r)$ is determined both by the function $\Delta f(\rho)$ and by

2.5 Critical Heterophase Fluctuations

the spatial dispersion of the equilibrium distribution $\rho_*(r)$. The potential V_{eff} is equal to zero on the spinodal and is negative in the unstable region.

Differentiating Eq. (2.39) with respect to r ($\kappa = \kappa(T)$), we have

$$\left[-\frac{1}{r^2} \frac{d}{dr} \left(r^2 \frac{d}{dr} \right) + \frac{2}{r^2} + \frac{1}{2\kappa} \frac{\partial^2 \Delta f}{\partial \rho^2} \right] \frac{d\rho_*(r)}{dr} = 0. \tag{2.49}$$

It follows from Eqs. (2.45) and (2.49) that the derivative $d\rho_*/dr$ is the eigenfunction of the operator $\hat{L}$ with $l = 1$ and the eigenvalue $e_{n,1} = 0$. According to the oscillation theorem of quantum mechanics [59], the circumstance that in the whole region of its definition, including the boundaries $r \to \infty$ and $r \to 0$, the derivative $d\rho_*/dr$ in a one-component system has no zeros, leads to $n = 0$ and

$$\psi_{0,1}(r) = \frac{d\rho_*}{dr}, \qquad e_{0,1} = 0. \tag{2.50}$$

The harmonic function $\psi_{0,1}(r)$ reflects a translational mode and describes the shift of the distribution of $\rho_*(r)$ as a whole. In the reference system which is connected with the center of the extremum distribution, the decomposition term in Eq. (2.48) with $n = 0, l = 1$ is equal to zero, and the translational mode does not change the energy of the system.

At small l, the harmonics $\psi_{0,l}(r)$ may be approximated by the function $d\rho_*/dr$ [60]. In this case, we get from Eqs. (2.45) and (2.49) the following result:

$$e_{0,l} \simeq \frac{l(l+1) - 2}{2R_*^2}, \tag{2.51}$$

where R_* is the effective radius of the extremum configuration. According to Eq. (2.51), the spectrum of the operator $\hat{L}$ contains only one negative eigenvalue $e_{0,0} = -R_*^{-2}$, which is responsible for the instability of the configuration. It means that, in the spherically symmetric case, all solutions of Eq. (2.39) with the boundary conditions (2.40) correspond to extremums of the surface $\Delta F[\rho]$ of the saddle-point type, i.e., they describe the critical nucleus. Any shift in the configurational space of the functions of local densities $\rho(r)$ from the point $\rho_*(r)$ in the direction specified by the numbers $n = l = m = 0$ leads to a decrease in the Helmholtz free energy. As in the region of low metastability, the eigenfunction $\psi_{0,0}(r)$ is localized mainly in a spherical layer of finite thickness with an effective radius R_* and its evolution is connected with an increase (decrease) in the nucleus size. The remaining eigenvalues of the operator $\hat{L}$ are positive. Modes with $n > 0$ and $l > 2$ describe changes of the nucleus shape retaining its volume, density profile, and the position of the center of mass. The set of states with $n = 1$ corresponds to perturbations at which a nucleus retains its spherical shape, but the density profile in the transition layer is distorted. At a certain value $n = n_*$, a continuous spectrum of nonlocalized eigenfuctions is found, which describes fluctuations in the volume of the liquid. As distinct from a metastable homogeneous phase, these fluctuations are "distorted" by the presence of a near-critical nucleus.

2.6
Relaxation Processes in Metastable Phases

A metastable state is a particular state of partial equilibrium. It is assumed that in a one-component system it is determined uniquely by specifying two thermodynamic variables. Changes in thermodynamic parameters cause a reconstruction of the configurations of the system leading to a transition from one equilibrium distribution to another. Such a transformation proceeds during a time τ_r, the characteristic relaxation time of the system. If the changes take place during times exceeding τ_r, a metastable system passes through a sequence of quasiequilibrium states. Otherwise the distribution of particles cannot follow the change of the state parameters and a frozen-in nonequilibrium state is obtained.

The process of relaxation of the system includes different mechanisms, each of which has its own characteristic time scales. Two kinds of stability may be correlated with two classes of space and time scales. The dynamic behavior of the system is determined by the utmost slowly relaxing parameters. In application to a one-component system, these parameters are the following hydrodynamic variables: local density, local velocity, and the local density of entropy. The characteristic scales of changes in hydrodynamic variables (modes), which will be represented by the variable, a, exceed the correlation radius, ζ, considerably. Let a_0 be the equilibrium value of a certain hydrodynamic variable, $a(\vec{r}, \tau)$, determined by the minimum value of the corresponding thermodynamic potential, E. The deviation from equilibrium causes a relaxation process—the value of $a(\vec{r}, \tau)$ tends in time to a_0. For a weakly nonequilibrium state, the rate of change of $a(\vec{r}, \tau)$ is proportional to the force conjugate to it, i.e., [61]

$$\frac{\partial a}{\partial \tau} = -\hat{\Gamma}\frac{\delta E}{\delta a}. \tag{2.52}$$

Here, the kinetic coefficient $\hat{\Gamma}$ is a linear operator. Its form is specified by the law of conservation of the parameter $a(\vec{r}, \tau)$. The law of conservation determines whether the perturbation will relax in a local way or, before disappearing, will propagate to the entire volume of the system. Regarding large-scale motions, we have [62]

$$\hat{\Gamma} = \Gamma_n - \Gamma_c \Delta, \tag{2.53}$$

2.6 Relaxation Processes in Metastable Phases

where Γ_n and Γ_c are smooth functions of temperature and density. If $a(\vec{r}, \tau)$ satisfies the law of conservation, $\Gamma_n = 0$ or $\hat{\Gamma} = \Gamma_n$ holds.

The process of relaxation differs qualitatively with respect to homophase and heterophase fluctuations with the presence or absence of a conservation law for the hydrodynamic variables. Let us examine a one-component isotropic system, whose state is characterized by a single mode—the density of the number of particles, $\rho(\vec{r}, \tau)$. The equilibrium value of ρ_0 is determined by the minimum of the Helmholtz free energy. Both homogeneous and inhomogeneous perturbations may evolve in the system, and in a metastable region the requirements of their infinitesimality are imposed on the latter, i.e., $|\delta\rho(\vec{r}, \tau)| = |\rho(\vec{r}, \tau) - \rho_0| < |\rho_m - \rho_0|$ (see Fig. 2.6c). Expanding the integrand in Eq. (2.33) in terms of $\delta\rho(\vec{r}, \tau)$ and keeping only terms in it not higher than of second order, we have

$$F[\rho] = \int \left[f(\rho_0) + \frac{1}{2}\frac{\partial^2 f}{\partial \rho^2}(\delta\rho)^2 + \kappa(\nabla\delta\rho)^2 \right] d\vec{r}. \tag{2.54}$$

If we represent the perturbation $\delta\rho(\vec{r}, \tau)$ in the form of a superposition of planar waves

$$\delta\rho(\vec{r}, \tau) = \sum_{\vec{q}} \rho_{\vec{q}}(\tau) \exp(i\vec{q}\vec{r}), \tag{2.55}$$

substitution of Eq. (2.55) into Eq. (2.54) results in

$$\Delta F[\rho] = F[\rho] - F_0 = \frac{1}{2}\sum_{\vec{q}} \chi^{-1}(\vec{q})|\rho_{\vec{q}}(\tau)|^2, \tag{2.56}$$

$$\chi^{-1}(\vec{q}) = \chi^{-1}(0) + 2\kappa q^2, \qquad \chi^{-1}(0) = \left(\frac{\partial^2 f}{\partial \rho^2}\right)_T = (\rho_0^2 \beta_T)^{-1}. \tag{2.57}$$

Here, $\chi(\vec{q})$ is the function describing the response of the system to perturbations with the wavelength $\lambda = 2\pi/q$, $\chi(0)$ being the long-wavelength limit of the response function. Summation in Eqs. (2.55) and (2.56) is performed with respect to wave numbers smaller than the inverse of the radius of the attractive interaction forces of the molecules. From the general condition of stability, $\Delta F[\rho] > 0$, it follows that $\chi^{-1}(\vec{q}) > 0$ holds for all values of $\vec{q}$. On the spinodal ($\chi^{-1}(0) = 0$), a one-component system loses its stabilizing reaction to long-wavelength perturbations retaining stability against short-wavelength density fluctuations. With deeper penetration into the unstable region, more and more short-wavelength perturbations lead to the disturbance of space homogeneity [64].

Let us assume that $\hat{\Gamma} = \Gamma_n$ (the local density) is a nonconserved parameter if the system has small velocity gradients causing viscous friction [65]). Then, as a consequence from Eqs. (2.52) and (2.56) for the regular growth rate of each of the Fourier components $\delta\rho(\vec{r},\tau)$, we have

$$\frac{\partial \delta\rho_{\vec{q}}}{\partial \tau} = -\Gamma_n \chi^{-1}(\vec{q})\delta\rho_{\vec{q}}. \tag{2.58}$$

Equation (2.58) gives the exponential law of change with time in the Fourier harmonics of density

$$\rho_{\vec{q}}(\tau) = \rho_{\vec{q}}(0)\exp(i\Omega_{\vec{q}}\tau), \tag{2.59}$$

where $\Omega_{\vec{q}}$ is the frequency of the collective mode, the reciprocal of the relaxation time

$$\Omega_{\vec{q}} = i\tau_{\vec{q}}^{-1} = i\Gamma_n \chi^{-1}(\vec{q}) = i\tau_0^{-1} + 2i\Gamma_n \kappa q^2, \qquad \tau_0^{-1} = \Gamma_n \chi^{-1}(0). \tag{2.60}$$

A decrease in the isothermal elasticity, observed as the spinodal is approached, results in an unlimited increase in the relaxation time of long-wavelength Fourier components of density. In this case, the relaxation times of the Fourier components with $\vec{q} \neq 0$ are finite and decrease with increasing $\vec{q}$.

In the spatially homogeneous case, the response function does not depend on the wave number, and the relaxation time increases proportional to the isothermal compressibility. The same nonequilibrium value of $\delta\rho$ in the whole system may be created only by an external action. These features are typical results of the mean-field approximation employed where spontaneous fluctuations are suppressed.

If the density is a conserved quantity, then $\hat{\Gamma} = -\Gamma_c \Delta$, and Eq. (2.52) is the continuity equation. In this case, instead of Eq. (2.60), we have

$$\Omega_{\vec{q}} = i\tau_{\vec{q}}^{-1} = i\Gamma_c \chi^{-1}(\vec{q}) q^2. \tag{2.61}$$

For hydrodynamic variables, the relation $\tau_{\vec{q}} \sim \Gamma_c \chi^{-1}(0) q^2$ holds. On the spinodal of a one-component system, the relaxation time of long-wavelength perturbations is infinitely large and finite for short-wavelength perturbations. Irrespective of the degree of vicinity to the spinodal, a homogeneous ($\vec{q} = 0$) variation of density in the whole volume will be retained as long as desired. It means that, accounting for the conservation law, weak density inhomogeneities will relax the slower, the smaller the density gradient.

Within the long-wavelength (hydrodynamic) limit, relaxation processes in one-component systems are connected with two collective modes: nonpropagating entropy fluctuations (the thermal diffusion mode) and pressure fluctuations (the acoustic mode) [66]. The time dependence of each of the collective modes looks like that presented in Eq. (2.59), where $\Omega_{\vec{q}}$ is a complex quantity. For the thermal diffusion and the acoustic modes, we obtain, respectively,

$$\Omega_{\vec{q}} = iD_T q^2, \tag{2.62}$$
$$\Omega_{\vec{q}} = \pm Cq + iD_s q^2. \tag{2.63}$$

Here, $D_T = \Lambda/\rho c_p$ is the thermal diffusivity, Λ is the thermal conductivity, $C = (\partial p/\partial \rho)_s^{1/2}$ is the low-frequency sound velocity, and D_s is the logarithmic damping coefficient of sound waves,

$$D_s = D_T \left(\frac{c_p}{c_v} - 1\right) + \frac{1}{\rho}\left(\frac{4}{3}\eta + \eta_v\right), \qquad (2.64)$$

where η and η_v are the shear and the bulk viscosity, respectively.

In the region of weak and moderate metastability of simple liquids ($T = 0.9 T_c$), the thermal diffusivity is $D_T \simeq (25\text{--}50) \times 10^{-9}\,\text{m}^2\,\text{s}^{-1}$, and the dispersion time of a temperature inhomogeneity at length scales, $l = 10^{-3}$ m, is $\sim (0.5\text{--}1)$ s. Under the same conditions the characteristic time of damping of acoustic excitations is approximately an order of magnitude smaller. The pressure in a liquid relaxes much more rapidly. For parameter values, $C = 400\,\text{m}\,\text{s}^{-1}$ and $l = 10^{-3}$ m, we obtain $\tau_r \simeq 4 \times 10^{-7}$ s.

As the stability boundary is approached, relaxation times of hydrodynamic variables increase. On the spinodal, at least one of the collective modes becomes unstable (soft)—the complex excitation frequency at a fixed $\vec{q}$ becomes zero. The conditions of damping of fluctuations may be regarded as conditions of stability for an equilibrium state with respect to infinitesimal perturbations [67, 68].

With regard to a one-component system the thermal diffusion mode is the soft one. The behavior of thermal diffusivity in the vicinity of the spinodal was examined by Zeldovich and Todes [69]. On the spinodal, $D_T = 0$ holds, and the thermal conductivity in this case is either a positive finite or a diverging (but slower than c_p) quantity. Beyond the spinodal $D_T < 0$, and exponentially increasing temperature differences appear in the system leading to the separation of the system into phases.

The velocity of a low-frequency sound is finite on a mechanical spinodal and becomes zero at the boundary of stability against adiabatic mechanical perturbations. The formulation of an answer to the question how the damping of the acoustic mode has to be described, is a more complicated task [67]. According to Eq. (2.64), the damping is infinitely large if all of the three kinetic coefficients D_T, η, and η_v at some supersaturation become zero.

In solutions, along with thermal diffusion and acoustic modes, there is also a diffusion mode responsible for the relaxation of composition fluctuations. Relaxation times of concentration inhomogeneities at small superheats in an argon–krypton solution are approximately five times as long as the time needed for equalization of temperature inhomogeneities. The diffusion mode becomes soft on the diffusion spinodal.

According to Eq. (2.61), relaxation times of hydrodynamic modes are proportional to the square of the wavelength. When a system is rapidly transferred into a metastable region, long-wavelength fluctuations are not excited.

In the absence of total thermodynamic equilibrium, equilibrium with respect to the interactions of the nearest neighbors is achieved rapidly. Such a system, which has relaxed by small-scale degrees of freedom, should be described by the mean-field approximation. This viewpoint was developed by Zeldovich [70]. It allows one to consider the equation of state, derived using the mean-field theory, as a certain zeroth-order approximation for the equation of state of a metastable system.

2.7
Dynamics of Heterophase Fluctuations

Let us examine a thermodynamic system in a metastable state. Infinitesimal perturbations cause a reversible modification of a metastable system, and after the lapse of time τ_0 an equilibrium distribution of small-scale fluctuations is established. Instability against finite perturbations leads to a spontaneous emergence of large-scale configurations of fields of hydrodynamic modes $a_i(\vec{r}, t)$, so-called heterophase fluctuations, in the system. Precritical heterophase fluctuations are not viable and are absorbed by the chaotic thermal motion. The decay of a metastable phase begins with the appearance of a supercritical heterophase fluctuation, the nucleus of the new phase. The distinguished linear scales of the problem are the correlation length ξ, the distance at which the effect of a local inhomogeneity on the distribution of density in its vicinity is realized and the size of a critical nucleus, R_*. The relation between ξ and the characteristic linear dimension of the critical nucleus, R_*, is established by the condition of extremum of the functional ΔE at the saddle point.

The characteristic expectation time of a critical nucleus is $\bar{\tau}$ (the relaxation time of large-scale degrees of freedom). For the characteristic times of two kinds of relaxation processes in the metastable phase τ_0 and $\bar{\tau}$ one can write [71]

$$\bar{\tau} \sim \tau_0 \exp\left(\frac{\Delta E(a_*)}{k_B T}\right) \gg \tau_0, \qquad (2.65)$$

where $\Delta E(a_*) = E(a_*) - E(a_A)$ is the difference of the energy values in the state of the metastable equilibrium (point A), $E(a_A)$, and at the saddle point of the potential barrier, $E(a_*)$.

Fluctuations of different sizes are constantly born and disappearing in systems of large numbers of particles due to chaotic thermal motion. In this case, the microscopic state of a medium changes many times faster than the macroscopic state of a system. Rapid spontaneous fluctuations of a medium may be regarded as a random process characterized by the correlation time τ_c, the memory time of a random process. In simple liquids, characteristic correlation times at the scales of the radius of interaction of the intermolecular forces r_0 are approximately equal to $2r_0/v_t \simeq 10^{-11}$ s, where v_t is the average velocity of thermal motion of the molecules.

The effect of small-scale fluctuations on the behavior of large-scale hydrodynamic modes can be taken into account by the introduction of a random force, $\zeta_i(\tau)$, as a source of emergence of noise in the ith hydrodynamic mode. The properties of the random force are determined under the assumption of the existence of an equilibrium distribution of small-scale (homophase) fluctuations in the metastable phase. The force $\zeta_i(\tau)$ has a Gaussian shape with a zero average value. The correlation between the values of the random force at two moments of time τ and τ' is different from zero only for time intervals $|\tau - \tau'| \sim \tau_c$, the time of interaction of small-scale and large-scale degrees of freedom. Since $\tau_c \sim \tau_0 \ll \bar{\tau}$, the limiting transition $\tau_c \to 0$ is usually performed considering the system as a medium without memory or with Gaussian white noise. The average value and the correlation function of such a noise will look like

$$\langle \zeta_i(\tau) \rangle = 0, \tag{2.66}$$

$$\langle \zeta_i(\tau) \zeta_j(\tau') \rangle = 2D_{ij}\delta_{ij}\delta(\tau - \tau'). \tag{2.67}$$

Here, D_{ij} is the intensity of the source of force $\zeta_i(\tau)$. Averaging is performed with respect to all realizations of the random force.

Taking into account the random force, the equations of motion for the fields of hydrodynamic modes read [62, 63]

$$\frac{\partial a_i}{\partial \tau} = -\sum_j \Gamma_{ij} \frac{\delta E}{\delta a_j} + \zeta_i(\tau). \tag{2.68}$$

Here, the elements of the matrix of kinetic coefficients, $\hat{\Gamma} = \Gamma_{ij}$, are linear operators, determined by Eq. (2.53).

According to the stochastic equation of motion (2.64), the hydrodynamic variables, $a_i(\vec{r}, \tau)$, are random quantities, and for them one may introduce the distribution function $P(\{a\}, \tau)$, where $\{a\}$ is the abbreviation for the set $a_i(\vec{r}, \tau)$ of variables. The theory of homogeneous random processes shows [72] that the function $P(\{a\}, \tau)$ which determines the density of probability of finding a system at the moment, τ, with the configuration of fields, $\{a\}$, satisfies the Fokker–Planck multidimensional equation [32, 60, 73, 74].

As was done in previous sections, we shall restrict our consideration to systems with one hydrodynamic mode, $\rho(\vec{r}, \tau)$. In the multidimensional space of functions $\rho(\vec{r})$ to each configuration of the density field there corresponds a certain phase point, and to the Helmholtz free energy, which is the functional of $\rho(\vec{r})$, a certain hypersurface. The relative minimum of the hypersurface, $\Delta F[\rho]$, corresponds to the metastable state, and the absolute minimum to the stable state. The system may leave the state of metastable equilibrium by overcoming the barrier formed by the hypersurface of the Helmholtz free energy around the point of the relative minimum. The most probable mechanical trajectory of a heterophase fluctuation will pass through the point to which the minimum barrier height (saddle point) corresponds.

In order to determine the flux of heterophase fluctuations, we write the Fokker–Planck equation as a continuity equation [60]

$$\frac{\partial P}{\partial \tau} = -\int \frac{\delta j}{\delta \rho(\vec{r})} d\vec{r}, \qquad (2.69)$$

where

$$j[\vec{r}] = -\hat{\Gamma}\left[\frac{\delta F}{\delta \rho(\vec{r})}P + k_B T \frac{\delta P}{\delta \rho(\vec{r})}\right] \qquad (2.70)$$

is the functional of $\rho(\vec{r})$, a component of the fluxes of probability in the infinite-dimensional space of functions $\rho(\vec{r})$. The first term in Eq. (2.70) determines the flux into the region of smaller values of the thermodynamic potential, the second one, into the direction of lower probability. According to Eq. (2.68), the coefficient to P is the rate of regular change of the variable $\rho(\vec{r}, \tau)$.

In the equilibrium state, the distribution function, $P = P_{eq}$, is determined by Eq. (2.18), where $W = \Delta F[\rho]$, the resultant flux, j, is equal to zero, and the Einstein relation follows from Eq. (2.70):

$$\Gamma_{ij} = \frac{D_{ij}}{k_B T}. \qquad (2.71)$$

In the critical region of the saddle point, defined as the region where $|F[\rho] - F[\rho_*]| \leq k_B T$, the quadratic approximation (2.41) is valid. A rearrangement of Eq. (2.43) transforms the quadratic form (2.41) into the normal representation given by Eq. (2.48). The variable $x_0(\vec{s} = 0)$, responsible for the instability of the critical configuration, gives the direction of the energetically most advantageous growth of a heterophase fluctuation at the hypersurface $\Delta F[\rho]$, which corresponds to the bottom of a valley going through the pass. In terms of the variables $x_{\vec{s}}$, the density of probability of fluxes, Eq. (2.70), takes the form

$$j_{\vec{s}} = \int \psi_{\vec{s}}(\vec{r}) j(\vec{r}) d\vec{r} = -\sum_{\vec{s}} \tilde{\Gamma}_{\vec{s}\vec{s}'}\left(\frac{\partial F}{\partial x_{\vec{s}'}}P + k_B T \frac{\partial P}{\partial x_{\vec{s}'}}\right). \qquad (2.72)$$

For purely dissipative systems, $\tilde{\Gamma}_{\vec{s}\vec{s}'}$ is a symmetric, positive definite matrix of the kinetic coefficients. If the variable $\rho(\vec{r})$ is retained, we have

$$\tilde{\Gamma}_{\vec{s}\vec{s}'} = \int (\nabla \psi_{\vec{s}}) \Gamma_c (\nabla \psi_{\vec{s}'}) d\vec{r}. \qquad (2.73)$$

2.7 Dynamics of Heterophase Fluctuations

Otherwise, we get

$$\tilde{\Gamma}_{\tilde{s}\tilde{s}'} = \int \psi_{\tilde{s}} \Gamma_n \psi_{\tilde{s}'} d\vec{r}. \tag{2.74}$$

In the stationary case, $\partial P/\partial \tau = 0$ holds, and from Eqs. (2.69) and (2.72) we have

$$\sum_{\tilde{s}} \frac{\partial j_{\tilde{s}}}{\partial x_{\tilde{s}}} = -\sum_{\tilde{s}\tilde{s}'} \frac{\partial}{\partial x_{\tilde{s}}} \tilde{\Gamma}_{\tilde{s}\tilde{s}'} \left(\frac{\partial F}{\partial x_{\tilde{s}'}} P + k_B T \frac{\partial P}{\partial x_{\tilde{s}'}} \right) = 0, \tag{2.75}$$

where $\tilde{\Gamma}_{\tilde{s}\tilde{s}'}$ is the transposed matrix of kinetic coefficients.

Following Langer [60], we find the solution of Eq. (2.75) in the form

$$P(\{x\}) = P_{eq}(\{x\}) \int_0^\infty \exp\left[-\frac{1}{2}\left(\varepsilon + \sum_{\tilde{s}} n_{\tilde{s}} x_{\tilde{s}} \right)^2 \right] d\varepsilon, \tag{2.76}$$

where $n_{\tilde{s}}$ is a constant vector connected with the direction of the flux of heterophase fluctuations through the pass. The boundary conditions for Eq. (2.75) are determined by the requirements of coincidence of the stationary and the equilibrium distribution functions close to the point of conventional minimum of the hypersurface, $\Delta F[\rho]$,

$$P(\{x\}) \simeq P_{eq}(\{x\}) \quad \text{for} \quad \{x\} \simeq \{x_A\} \tag{2.77}$$

and the finite total number of nuclei in the system

$$P(\{x\}) \simeq 0, \quad \text{when} \quad \{x\} \to \{\infty\}. \tag{2.78}$$

The latter condition means that beyond the well there is only a flux of nuclei going over the slope of the hypersurface $\Delta F[\rho]$ to infinity.

Substitution of Eq. (2.76) into Eq. (2.69) yields an equation for $n_{\tilde{s}}$. In the spectrum of the stability operator, Eq. (2.42), there is only one negative eigenvalue of e_0. The variable x_0 which corresponds to this eigenvalue is called the unstable variable. The distribution function in the unstable variable x_0 is found by integrating the full distribution function (2.76) over all stable variables $x_{\tilde{s}}(\tilde{s} \neq 0)$ at the condition $x_0 = \text{const}$,

$$P(x_0) = C_0 \exp\left(-\frac{\Delta F_*}{k_B T} \right) \int_0^\infty d\varepsilon \int \exp\left[-\frac{1}{2k_B T} \sum_{\tilde{s}} e_{\tilde{s}} x_{\tilde{s}}^2 - \right.$$

$$\left. -\frac{1}{2}(\varepsilon + \sum_{\tilde{s}} n_{\tilde{s}} x_{\tilde{s}})^2 \right] \prod_{s \neq 0} dx_{\tilde{s}}. \tag{2.79}$$

Here, C_0 is the normalization constant of the equilibrium distribution function with respect to the unstable variable.

The inner integral in Eq. (2.79) is a Gaussian integral. On determining the position of the maximum of the exponent of the exponential curve in terms of the variables $x_{\tilde{s}}(\tilde{s} \neq 0)$, after integration we have [75]

$$P(x_0) = P_{eq}(x_0) \left(\frac{|e_0|}{2\pi k_B T}\right)^{1/2} \int_{x_0}^{\infty} \exp\left(-\frac{|e_0|\varepsilon^2}{2k_B T}\right) d\varepsilon, \qquad (2.80)$$

where

$$P_{eq}(x_0) = C_0 \exp\left(-\frac{|e_0|x_0^2}{2k_B T}\right) \exp\left(-\frac{\Delta F_*}{k_B T}\right). \qquad (2.81)$$

The distribution with respect to the unstable variable does not depend on the kinetic coefficients, whereas the expression for the full stationary distribution function, Eq. (2.76), includes the components of the vector $n_{\tilde{s}}$ depending on $\tilde{\Gamma}_{\tilde{s}\tilde{s}'}$.

According to Eqs. (2.52) and (2.48), the rate of regular change of the hydrodynamic variable $x_{\tilde{s}}$ is described by the equation

$$\frac{\partial x_{\tilde{s}}}{\partial \tau} = -\sum_{\tilde{s}'} \tilde{\Gamma}_{\tilde{s}\tilde{s}'} e_{\tilde{s}'} x_{\tilde{s}'}. \qquad (2.82)$$

The eigenvalues of the matrix $\left(-\tilde{\Gamma}_{\tilde{s}\tilde{s}'} e_{\tilde{s}'}\right)$ are solutions to the characteristic equation

$$\det[\tilde{\Gamma}_{\tilde{s}\tilde{s}'} e_{\tilde{s}'} + \lambda \delta_{\tilde{s}\tilde{s}'}] = 0. \qquad (2.83)$$

Since the spectrum of the stability operator, Eq. (2.42), contains only one negative eigenvalue e_0, the matrix $\left(-\tilde{\Gamma}_{\tilde{s}\tilde{s}'} e_{\tilde{s}'}\right)$ can have only one positive eigenvalue, λ_0. This property results in the existence of one exponentially increasing solution to Eq. (2.82),

$$x_{\tilde{s}} = b_0 \exp(\lambda_0 \tau), \qquad (2.84)$$

where b_0 is the eigenvector of the matrix $\left(-\tilde{\Gamma}_{\tilde{s}\tilde{s}'} e_{\tilde{s}'}\right)$ corresponding to the eigenvalue λ_0. All the other eigenvalues λ_i ($i \neq 0$) are negative, and the solutions of Eq. (2.84) corresponding to them decay exponentially. The eigenvector $\vec{b}_0$ of the matrix $\left(-\tilde{\Gamma}_{\tilde{s}'} e_{\tilde{s}'}\right)$ corresponding to the eigenvalue λ_0 gives the direction of the probability flux through the saddle point of the pass. Owing to the nondiagonal character of the matrix of the kinetic coefficients, the most probable trajectory of the heterophase-fluctuation growth does not coincide generally with the energetically most advantageous path.

From Eq. (2.72) taking into account Eqs. (2.82), (2.83), and (2.84), after integration with respect to all $x_{\bar{s}}$ besides x_0, we have

$$j_{x_0} = \lambda_0 \left(x_0 P + \frac{k_B T}{|e_0|} \frac{\partial P}{\partial x_0} \right), \quad (2.85)$$

where $j_{x_0} = j(x_0)$ is the one-dimensional flux of heterophase fluctuations along the x_0-axis. In the steady state, $j(x_0)$ is independent of x_0. If the distribution (2.80) is established then all other components of the flux vanish. In this case, the value of j_{x_0} is not only a one-dimensional but also a total flux, i.e., it determines the number of viable nuclei emerging in a metastable system in a unit of time, J. Substitution of Eq. (2.80) into Eq. (2.85) yields

$$J = \lambda_0 \left(\frac{k_B T}{2\pi |e_0|} \right)^{1/2} P_{\text{eq}}(x_0) \mid_{x_0 = 0} = C_0 \lambda_0 \left(\frac{k_B T}{2\pi |e_0|} \right)^{1/2} \exp\left(-\frac{\Delta F_*}{k_B T} \right). \quad (2.86)$$

The !nucleation rate, J, is the main kinetic characteristic property of a relaxing metastable system. According to Eq. (2.86), the value of $\bar{\tau} \sim J^{-1}$ is determined to a large extent by a thermodynamic factor, namely the height of the activation barrier, $W_* = \Delta F_*$. The kinetic factors connected with the dynamics of growth of heterophase fluctuations affect $\bar{\tau}$ only through the parameter λ_0.

Gibbs [1] regarded W_* as the measure of stability for the metastable phase. The dimensionless complex in the power of the exponential function in Eq. (2.86) (the Gibbs number)

$$G_* = \frac{W_*}{k_B T} \quad (2.87)$$

determines the ratio of the height of the activation barrier to the average energy of thermal motion per degree of freedom. The Gibbs number is infinite on the binodal and is equal to zero on the spinodal. In the derivations of Eq. (2.86), we assumed that $G_* \gg 1$. Generally, the work of formation of a critical nucleus depends on the conditions how the process evolves and may be calculated employing thermodynamic [1, 76] or quasithermodynamic [58] methods.

2.8
Kinetic Nucleation Theory (Multiparameter Version)

The theory of nucleation mediated by thermal fluctuations was formulated in its basic features by Volmer and Weber [2], Farkas [3], Becker and Döring [4], Zeldovich [5], Frenkel [6], and others. It is commonly denoted as the classical nucleation theory. The classical theory describes heterophase fluctuations as microscopic fragments of a new phase (vapor bubble, liquid droplet).

This assumption limits the area of applicability to low supersaturations. Fragments of a new phase are assumed to be spherically symmetric pieces of the new phase. Their state is characterized by one distinguished (unstable) variable, the radius R or the number of molecules in a fragment. It is believed that formation and growth of the fragments are caused by uncompensated evaporation–condensation acts of single molecules. Such a point of view on the mechanism of growth of new-phase fragments introduced into nucleation theory by Farkas [3] ensures the slowness (diffuseness) of motion through a potential barrier and makes it possible to omit the term proportional to the acceleration in the stochastic equation (2.64) and the terms proportional to the rate and the derivatives of the rate in the Fokker–Planck equation, Eqs. (2.69) and (2.70). Besides, neglecting inertia effects in the one-parameter version of the nucleation theory ensures the establishment of an equilibrium size distribution of fragments in a metastable potential well.

However, in most cases of interest the state of a new-phase fragment is determined not only by one but by several variables. By this reason, nucleation is a multiparameter activation process. The multiparameter interpretation of nucleation considered in Refs. [60,75,77–79] gives rise to a broad variety of scenarios for the nucleation event, which is connected with a possible spread of time scales for different variables. In particular, there appear anomalous (different from classical) nucleation regimes: the bypassing of the saddle point of a potential relief by a nuclei flux, the disturbance of the Boltzmann equilibrium distribution for precritical aggregates, a nonequilibrium composition in a critical nucleus [80–82].

Let us now examine in more detail the multiparameter approach to the problem of nucleation. Among the parameters that characterize the state of a new-phase fragment we distinguish the most slowly relaxing (significant) variables. The variables that are not significant are included in a thermal ensemble which simulates small-scale fluctuations and is taken into account by random fields. The significant variables are counted off from their values for a critical nucleus and are denoted as a_i, where $i = 1, \ldots, N$ and N is the total number of significant variables. We suppose that in variables a_i the expression for the work of formation of a nucleus E in the vicinity of the saddle point has a quadratic form (without cross terms), i.e.,

$$E = E_* + \frac{1}{2}\sum_{ij} e_{ij} a_i a_j = E_* + \frac{1}{2}\sum_i e_i a_i^2. \tag{2.88}$$

The determination of variables for which the matrix $\widehat{E} = (e_{ij})$ is of diagonal form is reduced to the search for a respective transformation. Among the eigenvalues of the matrix $\widehat{E}$, one is negative.

In order to take into account inertia effects, we will introduce into consideration the rates of change of the variables a_i: $a_{N+i} \equiv da_i/d\tau \equiv \dot{a}_i$ ($i = 1, \ldots, N$).

In this case, the state of a new-phase fragment will be determined by a point in the 2N-dimensional phase space $\{a\} = a_1, \ldots, a_{2N}$. The energy of the system, E, in such a case is the Hamiltonian, H. In the vicinity of the saddle point, the Hamiltonian looks like

$$H = E(a_1, \ldots, a_N) + K(a_{N+1}, \ldots, a_{2N}) = E_* + \frac{1}{2} \sum_{i,j} h_{i,j} a_i a_j$$

$$= E_* + \frac{1}{2} \left(\sum_{i=1}^{N} e_i a_i^2 + \sum_{i=N+1}^{2N} m_{\text{eff}} a_i^2 \right), \quad (2.89)$$

where m_{eff} is the effective mass of the fragment in variables $\{a\}$.

The stochastic equations of motion, Eq. (2.64), retain their form taking into account the fact that $i, j = 1, \ldots, 2N$ and $E = H$. In dynamical systems the matrix $\hat{\Gamma}$ contains, besides the symmetric kinetic part, $\hat{\Gamma}^+$ ($\Gamma_{ij}^+ \neq 0$ only for the subscripts i and j corresponding to the rates), an antisymmetric dynamic part, $\hat{\Gamma}^-$. It is connected with the fact that the variables a_i ($i = 1, \ldots, N$) (coordinates) remain unchanged with changes of the sign of time, whereas the variables a_{N+i} (rates) change sign [20].

The evolution in time of the distribution function of new-phase fragments in variables $\{a\}$ obeys a Fokker–Planck multidimensional equation which may be represented as [73, 74]

$$\frac{\partial P(\{a\}, \tau)}{\partial \tau} = \sum_{ij}^{2N} \frac{\partial}{\partial a_i} \left[D_{ij} \left(\frac{1}{k_B T} \frac{\partial H}{\partial a_j} + \frac{\partial}{\partial a_j} \right) \right] P(\{a\}, \tau), \quad (2.90)$$

where D_{ij} is the matrix of diffusion coefficients in the space $\{a\}$ related to the matrix of kinetic coefficients, Γ_{ij}, by relation (2.71). We are interested in the stationary solution of Eq. (2.90) corresponding to the flux of fragments of an ambient phase from the region of heterophase fluctuations where equilibrium is considered to be established, through the saddle to the other side into the region of two-phase states. In the stationary case, Eq. (2.90) looks after being linearized in the vicinity of the saddle point like

$$\sum_{ij} \frac{\partial}{\partial a_i} D_{ij} \left(\frac{1}{k_B T} \sum_{\kappa} h_{j\kappa} a_\kappa + \frac{\partial}{\partial a_j} \right) P(\{a\}) = 0. \quad (2.91)$$

The boundary conditions for this equation presuppose that equilibrium is disturbed only in the vicinity of the saddle point. Outside this region, Eqs. (2.77) and (2.78) are fulfilled. If the equilibrium distribution function $P_{\text{eq}}(\{a\})$, being a solution of Eq. (2.91), is established, fluxes of fragments along the axes a_i vanish. This equilibrium distribution is given by the expression

$$P_{\text{eq}}(\{a\}) = C \exp\left[-\frac{H(\{a\})}{k_B T} \right]. \quad (2.92)$$

By analogy with the one-dimensional case [83], we search for the stationary solution of Eq. (2.90) in the form

$$P(\{a\}) = w(\{a\})P_{eq}(\{a\}). \tag{2.93}$$

Substitution of Eq. (2.93) into Eq. (2.91) gives

$$\sum_{ij} D_{ij} \left(\frac{1}{k_B T} \sum_{\kappa} h_{j\kappa} a_{\kappa} + \frac{\partial}{\partial a_j} \right) \frac{\partial}{\partial a_i} w(\{a\}) = 0. \tag{2.94}$$

With the boundary conditions (2.77) and (2.78), the solution of this equation is

$$w(\{a\}) = \frac{1}{(2\pi k_B T)^{1/2}} \int_u^\infty \exp\left(-\frac{z^2}{2k_B T}\right) dz. \tag{2.95}$$

Here, the value of u is determined by deviations from the saddle point, i.e.,

$$u = \sum_i n_i a_i. \tag{2.96}$$

In Eq. (2.96), $\vec{n}$ is a constant vector which points into the direction of the fastest descent from the pass. Substitution of Eq. (2.95) into Eq. (2.94) leads to the following equation for determining $\vec{n}$:

$$\sum_{ij} n_i \widetilde{D}_{ij} h_{j\kappa} + \left(\sum_{ij} n_i D_{ij} n_j \right) n_\kappa = 0, \tag{2.97}$$

where $\widetilde{D}_{ij}$ is the transposed matrix to D_{ij}. Equation (2.97) leads to the characteristic equation

$$-\sum_{ij} n_i \widetilde{D}_{ij} h_{j\kappa} = k_B T \lambda n_\kappa \tag{2.98}$$

for determining the eigenvalues of λ and the eigenvectors $\vec{n}$ of the matrix $-\widehat{\widetilde{D}}\widehat{H}$ and the condition of normalization of the eigenvector $\vec{n}$,

$$\sum_{ij} n_i D_{ij} n_j = k_B T \lambda. \tag{2.99}$$

The latter relation may be written in an equivalent form as [74]

$$\sum_{ij} n_i h_{ij}^{-1} n_j = -1, \tag{2.100}$$

where $\widehat{H}^{-1} = (h_{ij}^{-1})$ is the matrix inverse to $\widehat{H}$.

2.8 Kinetic Nucleation Theory (Multiparameter Version)

We assume that in the diagonal representation of the Hamiltonian (2.89) to the unstable variable, denoted as a_0, a certain, formally introduced subscript $i = 0$ is assigned to one of the $2N$ subscripts of variables a_i. In this case, $h_0 < 0$ holds. In accordance with the presence of one unstable variable the matrix $-\widetilde{D}\widehat{H}$ has one positive eigenvalue of λ_0, and all other λ_i possess a negative real part. Since the matrix $-\widetilde{D}\widehat{H}$ is real, its eigenvector $\vec{n}_0$ with a positive eigenvalue may also be considered real.

In accordance with Eq. (2.98), the eigenvalue λ_0 is determined as the root of the characteristic equation (in the matrices $-\widetilde{D}\widehat{H}$ and $-\widehat{D}\widehat{H}$ the eigenvalues λ are the same, as distinct from the eigenvectors)

$$\det(\widehat{D}\widehat{H} + k_B T \lambda \widehat{I}) = 0. \tag{2.101}$$

Far away from the saddle point, fluctuations are insignificant, and fragments that have passed through the saddle move, as follows from Eq. (2.64), according to the equation

$$\frac{da_i}{d\tau} = -\frac{1}{k_B T} \sum_j D_{ij} h_j a_j. \tag{2.102}$$

The general solution of Eq. (2.102) taking into account Eq. (2.98) is

$$a_i = A_0 n_{0i} \exp(\lambda_0 \tau) + \sum_{j=1}^{2N-1} A_j n_{ij} \exp(\lambda_j \tau), \tag{2.103}$$

where the coefficients $A_0, A_1, \ldots, A_{2N-1}$ are determined by the initial values of a_i.

From Eq. (2.103) it follows that at $\tau \to \infty$ the value of $|a_0| \to \infty$, and all a_i where $i \neq 0$ become zero. Thus, the variable a_0 is unstable not only thermodynamically determined by the work of nucleus formation, but also kinetically by the macroscopic equations of motion, Eq. (2.102).

The distribution function with respect to the variable a_0 is found by integration of the complete distribution function, Eqs. (2.93) and (2.95), over all stable variables a_i ($i \neq 0$) at the condition that $a_0 = \text{const}$,

$$P(a_0) = \frac{C}{(2\pi k_B T)^{1/2}} \exp\left(-\frac{W_*}{k_B T}\right) \exp\left(-\frac{1}{2 k_B T} \sum_i h_i a_i^2\right)$$

$$\times \int_u^\infty \exp\left(-\frac{z^2}{2 k_B T}\right) dz \prod_{i \neq 0} da_i. \tag{2.104}$$

Taking into consideration Eqs. (2.98) and (2.99), this procedure leads to a result coinciding with Eqs. (2.80) and (2.81), i.e., $e_0 = h_0$.

In the vicinity of the saddle point, the multidimensional equation of nucleation kinetics, Eq. (2.90), may be written in the form of a continuity equation for the flux of states of the system under consideration directed from the region of heterophase fluctuations into the two-phase region (see Eqs. (2.90) and (2.91)). For the density of the fluxes of states along the variable a_i according to Eq. (2.90), we have

$$j_i(\{a\},\tau) = -\sum_j D_{ij}\left(\frac{1}{k_BT}\frac{\partial H}{\partial a_j} + \frac{\partial}{\partial a_j}\right)P(\{a\},\tau). \tag{2.105}$$

Hence, taking into account Eqs. (2.92), (2.95), and (2.96), we obtain

$$j_i(\{a\}) = \frac{1}{(2\pi k_BT)^{1/2}}\sum_j D_{ij}n_j P_{eq}(\{a\})\exp\left(-\frac{u^2}{2k_BT}\right). \tag{2.106}$$

In order to find the nucleation rate it is necessary to integrate the complete flux $\sum_i j_i(\{a\})$ going through the pass with respect to its cross-section. Since the divergence of a complete stationary flux is equal to zero, the result of integration is independent of the orientation of the hypersurface, with respect to which the integration is performed, relative to the flux direction. Determining the surface of integration by the condition $u = 0$, we have

$$J = \sum_i \int_{u=0} j_i(\{a\})ds_i. \tag{2.107}$$

Substitution of Eq. (2.106) into Eq. (2.107), taking into account the normalization condition (2.99), yields

$$J = C\frac{\lambda_0}{2\pi}\left[\left|\det\left(\frac{\widehat{H}}{2\pi k_BT}\right)\right|\right]^{-1/2}\exp\left(-\frac{W_*}{k_BT}\right), \tag{2.108}$$

where C is the normalization constant of the complete equilibrium distribution function given by Eq. (2.92).

Separating the distribution function with respect to the unstable variable from Eqs. (2.99) and (2.89) we have

$$P_{eq}(\{a\}) = P_{eq}(a_0)\prod_{i\neq 0}P_{eqi}(\{a\})$$

$$= P_{eq}(a_0)\prod_{i\neq 0}C_i\exp\left(-\frac{1}{2k_BT}\sum_{i\neq 0}^{2N}h_ia_i^2\right). \tag{2.109}$$

The conditions of normalization for the functions $P_{eqi}(\{a\})$ describing the equilibrium distribution of fragments over the variables a_i ($i \neq 0$) are

$$C_i\int_{-\infty}^{+\infty}\exp\left(-\frac{1}{2k_BT}\sum_{i\neq 0}^{2N}h_ia_i^2\right) = 1. \tag{2.110}$$

It follows from Eqs. (2.109) and (2.110) that

$$C = C_0 \prod_{i \neq 0} C_i = C_0 \prod_{i \neq 0} \left(\frac{h_i}{2\pi k_B T} \right)^{1/2}. \tag{2.111}$$

Substitution of Eq. (2.111) into Eq. (2.108) gives Eq. (2.86). Equations (2.108) and (2.86) are the final result of the multidimensional steady-state nucleation theory. With respect to their form, they coincide with the equation obtained in one-dimensional nucleation theory [5,6]. All the information on the effect of multiparametricity on the rate of passage of new-phase fragments through the saddle point of the potential barrier is contained in the increased decrement of the unstable variable, λ_0.

We rewrite, now, Eq. (2.108) in the form which is usually employed when compared to experiment [9, 10]

$$J = \rho z_0 z_1 \lambda_0 \left(\frac{k_B T}{2\pi |e_0|} \right)^{1/2} \exp\left(-\frac{W_*}{k_B T} \right) = \rho z_0 B \exp(-G_*), \tag{2.112}$$

where ρ is the number of particles per unit volume of the metastable phase, z_0 is the factor that corrects the normalization of the equilibrium distribution function, z_1 is the factor that determines the transition from the distribution function with respect to the number of molecules in the nuclei, and B is the kinetic prefactor. The value of B is proportional to the rate of passage of fragments through the critical size.

2.9
Approximations and Limitations of Classical Nucleation Theory

We examine, in the present section, the hierarchy of times given by the stochastic equation of motion, Eq. (2.64). For this purpose we restrict ourselves to the simplest case, where the state of a new-phase fragment is determined by only two variables, the size, R, and the rate of its change, $\dot{R}$. As was done before, by choosing the origin of the system of coordinates at the saddle point of the potential barrier, we have $a_1 = R - R_* = x$ and $a_2 = \dot{R} = v$. From Eq. (2.64) we get

$$\dot{x} = -\Gamma_{xx} \frac{\partial H}{\partial x} - \Gamma_{xv} \frac{\partial H}{\partial v},$$

$$\dot{v} = -\Gamma_{vx} \frac{\partial H}{\partial x} - \Gamma_{vv} \frac{\partial H}{\partial v} + \zeta_v(\tau). \tag{2.113}$$

In the vicinity of the saddle point, the Hamiltonian has the form of Eq. (2.88). By differentiating Eq. (2.88) with respect to x and v and substituting the obtained result into Eq. (2.113), we have

$$\dot{x} = \Gamma_{xx}|e|x - \Gamma_{xv}m_{\text{eff}}v, \tag{2.114}$$

$$\dot{v} = \Gamma_{vx}|e|x - \Gamma_{vv}m_{\text{eff}}v + \zeta_v(\tau). \tag{2.115}$$

From the identity $\dot{x} = v$ and the principle of symmetry of kinetic coefficients for the dynamic part of the matrix $\hat{\Gamma}$ it follows that $\Gamma_{xx} = 0$ and $\Gamma_{vx} = -\Gamma_{xv} = 1/m_{\text{eff}}$, where m_{eff} is the effective mass of the fragment in variables (x,v). In this case, the system of equations (2.114) and (2.115) takes the form

$$\dot{x} = v, \tag{2.116}$$

$$\dot{v} = \frac{|e|}{m_{\text{eff}}}x - \Gamma_{vv}m_{\text{eff}}v + \zeta_v(\tau), \tag{2.117}$$

where the random force $\zeta_v(\tau)$ is determined by the expressions

$$\langle \zeta_v(\tau) \rangle = 0, \tag{2.118}$$

$$\langle \zeta_v(\tau)\zeta_v(\tau') \rangle = 2D_{vv}\delta(\tau-\tau') = 2k_B T \Gamma_{vv}\delta(\tau-\tau'). \tag{2.119}$$

Equation (2.117) may be represented as

$$\left(\frac{dv}{dx} + \Gamma_{vv}m_{\text{eff}}\right)v = \frac{|e|}{m_{\text{eff}}}x + \zeta_v(\tau). \tag{2.120}$$

Hence it follows that inertia effects in the evolution of the fragments may be neglected if

$$\frac{d\dot{x}}{dx} \ll \Gamma_{vv}m_{\text{eff}}. \tag{2.121}$$

In this case, the system of equations (2.116) and (2.117) is reduced to one stochastic equation

$$\dot{x} = \frac{|e|}{\Gamma_{vv}m_{\text{eff}}^2}x + \frac{1}{\Gamma_{vv}m_{\text{eff}}}\zeta_v(\tau) = \frac{|e|}{\Gamma_{vv}m_{\text{eff}}^2}x + \zeta_x(\tau), \tag{2.122}$$

where

$$\langle \zeta_x(\tau) \rangle = 0, \tag{2.123}$$

$$\langle \zeta_x(\tau)\zeta_x(\tau') \rangle = 2D_{xx}\delta(\tau-\tau'). \tag{2.124}$$

Here, the notation D_{xx} is introduced for the diffusion coefficient in the space of fragment sizes. Equation (2.119) includes the diffusion coefficient in the space of velocities D_{vv}. According to Eqs. (2.119) and (2.122), we have

$$D_{xx} = \frac{k_B T}{\Gamma_{vv}m_{\text{eff}}^2}. \tag{2.125}$$

In application to the case of a purely deterministic motion, Eq. (2.122) can be rewritten as

$$\frac{d\dot{x}}{dx} = \frac{|e|}{\Gamma_{vv} m_{\text{eff}}^2}. \tag{2.126}$$

In this case, the condition of neglecting inertia effects, Eq. (2.121), takes the form

$$\frac{|e|}{\Gamma_{vv} m_{\text{eff}}^2} \ll \Gamma_{vv} m_{\text{eff}}. \tag{2.127}$$

According to Eqs. (2.116) and (2.117), three characteristic time scales can be distinguished with respect to the fragment evolution. The first characteristic is the time of relaxation of the velocity, τ_v, due to the action of the frictional force determined by the second term on the right-hand side of Eq. (2.117). If any other forces are absent, it follows from Eq. (2.117) that

$$\tau_v = \frac{1}{\Gamma_{vv} m_{\text{eff}}}. \tag{2.128}$$

The potential barrier $E(x)$ restricts the motion of fragments, and there appears a second time scale, τ_D, which is the characteristic diffusion time, i.e., the time required for a fragment to pass the critical region of the potential barrier,

$$\tau_D = \frac{x_+^2}{D_{xx}}. \tag{2.129}$$

According to Eq. (2.88), $x_+^2 \simeq k_B T / |e|$ holds. With Eq. (2.125), from Eq. (2.129) for τ_D we have the relation

$$\tau_D = \frac{\Gamma_{vv} m_{\text{eff}}^2}{|e|}. \tag{2.130}$$

The third term in Eq. (2.117) is the random (Langevin) force that takes into account the atomic structure and the fluctuations of the metastable phase. In view of the δ-correlation of the Langevin source, Eq. (2.67), the correlation time equals $\tau_c = 0$. Thus, only two time parameters τ_v and τ_D are different from zero. With Eqs. (2.128) and (2.130), the condition of neglecting inertia effects, Eq. (2.127), takes the form

$$\tau_v \ll \tau_D. \tag{2.131}$$

Inequality (2.131) implies that if a nucleation theory is developed in the configurational space, it examines only probabilities of transitions in periods exceeding the relaxation time, τ_v, considerably. In the theory of Brownian motion, such regime is known as the regime of strong friction [80]. It presupposes that the Maxwellian velocity distribution is instantaneously established for precritical fragments, and the formation of supercritical nuclei results from the slow (on the scale of the problem under consideration) motion of fragments via the potential barrier along the size axis.

In variables (x, v) the time-independent Fokker–Planck equation (2.91) linearized in the vicinity of the saddle point has the form

$$\left[-\frac{\partial}{\partial x} v - \frac{\partial}{\partial v} \left(\frac{|e|}{m_{\text{eff}}} x - D_{vv} \frac{m_{\text{eff}}}{k_B T} v \right) + D_{vv} \frac{\partial^2}{\partial v^2} \right] P(x, v) = 0. \tag{2.132}$$

The boundary conditions for this equation are

$$P(x, v) = \begin{cases} P_{\text{eq}}(x, v), & x = -\infty \\ 0, & x = +\infty \end{cases} \tag{2.133}$$

In accordance with the procedure of solving the Fokker–Planck multidimensional equation described in the previous section we determine the positive eigenvalue λ_0 and the eigenvector $\vec{n}_0$ of the matrix $-\widehat{D}\widehat{H}$ corresponding to it. From Eq. (2.96) the following system of equations can be derived:

$$-\frac{|e|}{m_{\text{eff}}} n_v = \lambda n_x, \tag{2.134}$$

$$-n_x - \frac{D_{vv} m_{\text{eff}}}{k_B T} n_v = \lambda n_v, \tag{2.135}$$

with the solution

$$\lambda_0 = -\frac{1}{2} \frac{D_{vv} m_{\text{eff}}}{k_B T} + \sqrt{\left(\frac{D_{vv} m_{\text{eff}}}{2 k_B T} \right)^2 + \frac{|e|}{m_{\text{eff}}}}. \tag{2.136}$$

The eigenvector corresponding to this eigenvalue has the form

$$\vec{n} = \chi \left(1, -\frac{\lambda_0 m_{\text{eff}}}{|e|} \right). \tag{2.137}$$

Thus, the direction of the vector $\vec{n}$, determining the direction of the fastest descent from the pass, through λ_0 depends on all the physical parameters of the problem. In the limiting case that inertia effects may be neglected (Eq. (2.127)), from Eqs. (2.137) and (2.125) we have

$$\lambda_0 \simeq \frac{|e| k_B T}{D_{vv} m_{\text{eff}}^2} = k_B T |e| D_{xx}. \tag{2.138}$$

At $D_{vv} \to \infty$, we have $\lambda_0 \to 0$ and $\vec{n} \to \chi(1, 0)$, i.e., the direction of the vector $\vec{n}$ coincides with the x-axis.

2.9 Approximations and Limitations of Classical Nucleation Theory

From the normalization conditions (2.99) and (2.100), we find a value of χ equal to

$$\chi = \left(\frac{|e|^2 k_B T}{D_{vv} m_{\text{eff}}^2 \lambda_0} \right)^{1/2}. \tag{2.139}$$

By substitution of Eqs. (2.136), (2.137), and (2.139) into Eqs. (2.93) and (2.95) for the steady-state distribution function of new-phase fragments along the variables x and v we have

$$\frac{P(x,v)}{P_{\text{eq}}(x,v)} = \left(\frac{|e|^2}{2\pi D_{vv} m_{\text{eff}}^2 \lambda_0} \right)^{1/2} \int_{x-\lambda_0 m_{\text{eff}} v/|e|} \exp\left(-\frac{|e|^2 z^2}{2 D_{vv} m_{\text{eff}}^2 \lambda_0} \right) dz. \tag{2.140}$$

When inertia effects are neglected, λ_0 is determined by Eq. (2.138). The substitution of this relation into Eq. (2.140) at $D_{vv} \to \infty$ gives

$$\frac{P(x)}{P_{\text{eq}}(x)} = \left(\frac{|e|}{2\pi k_B T} \right)^{1/2} \int_x \exp\left(-\frac{|e| z^2}{2 k_B T} \right) dz. \tag{2.141}$$

This function in dependence on the unstable variable $P(x)$ may also be obtained by integrating the complete distribution function (2.140) along the stable variable v provided that $x = \text{const}$. As is evident from Eq. (2.141), the stationary distribution along the unstable variable x does not depend on the kinetic coefficients whereas the expression for the complete distribution function (2.140) includes elements of the matrix of the generalized diffusion coefficient. Expressions (2.136), (2.140), and (2.86) are the final result of solving the stationary nucleation problem.

Let us examine the limits of applicability of the solutions to Eqs. (2.140) and (2.141). The integral in Eq. (2.140) becomes a constant if

$$\frac{\lambda_0 m_{\text{eff}} v}{|e|} - x \gg \left(\frac{\lambda_0 D_{vv} m_{\text{eff}}^2}{|e|^2} \right)^{1/2}. \tag{2.142}$$

In this region of the parameters (x,v), the stationary distribution function $P(x,v)$ coincides with the equilibrium distribution function $P_{\text{eq}}(x,v)$. Condition (2.142) is fulfilled only for energies smaller than the work of formation of a critical nucleus when

$$\frac{m_{\text{eff}} v^2}{2} - \frac{|e| x^2}{2} < 0. \tag{2.143}$$

Substitution of Eq. (2.143) into Eq. (2.142) gives

$$-x \gg \left(\frac{\lambda_0 D_{vv}}{|e|} \right)^{1/2} \frac{m_{\text{eff}}}{\left(|e|^{1/2} - \lambda_0 m_{\text{eff}}^{1/2} \right)}. \tag{2.144}$$

This condition determines the range of sizes for new-phase fragments where their equilibrium distribution is distorted only slightly by transitions over the potential barrier. If inertia effects in the motion of fragments can be neglected (Eq. (2.127)), then, taking into account Eq. (2.138), from Eq. (2.144) we get

$$-x \gg \left(\frac{k_B T}{|e|}\right)^{1/2}. \tag{2.145}$$

It means that the function $P(x,v)$ differs from the equilibrium distribution function $P_{eq}(x,v)$ only in the critical region of the potential barrier, where $E_* - |E(x)| \simeq k_B T \ll E_*$. As is evident from Eq. (2.141), beyond this region the stationary distribution coincides with respect to the unstable variable x with the equilibrium distribution $P_{eq}(x)$.

Otherwise, if Eq. (2.127) and

$$\frac{|e| k_B T}{D_{vv} m_{eff}^2} \gg \frac{D_{vv} m_{eff}}{k_B T} \tag{2.146}$$

hold, it follows from Eqs. (2.136) and (2.86) that

$$\lambda_0 = \left(\frac{|e|}{m_{eff}}\right)^{1/2}, \tag{2.147}$$

$$J = C_0 \left(\frac{k_B T}{2\pi m_{eff}}\right)^{1/2} \exp\left(-\frac{W_*}{k_B T}\right). \tag{2.148}$$

Formula (2.148) corresponds to the theory of absolute reaction rates [74], where inertia effects in the evolution of a nucleus in the absence of "friction" lead to the independence of the flux over the barrier from the shape of the barrier.

In fulfilling inequality (2.146), condition (2.144) is reduced to

$$-x \gg \left(\frac{k_B T}{|e|}\right)^{1/2} \left(\frac{|e|^{1/2} k_B T}{D_{vv} m_{eff}^{3/2}}\right)^{1/2} \tag{2.149}$$

and the function $P(x,v)$ is different from the equilibrium distribution in a much wider range than that given by inequality (2.145). When $D_{vv} \to 0$, the quadratic approximation for the top of the potential barrier used in solving Eq. (2.132) is violated. This property determines the limits of applicability of Eqs. (2.136), (2.86), and (2.148). In a first approximation, the width of the potential barrier (as well as of the potential well) is equal to $(W_*/|e|)^{1/2}$. Then from Eqs. (2.146) and (2.149) it follows that

$$\frac{k_B T |e|^{1/2}}{m_{eff}^{3/2}} \gg D_{vv} \gg \frac{(k_B T)^2 |e|^{1/2}}{W_* m_{eff}^{3/2}}. \tag{2.150}$$

This inequality determines the range of applicability of Eqs. (2.136), (2.86), and (2.148). Since homogeneous nucleation is a purely fluctuation phenomenon, in the limit $D_{vv} \to 0$ we have to expect $J = 0$. Equations (2.136), (2.86), and (2.148) do not satisfy such asymptotics.

2.10
Nucleation at a High Degree of Metastability

The classical nucleation theory is developed based on the assumption that the characteristic passage time of a new-phase fragment through the critical region of a potential barrier, τ_D, considerably exceeds the relaxation time, τ_v, connected with the dynamics of interaction between the fragment and the medium ($\tau_D \gg \tau_v$). In this case, the Maxwellian fragment velocity distribution is established immediately, and the "slow" motion of fragments over the potential barrier along an unstable variable ensures the presence of the Boltzmann equilibrium distribution for precritical fragments.

An illustrative model of the evolution of the new phase, providing diffusive motion along an unstable variable, is the Szilard schema. According to this model, a chain of alternating acts of attachment and detachment of single molecules at the surface of a fragment determines the evolution. This approach is introduced into nucleation theory by Farkas [3]. This schema was first used in describing the condensation of supersaturated vapors, where the formation of a nucleus by the mechanism of elementary evaporation and condensation acts seems to be evident. For a superheated liquid, such a nucleation mechanism may be expected to occur in a region of weak metastability, at temperatures that are not far from the critical point. It is less evident at cavitation in a stretched liquid or destruction of a stretched crystal. Here, critical nuclei are voids. The use of the traditional schema in this case requires the introduction into consideration of a virtual particle, a hole [9].

The classical scenario of formation of a new-phase nucleus is the most probable, but not the only possible one. At high supersaturations, when a system is in a state in the vicinity of the spinodal (in a region of high metastability), the appearance of a nucleus may not be preceded by a sequence of single acts of attachment and detachment of molecules. The possibility of such different cooperative mechanism of nucleation was first suggested by Skripov [9]. The cooperative decay of metastable systems under high supersaturations is confirmed by the results of computer experiments [84, 85]. A nucleation theory built on the assumption of a the cooperative mechanism is examined in [86–88].

We will evaluate the characteristic times of processes assuming that nucleation is initiated by the decay of unstable regions that form in the vicinity of the spinodal as a result of chaotic thermal motion [87]. The dynamics of such a process is different from the condensation–evaporation mechanism, and the condition $\tau_D \gg \tau_v$ may be violated. Following Zeldovich and Todes [69] we assume that the rate of decay of an unstable state in a one-component system is limited by the heat exchange between the expanding and the contracting elements of the volume. In this case, the decrement of the increase of density inhomogeneities (a quantity inversely proportional to τ_v) will be determined by the thermal diffusivity of the medium, D_T. Perturbations with wavelengths exceeding a certain critical value, l_*, are unstable. The value of l_* is determined by the isothermal compressibility, $l_* \sim |\beta_T|^{1/2}$ [89]. At the decay of unstable regions, nucleus formation is possible if $l_* \geq R_*$. In a linear approximation, $\tau_v \sim l_*^2/D_T$. In the vicinity of the spinodal $\beta_T \sim (\rho - \rho_{sp})^{-1}$ holds. If the heat conductivity on the spinodal is finite, then the relations $D_T \sim (\rho - \rho_{sp})$ and $\tau_v \sim (\rho - \rho_{sp})^{-2}$ are fulfilled. Assuming a quadratic potential barrier, the time τ_D is evaluated by data on the effective nucleus mass and the surface tension and depends only slightly on the distance from the spinodal [87]. Thus, in the vicinity of the spinodal it is possible to realize the condition $\tau_v \gg \tau_D$, which is characterized by a weak interaction of a new-phase fragment with the parent medium. The term "weak" means that the interaction with the medium has a weak effect on the motion of a precritical fragment and is not sufficient for the maintenance of the Boltzmann equilibrium distribution in a metastable state, which is disturbed when new-phase fragments leave it. The diffusion variable of the fragment in this case is not its size, but the energy, which is the most slowly-changing variable [87, 88]. If the energy generated during the decay of an unstable region proves to be insufficient for the formation of a viable new-phase fragment (nucleus), the originating density inhomogeneity disperses. At a subsequent appropriate fluctuation, another unstable region is formed and through its decay the system with a new amount of energy makes another attempt to overcome the barrier. Thus, when the interaction between a new-phase fragment and the medium is weak, the kinetics is determined by the energy diffusion from the depth of the potential well to the value W_* equal to the height of the potential barrier. In this case, thermal equilibrium is absent for precritical fragments.

The motion of a Brownian particle, representing to some extent a certain analog of a new-phase fragment, at a very weak interaction with the medium, was first examined by Kramers [83]. For the case of two independent variables, the characteristic size of a fragment, $a_1 = R - R_* = x$, and the rate of its change, $a_2 = \dot{x} = v$, the kinetic equation of nucleation in the regime of energy diffusion may be derived from the Fokker–Planck equation, Eq. (2.90), when going over in the latter from the variables (x, v) to the variables angle, ϕ, and total energy, H,

$$H = \frac{m_{\text{eff}} v^2}{2} + E \qquad (2.151)$$

2.10 Nucleation at a High Degree of Metastability

or angle, ϕ, and action, I,

$$I = \oint \{2m_{\text{eff}}[H - E(x)]\}^{1/2} dx. \tag{2.152}$$

Here, as before, m_{eff} is the effective mass of a new-phase fragment. Along the action variable, the kinetic equation has the form [83, 87]

$$\frac{\partial P(I,\tau)}{\partial \tau} = \frac{\partial}{\partial I}\left\{D(I)\left[\frac{1}{k_B T}\frac{dH}{dI} + \frac{\partial}{\partial I}\right]\right\} P(I,\tau), \tag{2.153}$$

where $D(I)$ is the diffusion coefficient along the coordinate I, and $P(I,\tau)$ is the function of action distribution of new-phase fragments. For the energy distribution function, we have

$$\frac{\partial P(H,\tau)}{\partial \tau} = \frac{\partial}{\partial H}\left\{D(H)\left[\frac{1}{k_B T} + \frac{\partial}{\partial H}\right]\right\} \frac{\partial H}{\partial I} P(H,\tau), \tag{2.154}$$

where $D(H)$ is the diffusion coefficient along the variable H. The diffusion coefficients $D(I)$ and $D(H)$ are connected via

$$D(H) = D(I)\frac{\partial H}{\partial I}. \tag{2.155}$$

In the stationary case from Eqs. (2.152) and (2.154) for the nucleation rate in the space of energies we have

$$J = -D(H)\left[\frac{1}{k_B T} + \frac{\partial}{\partial H}\right] P(I). \tag{2.156}$$

This expression may be presented as

$$J = -D(H)\exp\left(-\frac{H}{k_B T}\right) \frac{\partial}{\partial H}\left[P(H)\exp\left(\frac{H}{k_B T}\right)\right]. \tag{2.157}$$

Integrating Eq. (2.157) with respect to energy from a value close to the value of H at the bottom of the potential well (point A) to the value of H corresponding to a certain point B removed from the top of the potential barrier, we get

$$J = \frac{\left. P\exp\left(\frac{H}{k_B T}\right)\right|_{\text{near}A} - \left. P\exp\left(\frac{H}{k_B T}\right)\right|_B}{\int\limits_{\text{near}A}^{B} \frac{1}{D(H)}\exp\left(\frac{H}{k_B T}\right) dH}. \tag{2.158}$$

Owing to the integral divergence in Eq. (2.158) at $H = 0$, integration should be performed starting with an energy of the order of the thermal energy, i.e., $k_B T$. Moreover, assuming that the new-phase fragments that have passed over the potential barrier do not return, the term $P \exp(H/k_B T)|_B$ in Eq. (2.158) may be set equal to zero, and the upper limit of the integral in Eq. (2.158) may be considered equal to W_*. Thus, we obtain

$$J \simeq P \exp\left(\frac{H}{k_B T}\right)\bigg|_{nearA} \left[\int_{k_B T}^{W_*} \frac{1}{D(H)} \exp\left(\frac{H}{k_B T}\right)\right]^{-1}. \tag{2.159}$$

In a first approximation, we may assume

$$P \exp\left(\frac{H}{k_B T}\right)\bigg|_{nearA} = C_E, \tag{2.160}$$

where C_E is the normalization constant of the energy distribution function of fragments in the potential well.

At $W_* \gg k_B T$, the main contribution to the integral in Eq. (2.159) is made by the values of H close to W_*. Taking $D(H) \simeq D(W_*)$, we get

$$J \simeq C_E \frac{D(W_*)}{k_B T} \exp\left(-\frac{W_*}{k_B T}\right) \tag{2.161}$$

or

$$J \simeq C_I \left(\frac{\partial H}{\partial I}\right)_A \frac{D(I_*)}{k_B T} \exp\left(-\frac{W_*}{k_B T}\right). \tag{2.162}$$

Here, I_* is the value of the action at the peak of the potential barrier, C_I is the normalization constant of the action distribution function, and the subscript A refers to the potential well.

It follows from Eqs. (2.86) and (2.161) that two qualitatively different channels of spontaneous formation of new-phase nuclei lead to formally equivalent expressions for the stationary nucleation rate. In both cases, the exponential term, $\exp(-W_*/k_B T)$, dominates. The differences are connected with the value and the temperature dependence of the pre-exponential factor. The problem of determining the coefficients $D(W_*)$ and $D(I_*)$ in the regime of energy diffusion is complicated by the absence of an equilibrium distribution for precritical fragments.

In the limit $D(W_*) \to 0$, the nucleation rate tends to zero, i.e., $J = 0$. This property distinguishes Eq. (2.161) from another asymptotic formula (2.148), which implies that nucleation is also possible in the absence of fluctuations. This result is a consequence of the assumption of the presence of equilibrium in the ensemble of precritical fragments made in deriving Eq. (2.148).

In conclusion we shall note that cooperative processes of formation of new-phase nuclei, as an alternative to the Szilard schema of single acts of condensation and evaporation, may also proceed without the participation of the unstable phase. A rapid formation of local inhomogeneities with properties of the competing phase may also take place in the process of chaotic thermal motion.

2.11
Nucleation Bypassing the Saddle Point

Both in classical homogeneous nucleation theory (Section 2.7) and in its nonclassical version (Section 2.8) it is assumed that the most probable trajectory of motion for new-phase fragments always passes through the saddle point of a potential barrier. The possibility of an anomalous regime, where the equilibrium distribution for precritical fragments is not established and nucleation proceeds bypassing the saddle point, was discussed by Trinkaus [80]. Using two-dimensional nucleation as an example, he showed the possibility of such an anomaly for the case, if one of the elements of the matrix of the generalized diffusion coefficients is much smaller than the others, and suggested a theory for such a nucleation regime. Later on, the nucleation regime bypassing the saddle point was examined in [81, 82, 90]. This problem is most important in nucleation in multicomponent systems, where situations may arise easily that the rate of supply to a growing fragment of one of the mixture components will prove to be lower than that of the others by many orders of magnitude.

Here, we examine multidimensional nucleation without taking into account inertia effects. As we have done before, we represent the state variables of fragments of the newly evolving phase by a_i ($i = 1, \ldots, N$) and the set of these variables by $\{a\}$. The distribution function of fragments $P(\{a\}, \tau)$ is found by solving Eq. (2.90). It is believed that the potential relief in the space of $\{a\}$ has a saddle point and is described by the matrix $\hat{E}$, which has one negative eigenvalue and $N - 1$ positive eigenvalues. To simplify the subsequent calculations, we also assume that the matrix of the generalized diffusion coefficients, $\hat{D}$, is diagonal with one of the elements of the matrix $\hat{D}$, let it be D_{11}, considerably smaller than the others.

The small value of D_{11} implies that the motion along the variable a_1 is retarded, all the other variables $a_2, \ldots, a_N$ having sufficient time for adiabatic adjustment to this motion. In a multidimensional space, all other variables adjust to the slow variable a_1 being determined via

$$a_2^0(a_1), \ldots, a_N^0(a_1), \qquad (2.163)$$

and resulting from the conditions ($i \neq 1$)

$$\left(\frac{\partial E}{\partial a_i}\right)_{a_1} = 0, \quad \left(\frac{\partial^2 E}{\partial a_i^2}\right)_{a_1} > 0. \qquad (2.164)$$

In this case, the distribution function $P(\{a\}, \tau)$ reads

$$P(\{a\}, \tau) = P_0(a_1, \tau) \phi_{eq}(\{a'\}), \tag{2.165}$$

$$\phi_{eq}(\{a'\}) = \exp\left(-\frac{1}{2k_B T} \sum_{i,j} q_{ij} \Delta a_i \Delta a_j\right), \tag{2.166}$$

$$q_{ij} = \left(\frac{\partial^2 E}{\partial a_i \partial a_j}\right)_{\substack{a_1 = \text{const} \\ a_i = a_i^0(a_1)}}, \quad \Delta a_i = a_i - a_i^0(a_1), \tag{2.167}$$

i.e., it becomes of Boltzmann type with respect to the variables $\{a'\}$. Here, the notation $\{a'\}$ means that the variable a_1 is excluded from the set of variables $\{a\}$.

Let us exclude the fast variables $\{a'\}$ from Eq. (2.90). For this purpose, we will introduce the distribution function of the slow variable

$$P(a_1, \tau) = \int P(\{a\}, \tau) d\{a'\} \tag{2.168}$$

and integrate Eq. (2.90) with respect to the fast variables $\{a'\}$. As a result we have

$$\frac{\partial P(a_1, \tau)}{\partial \tau} = -\frac{\partial J_1}{\partial a_1} - j(a_1). \tag{2.169}$$

Here J_1 is the total flux of fragments at a fixed value of a_1, and $j(a_1)$ is the total flux in the space of $\{a'\}$ at a given value of a_1. If we assume that in the vicinity of the line $a_i^0(a_1)$ the element D_{11} of the matrix of the generalized diffusion coefficients depends only weakly on the variables $\{a\}$, and that $P(\{a\}, \tau)$ has the form of Eq. (2.165), then the flux J_1 may be presented as

$$J_1 = -D_{11} \exp\left(-\frac{E_{\text{eff}}}{k_B T}\right) \frac{\partial}{\partial a_1} \exp\left(\frac{E_{\text{eff}}}{k_B T}\right) P(a_1, \tau), \tag{2.170}$$

where

$$E_{\text{eff}} = -k_B T \ln \int \exp\left(-\frac{E}{k_B T}\right) d\{a'\} \tag{2.171}$$

is the effective potential, which is a potential of the average force along the coordinate a_1. The value of D_{11} in Eq. (2.170) is calculated on the line $a_{i0}(a_1)$ and is a function of a_1, exclusively.

Next we will examine the stationary case assuming that $\partial P/\partial \tau = 0$. From Eq. (2.169) taking into account Eqs. (2.165)–(2.168), (2.170) a time-independent equation for the determination of the function $P(a_1)$ follows as

$$-\frac{d}{da_1} D_{11} \left(\frac{P}{k_B T} \frac{dE_{\text{eff}}}{da_1} + \frac{dP}{da_1}\right) + j(a_1) = 0, \tag{2.172}$$

$$E_{\text{eff}} = E_0 + \frac{k_B T}{2} \ln \omega, \qquad E_0 = E(a_i^0(a_1)). \tag{2.173}$$

The quantity $\omega(a_1)$ in Eq. (2.173) has been obtained as a result of integrating Eq. (2.166) with respect to the variables $\{a'\}$ as

$$\omega(a_1) = \frac{\det \widehat{q}(a_1)}{(2\pi k_B T)^{N-1}}, \tag{2.174}$$

$$\int \phi_{\text{eq}} d\{a'\} = \omega^{-1/2}(a_1), \quad P(a_1) = P_0(a_1)\omega^{-1/2}(a_1). \tag{2.175}$$

In the stationary case, Eq. (2.169) reads

$$\frac{\partial J_1}{\partial a_1} + j(a_1) = 0. \tag{2.176}$$

The second term, $j(a_1)$, in Eq. (2.176) determines the withdrawal of new-phase fragments. If there is no saddle point in the subspace $\{a'\}$, then $j(a_1) = 0$ holds and a withdrawal does not occur. In this case, Eq. (2.176) takes the form

$$\frac{\partial J_1}{\partial a_1} = \frac{\partial J}{\partial a} = 0 \tag{2.177}$$

and describes ordinary one-dimensional nucleation.

If in the space $\{a'\}$ there is a saddle point, then nucleation is possible at a fixed value of a_1, the process scenario being determined by the outflow term. If, despite the presence of $j(a_1)$, fragments of a new phase moving along the line given by Eq. (2.163) grow to values exceeding $a_{1*} = 0$, the process will be determined by the passage of the saddle-point region. If the run-off inhibits the growth of nuclei at a certain level $a_1 < 0$, a Boltzmann-type equilibrium distribution close to the critical region of the potential barrier will not be achieved, and the line of the flux of new-phase fragments will bypass the saddle point.

An analytical solution to the diffusion problem with a run-off can be obtained only in the limiting cases of a strong and a weak run-off. For an approximate solution of Eq. (2.176), we will use the approach suggested in Refs. [82, 90]. The Gibbs thermodynamic potential at the initial point on the line given by Eq. (2.163) is $E_0(a_1)$. The position of the saddle point in the subspace $\{a'\}$ is shifted with respect to $\{a_*\}$ of the saddle point in the space $\{a\}$. If we assume that the saddle point of the subspace $\{a'\}$ is in the critical region where the expansion Eq. (2.88) is valid, then for the thermodynamic potential at the shifted saddle point we can write

$$E_+ = E_* + \frac{1}{2}\frac{\det \widehat{E}}{\det \widehat{E}_1} a_1^2, \tag{2.178}$$

where the matrix $\widehat{E}_1$ is the minor of the matrix $\widehat{E}$ obtained by deletion of the column and the line with numbers 1. Since at the saddle points the determinants of the matrices $\widehat{E}$ and $\widehat{E}_1$ are negative, the second component in Eq. (2.178) is positive.

According to Eq. (2.178), the activation energy for passing through the shifted saddle point in the subspace $\{a'\}$ is

$$W' = \Delta E'(a_1) = E_+ - E_0 = E_* + \frac{1}{2}\frac{\det \widehat{E}}{\det \widehat{E}_1}a_1^2 - E_0(a_1). \qquad (2.179)$$

The run-off term $j(a_1)$ may be calculated according to Eqs. (2.101) and (2.108) with the substitutions $C \to P_0(a_1)$, $\widehat{D} \to \widehat{D}_1(a_1)$, $\widehat{H} \to \widehat{E}_1(a_1)$, and $W_* \to W'$ as

$$j(a_1) = \left\{ P_0(a_1)\frac{\lambda_1}{2\pi}\left[\frac{|\det \widehat{E}_1(a_1)|}{2\pi k_B T}\right]^{1/2} \right\} \exp\left(-\frac{W'}{k_B T}\right), \qquad (2.180)$$

where λ_1 is the positive root of the equation

$$\det\left[\lambda_1 \widehat{I} + \widehat{D}_1(a_1)\widehat{E}_1(a_1)/k_B T + \lambda_1\right] = 0. \qquad (2.181)$$

Now moving from the function $P_0(a_1)$ to the function $P(a_1) = P_0(a_1)\omega^{-1/2}$, for the run-off term we get

$$j(a_1) = \gamma P(a_1), \qquad (2.182)$$

$$\gamma(a_1) = \frac{\lambda_1}{2\pi}\left[\frac{\det \widehat{q}(a_1)}{\det \widehat{E}_1(a_1)}\right]^{1/2} \exp\left(-\frac{W'}{k_B T}\right). \qquad (2.183)$$

According to Eqs. (2.172) and (2.182), Eq. (2.176) can be presented as

$$\frac{d}{da_1}D_{11}(a_1)\exp\left(-\frac{E_{\text{eff}}}{k_B T}\right)\frac{d}{da_1}\exp\left(\frac{E_{\text{eff}}}{k_B T}\right)P(a_1) = \gamma(a_1)P(a_1). \qquad (2.184)$$

The diffusion equation with a run-off describes the competition of two processes: diffusion relaxation leading to the establishment of an equilibrium distribution and superbarrier transitions in the motion in the subspace $\{a'\}$ resulting in the withdrawal of particles from the critical region of the saddle point.

Expression (2.179) is valid only in the critical region of the saddle point. However, even as we move away from the saddle point, the function $\Delta E'(a_1)$ retains its main property; it decreases in the interval from $a_1 = a_A$ to $a_1 = 0$. Therefore, the run-off term $j(a_1)$ "is turned on" at quite a great distance from the bottom of the well ($a_1 = a_A$). Until the run-off force has not reached a certain critical value, the distribution function $P(a_1)$ is close to equilibrium $P_{eq}(a_1)$. Then, the run-off is exponentially rapidly established. If we accept the indicated qualitative pattern, we may substitute the solution of the problem with a run-off by a solution neglecting it, but with an absorbing boundary condition at a certain point $a_1 = a_\alpha$. Then, instead of Eq. (2.184), we have to look for the solution of equation

$$\frac{d}{da_1} D_{11} P_{eq}(a_1) \frac{d}{da_1} \frac{P(a_1)}{P_{eq}(a_1)} = 0, \qquad P(a_\alpha) = 0. \tag{2.185}$$

Integrating Eq. (2.184) on the interval (a_0, a_α), we get

$$J_1(a_\alpha) - J_1(a_0) + \int_{a_0}^{a_\alpha} \gamma(a_1) P(a_1) da_1 = 0, \tag{2.186}$$

where, according to Eq. (2.170), $J_1(a_1)$ is

$$J_1(a_1) = -D_{11}(a_1) P_{eq}(a_1) \frac{d(P/P_{eq})}{da_1}. \tag{2.187}$$

If the point a_α is at the bottom of the well, then $J_1(a_0) \to J_+$, where J_+ is the nucleation rate sought. As a result, according to Eq. (2.186), we have

$$J_+ = J(a_\alpha) + \int_0^{a_\alpha} \gamma(a_1) P_{eq}(a_1) da_1. \tag{2.188}$$

Here, it is assumed that equilibrium is established up to the nearest vicinity of the point a_α where the absorbing boundary condition is imposed. The position of the point a_α at the hypersurface of the potential relief may be evaluated from the condition of equality of the rates of the processes of diffusion relaxation and superbarrier transitions [82]

$$D_{11} \left[\frac{E'_{\text{eff}}(a_\alpha)}{k_B T} \right] = \gamma(a_\alpha). \tag{2.189}$$

Equation (2.189) follows from Eqs. (2.184), (2.185), and (2.187) if we assume that at the point a_α the function P/P_{eq} changes more rapidly than P_{eq} and D_{11}. The run-off term $\gamma(a_\alpha)$ increases away from the bottom of the potential well. Therefore, the closer the point a_α to the bottom, the smaller the value of D_{11}. The integral in Eq. (2.188) can also be calculated for an arbitrary position of the point a_α, as long as it is located within the limits of the critical region of the saddle point. In such cases, we get [90]

$$J_+ = J \frac{[1 + \text{erf}(\Delta)]}{2} + J_1(a_\alpha), \tag{2.190}$$

$$\Delta = \left(\frac{\det \widehat{E}}{2k_B T \det \widehat{E}_1} \right)^{1/2} a_\alpha. \tag{2.191}$$

If $\Delta \gg 1$, then $\mathrm{erf}(\Delta) \approx 1$, $J_1(a_1) \to 0$ and $J_+ \simeq J$, i.e., Eq. (2.190) describes the ordinary nucleation regime through the saddle point. If $a_\alpha < 0$ and besides $|\Delta| \gg 1$, then

$$J_+ = J \frac{\exp(-\Delta^2)}{2\pi^{1/2}|\Delta|} + J_1(a_\alpha). \tag{2.192}$$

When condition (2.189) is used for finding the point a_α, the expression for the nucleation rate, Eq. (2.192), may be presented as

$$J_+ = J_1(a_\alpha) \left[1 + \frac{E'_{\mathrm{eff}}(a_\alpha) \det \widehat{E}_1}{|a_\alpha| \det \widehat{E}} \right]. \tag{2.193}$$

Hence it follows that as the point a_α departs from the saddle point, the contribution of the first term in Eq. (2.190) decreases. The one-dimensional growth of a new-phase nucleus proceeds up to the point a_α, whereupon the formation of a critical nucleus via the saddle point in the subspace $\{a'\}$ can be observed, i.e., with bypassing the main saddle point. In this case, $J_+ \approx J_1(a_\alpha)$ holds. Generally, the nucleation rate proves to be smaller than that one would calculate by the commonly employed formula, Eq. (2.108).

In the following, we examine the changes in the mechanism of growth of new-phase nuclei due to changes in the element of the matrix D_{11} of the generalized diffusion coefficient. As long as the value of D_{11} is comparable with the other elements of the matrix $\widehat{D}$, the growth proceeds through the saddle point. The nucleation rate is determined by the general expression (2.108), which includes the barrier parameters. As the element D_{11} decreases, a transition to the asymptotics of this expression can be observed which differs only by the determination of the eigenvalue λ. Instead of Eq. (2.101) we have the equation

$$\det \left(\frac{\widehat{D}\widehat{E}}{k_B T} + \lambda \widehat{I} \right)_1 = 0. \tag{2.194}$$

Here, the subscript 1 means that we take the minor of the matrix obtained by deleting the line and the column with numbers 1. The pre-exponential factor in Eq. (2.108) no longer depends on the element D_{11} of the matrix of the generalized diffusion coefficient. However, the classical nucleation theory remains valid as long as the condition $a_\alpha \gg 0$ is fulfilled. With a further decrease of D_{11}, when point a_α approaches the bottom of the well, a new growth mechanism develops characterized by a different dependence of the nucleation rate, Eq. (2.193), on the parameters of the problem under consideration, the absence of equilibrium for precritical aggregates, the bypassing of the saddle point, and a different composition of the critical nucleus. The process rate in this case is controlled by an element of the matrix D_{11} of the generalized diffusion coefficient, the slow variable a_1. Equation (2.190) ensures with the change of a_α a smooth transition from nucleation through the saddle point to nucleation bypassing it.

Apart from the value of D_{11}, the boundary of transition to the anomalous regime of nucleation depends on the height of the activation barrier (Eqs. (2.190), (2.193)). If the activation energy is high, the region of the anomalous regime may prove to be unattainable. At the same time, at low barriers ($\leq 10k_BT$) the transition to the regime with bypassing the saddle point proves to be less critical for the value of D_{11}.

2.12
Some Comments on Nucleation Theory

A solution to the problem of the theoretical description of the nucleation kinetics in the form of Eqs. (2.86) and (2.108) has been obtained for the stationary state. It is the distribution function of new-phase fragments in chosen variables $P(\{a\})$ that is stationary. A nonstationary solution to the multidimensional kinetic equation, Eq. (2.90), is considered in Refs. [77,78]. It is assumed that since the stable variables a_i ($i \neq 0$) are not connected with the process of overcoming the activation barrier, they are faster than the unstable variable, a_0. The distribution function of fragments in the unstable variable is written then as [78]

$$P(a_0, \tau) = P(a_0) + \sum_{j=1} H_j(a_0) \exp(-j\lambda_0 \tau), \quad (2.195)$$

where the coefficients $H_j(a_0)$ are expressed in terms of the Hermite polynomials. The stationary distribution is represented in Eq. (2.195) by the first term. The value of $j\lambda_0$ ($j = 1, 2, \ldots$) in the exponent of the exponential function gives the spectrum of inverse times for the establishment of the stationary distribution, in which the dominating term is that with $j = 1$. In a first approximation, we have

$$P(a_0, \tau) = P(a_0) + [P_{\text{eq}}(a_0) - P(a_0)] \exp(-\lambda_0 \tau). \quad (2.196)$$

After the time $\tau_l \simeq \lambda_0^{-1}$, supercritical fragments, irrespective of their initial distribution, begin to form at a regular rate, J.

The nonstationary evolution of a flux of new-phase fragments in the critical region of the potential barrier is presented by Zeldovich [5] as

$$J'(\tau) = J \exp\left[-\frac{e_0 a_+^2}{4 k_B T \lambda_0 \tau}\right], \qquad (2.197)$$

where a_+ is the value of an unstable variable at the lower boundary of the critical region. The time of relaxation, τ_l, of the flux, J', to the stationary value of J may be evaluated as the time of diffusive motion of fragments across the critical region

$$\tau_l \simeq \frac{e_0 a_+^2}{4 k_B T \lambda_0} = \frac{1}{2\lambda_0}. \qquad (2.198)$$

In the order of magnitude, this estimate of the period of nonstationarity coincides with other, more rigorous approaches [91–95].

An important task in homogeneous nucleation theory is the determination of the equilibrium distribution function of fragments with respect to the unstable variable. The presence of a stationary flux of nuclei deforms the distribution function $P_{eq}(a_0)$ at $a_0 < 0$ only slightly. The factor C_0 which appears in the final result of the stationary nucleation theory (2.86) is the normalization constant of the equilibrium distribution function in the unstable variable. This factor cannot be expressed exclusively in terms of macroscopic characteristics [61]. A correct solution to the problem of determination of the distribution function, $P_{eq}(a_0)$, presupposes the consideration of the Gibbs grand canonical ensemble [79, 96–98]. In the phase space of a system it is necessary to distinguish regions pertaining to the microheterogeneous ("droplet") states of a substance. It is furthermore necessary to sum up the distribution function of the ensemble with respect to all states leaving only the distinguished fragment variables free. However, such a procedure cannot be realized up to a certain completion. Therefore, the problem is solved employing a number of assumptions. For a qualitative evaluation of C_0 it may be assumed that this factor is proportional to the density, ρ, of the number of particles in the metastable phase and the derivative dN/da_0, where N is the number of particles in a new-phase nucleus.

The classical nucleation theory does not take into account the effects of the possible memory of the medium. Since the mass of new-phase fragments considerably exceeds the mass of a molecule, correlation times in the subsystem of nuclei are much larger than those characterizing the adjustment of the medium. As shown in Ref. [99], the non-Markovian dynamics can always be presented as a Markovian process in a system with a large number of variables.

The homogeneous nucleation theory presupposes the presence of a single source of heterophase fluctuations in the metastable system, which is the chaotic thermal motion. Such a mechanism of formation of viable new-phase centers is known as homogeneous nucleation. Purely homogeneous nucleation may be observed only in homogeneous systems which are not exposed to initiating external actions and do not contain any foreign inclusions. In practice, there are always different impurities, which may initiate the appearance of new-phase nuclei. Such "impurities" in superheated liquids are dissolved gases, gas-saturated solid particles, and cracks in vessel walls, places with reduced wettability, etc. Nucleation on foreign inclusions is called heterogeneous. Since it is impossible to create ideally pure conditions in an experiment, there are always some doubts as to whether the homogeneous mechanism of nucleation is realized.

For constructing a theory of heterogeneous nucleation, an approach similar to that used in the consideration of homogeneous nucleation can be employed. The main difficulty here is connected with the lack of data on the activity of foreign centers and their distribution in the volume of the metastable phase. When centers of fluctuational growth are precritical new-phase inclusions, for instance electronic bubbles in a quantum liquid, the stationary rate of heterogeneous nucleation can be written as [100, 101]

$$J_{het} \simeq BP_{het}(a_0) \simeq C'B \exp\left[-\frac{(W_* - W_0)}{k_B T}\right], \qquad (2.199)$$

where $P_{het}(a_0)$ is the distribution function of inclusions in the variable a_0 and W_0 is equal to the minimum work of formation of an inclusion.

The exponential factor in Eq. (2.199) is smaller than in Eq. (2.86), and therefore heterogeneous nucleation manifests itself earlier than homogeneous nucleation. Since $\rho_{het} \ll \rho$ and the value of the kinetic factor at heterogeneous nucleation is limited from above by the value of B for the homogeneous nucleation mechanism (in Eq. (2.199), B_{het} is taken equal to B), $\rho_{het} B_{het} \ll \rho B$ and the rate of heterogeneous nucleation is a weaker temperature function than that of homogeneous nucleation. At a certain superheating, $\Delta T = \Delta T_+$, the rate J_{het} is equal to J, and for $\Delta T > \Delta T_+$ homogeneous nucleation will be dominant even with a comparatively small number of boiling centers in the system. The great steepness of the temperature dependence of J as compared to J_{het} makes it possible to experimentally distinguish the effects of heterogeneous and homogeneous nucleation.

3
Attainable Superheating of One-Component Liquids

3.1
Two Approaches to the Determination of the Work of Formation of a Critical Bubble

The minimum work, W, required in order to create a fragment of a new phase of given size in a metastable system depends on the mechanism and the conditions with regard to which the process takes place. Let us consider the process of homogeneous formation of a vapor phase in a superheated liquid at constant external pressure, p', and temperature, T. In a one-component system, the constancy of p' and T implies also constancy of the chemical potential, μ', of the liquid. The difference in the values of the Gibbs thermodynamic potential for the initial state, which consists of a homogeneous metastable liquid, and the final state, which includes a vapor bubble of radius, R, and a metastable liquid, determines the work of formation of a certain amount of the vapor phase via

$$W = \Delta\Phi = (p' - p'')V + \sigma A + (\mu'' - \mu')N''. \tag{3.1}$$

Here, σ is the surface tension, N'' is the number of molecules, $V = \frac{4}{3}\pi R^3$, is the volume and $A = 4\pi R^2$ is the surface area of the bubble. All of these quantities will be determined employing the surface of tension as the dividing surface. One prime in Eq. (3.1) indicates that the respective parameter describes the state of the liquid, whereas two primes specify the intensive state parameters of the vapor.

The internal equilibrium of a bubble as an isolated complex is characterized by [76]

$$A d\sigma + S'' dT - V dp'' + N'' d\mu'' = 0. \tag{3.2}$$

Differentiation of Eq. (3.1) taking into account Eqs. (3.2) employing the conditions of constancy of T, p', μ' and the relation $N = N' + N''$ yields

$$d\Delta\Phi = \left[(p' - p'') + \sigma \frac{dA}{dV}\right] dV + (\mu'' - \mu')dN''. \tag{3.3}$$

Explosive Boiling of Superheated Cryogenic Liquids. Vladimir G. Baidakov
Copyright © 2007 WILEY-VCH Verlag GmbH & Co. KGaA, Weinheim
ISBN: 978-3-527-40575-6

In a thermodynamic equilibrium state, the function $\Delta\Phi(V, N'')$ has an extremum, i.e., $(d\Delta\Phi)_* = 0$, and from Eq. (3.3) the conditions are obtained for mechanical

$$p''_* - p' = \sigma \left.\frac{dA}{dV}\right|_* = \frac{2\sigma}{R_*} \tag{3.4}$$

and diffusion

$$\mu''(p''_*, T) = \mu'(p', T) \tag{3.5}$$

equilibrium of a vapor bubble in a superheated liquid.

At the extremum of $\Delta\Phi$, we obtain with Eqs. (3.4) and (3.5) the following expression:

$$(d^2\Delta\Phi)_* = \sigma \left.\frac{d^2 A}{dV^2}\right|_* (dV)^2 + d\sigma dA - dp'' dV + d\mu'' dN''. \tag{3.6}$$

Assuming that the pressure at any time is connected with the number of particles in a bubble by the equation of state of the ideal gas, and the surface tension at the vapor bubble/liquid interface is a function only of temperature, Eq. (3.6) can be presented as

$$\begin{aligned}(d^2\Delta\Phi)_* &= -\left(\frac{2\sigma A_*}{9V_*^2} + \frac{N''_* k_B T}{V_*^2}\right)(dV)^2 - 2\frac{k_B T}{V_*}dVdN'' + \frac{k_B T}{N''}(dN''_*)^2 \\ &= -\frac{2\sigma A_*}{9V_*^2}(dV)^2 + \frac{V_*}{p''_*}(dp'')^2. \end{aligned} \tag{3.7}$$

It follows from Eq. (3.7) that in variables V and p'', as distinct from the set of variables V and N'', the quadratic form $d^2\Delta\Phi$ contains no cross terms. The different signs of the coefficients in the squares of the differentials dV (or dR) and dp'' indicate that the surface of the Gibbs thermodynamic potential in the vicinity of the extremum point is a hyperbolic paraboloid, and the extremum point itself is a saddle point (Fig. 3.1). The line along the ridge represents a water shed between the region of a relative (metastable) minimum and the region containing the absolute (stable) minimum of the thermodynamic potential. The line tangent to the water-shed line at the saddle point coincides with the pressure axis and passes the thermodynamic potential surface at a height lower than the water-shed line, i.e., the variable $(p'' - p''_*)$ is stable. The line along the valley going through the pass coincides with the axis of the bubble volume. The equilibrium of a bubble within a metastable liquid with respect to the variable $(V - V^*)$ is unstable. The saddle point is the point of intersection of the lines of the water shed and the path along the valley, and also of the lines of mechanical (Eq. (3.4)) and diffusion (Eq. (3.5)) equilibrium of a vapor bubble. The bubble in an unstable equilibrium, corresponding to the saddle point, is the critical nucleus in the considered process of boiling.

3.1 Two Approaches to the Determination of the Work of Formation of a Critical Bubble

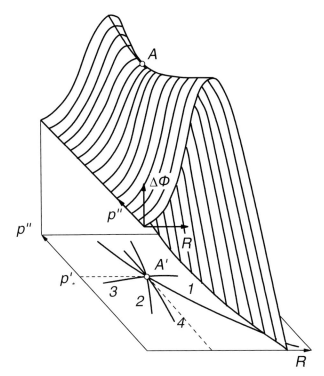

Fig. 3.1 The surface of the Gibbs thermodynamic potential and its projection onto the plane (R, p''). (1, 2): lines of mechanical and diffusion equilibrium of a bubble; (3, 4): lines of spillway and water shed of the saddle surface; A: saddle point.

Substitution of Eqs. (3.4) and (3.5) into Eq. (3.1) results in the following expression for the work of formation of a critical nucleus [1]:

$$W_* = \min \max \Delta \Phi = \frac{1}{3}\sigma A_* = \frac{1}{2}(p''_* - p')V_* = \frac{16\pi \sigma^3}{3(p''_* - p')^2}. \tag{3.8}$$

In the range of weak metastability, $(G_* = W_*/k_B T \gg 1)$, a critical nucleus contains a large number of molecules, and the pressure and density in the critical nucleus are close to the equilibrium values of the state parameters of the co-existing macroscopic phases. The surface tension of the critical bubble/liquid interface may be taken in a first approximation as equal to its value at a planar interface. Such assumption is commonly denoted as the macroscopic or capillarity approximation. The difference of pressures $(p'' - p')$ at small degrees of superheating can be approximated with a good accuracy via [9]

$$p''_* - p' = (p_s - p')\left(1 - \frac{\rho''_s}{\rho'_s}\right), \tag{3.9}$$

where p_s is the saturation pressure at a planar interface, and ρ'_s and ρ''_s are the orthobaric densities. All the quantities in Eq. (3.8) can be experimentally determined in this case.

Penetrating deeper into the metastable region, the radius of the surface of tension, R_*, of the critical nucleus decreases, and the compressibility of the liquid and the correlation radius, ξ, increase. Since the correlation length, ξ, is directly proportional to the thickness of the interface, at a certain supersaturation the appearance of totally inhomogeneous nuclei can be expected. Some authors [56, 102] suppose that this case determines the limit of applicability of the Gibbs method. In discussing this problem, Gibbs writes [1]: "This, however, will cause no difficulties if we regard the inner-mass phase as the one determined by the same relations" (Eqs. (3.4), (3.5) in the present book, V.B.) "connecting it with the outer mass as in all other cases." And further, in explaining his thoughts, in a note on this paragraph, he continues: "When applying our formulae to a microscopic ball of water in vapor, we should regard the density and the pressure of the inner mass not as the actual density or pressure at the center of the ball, but as the density of liquid water, which has the temperature and the potential of vapor." With such an approach to the determination of the nucleus properties, the thermodynamic method will be valid down to values of the work of critical cluster formation equal to $W_* = 0$.

The value of the surface tension will be, in general, not equal to the respective value for an equilibrium coexistence of both phases at planar interfaces. The conclusion about the dependence of the surface tension of a critical nucleus on the radius of curvature of the separating surface follows logically from Gibbs' theory of capillarity. Since the value of W_* does not depend on the choice of the separating surface, and, as distinct from σ and R_*, may be determined beyond the Gibbs method, Eqs. (3.4) and (3.8) may be regarded as equations determining the function $\sigma(R_*)$ [57, 103].

An approach to compute the work of formation of a critical nucleus different from that of Gibbs was suggested by Cahn and Hilliard [58]. It does not require the introduction of a parameter of the nucleus such as surface tension. The Cahn–Hilliard method is based on van der Waals' theory of capillarity [7]. In the van der Waals approach, the Helmholtz excess free energy, $\Delta F[\rho]$, of an inhomogeneous system is expressed as a functional of the local density, $\rho(\vec{r})$, and is written in the form of Eq. (2.36). For an incompressible liquid of infinite size, the density distribution in a critical bubble is found by solving Eq. (2.39) with the boundary conditions, Eq. (2.40). The function $\rho_*(\vec{r})$ corresponds to the saddle point of the potential barrier, $\Delta F[\rho]$, separating in the configurational space the state of metastable equilibrium from an absolutely stable state,

$$W_* = \min \max \Delta F[\rho]. \tag{3.10}$$

Excluding the term $\kappa(\nabla\rho)^2$ with the help of Eq. (2.39) from Eq. (2.36), we have

$$W_* = 4\pi \int_0^\infty \left\{\Delta f(\rho) - \frac{1}{2}(\rho - \rho_0)[\mu(\rho) - \mu_0]\right\} r^2 dr. \tag{3.11}$$

In the vicinity of the line of phase equilibrium, the effective radius of the nucleus is large compared with the thickness of the transition layer, and integration of Eq. (2.36) yields the Gibbs expression for the work of formation of a nucleus, i.e.,

$$W = \Delta F \simeq 4\pi R^2 \sigma + \frac{4}{3}\pi R^3 (p_0 - p_s), \tag{3.12}$$

where

$$4\pi R^2 \sigma \simeq 4\pi \kappa \int_0^\infty (\nabla \rho)^2 r^2 dr, \tag{3.13}$$

$$p_0 - p_s \simeq \Delta f(\rho', \rho''). \tag{3.14}$$

From Eqs. (3.11) and (3.12), we obtain the formula for the work of formation of a critical nucleus, Eq. (3.8).

The Cahn–Hilliard approach requires the knowledge of the coefficient, κ, and the density of free energy of a homogeneous system, Δf, in the metastable and unstable regions. Such information may be obtained on the basis of experimental data on the first- and the second-order derivatives of the thermodynamic potential and the surface tension. Another way of finding the unknown parameters of the functional of the Helmholtz free energy is connected with the application of the statistical theory of the liquid state.

3.2
Boiling-Up Kinetics of Superheated Liquids

The problem of boiling-up of a superheated one-component liquid is a multiparameter problem of the kinetics of first-order phase transitions. Even in an isothermal approximation, the thermodynamic state of a vapor-phase nucleus is characterized by, at least, two macroscopic variables (see Eq. (3.7)). Taking into account the nonisothermality of the nucleation process, the inertial forces, the nonspherical shape of a bubble, and some other factors, the number of parameters determining the state of the bubble increases.

The stationary nucleation process in a superheated one-component liquid was first examined by Döring [104]. It was assumed by him that the growth of vapor bubbles is exclusively determined by the rates of evaporation and condensation of molecules. The condition of mechanical equilibrium, Eq. (3.4), was considered to be fulfilled for the whole ensemble of precritical bubbles. A boiling-up liquid was assumed to be ideal and inertialess, and a vapor bubble was characterized by only one variable, the number of molecules in the bubble. Employing the above-mentioned approximations, an expression for the nucleation rate was obtained. In the literature, it is known as the Döring–Volmer formula [104–107]. The kinetic factor of the Döring–Volmer formula takes the form (see Eq. (2.112))

$$B = B_1 = \left[\frac{6\sigma}{\pi m(3-b)}\right]^{1/2}, \qquad (3.15)$$

and the factor that corrects the normalization of the equilibrium distribution function can be written as

$$z_0 = \exp\left(-\frac{l}{k_B T}\right). \qquad (3.16)$$

In Eqs. (3.15) and (3.16), m is the molecular mass, $b = 1 - p'/p''_* = 2\sigma/(p''_* R_*)$, and l is the heat of evaporation per molecule.

Stranski [106, 107] supposed the introduction of the factor z_0 of Eq. (3.16) into the function of bubble size distribution by Volmer [105] to be incorrect and suggests to take the value $z_0 = 1$ instead. Zeldovich [5] examined cavitation in a viscous liquid deliberately excluding processes of evaporation of molecules into a bubble and heat relaxation from consideration. As distinct from Döring, who regarded the evolution of bubbles that originated by means of fluctuation as a sequence of condensation and evaporation acts of single molecules, Zeldovich, making use of the theory [83], presented the fluctuational growth of precritical bubbles as a one-dimensional diffusion process in the space of their sizes.

The most rigorous and complete solution to the problem of boiling-up of a pure liquid was obtained by Kagan [108] who took into account all main factors determining the bubble growth: viscous and inertial forces, the rate of evaporation of molecules into the cavity and the rate of heat supply to the bubble. As was supposed by Döring [104] and Zeldovich [5], the state of a bubble was described by one variable, its characteristic size. Deryagin, Prokhorov, and Tunitsky [79] analyzed the growth of precritical bubbles in a superheated liquid as a process of their diffusion in the space of two variables, the volume, V, and the pressure, p'', of the vapor in a bubble. The problem was solved without taking into account the effect of the inertial forces and heat relaxation at the bubble–liquid interface.

Let us examine the boiling-up kinetics of a one-component liquid employing the following approximations: (i) Moderate stretching of the liquid or degrees of superheating which result in a sufficiently large nucleation barrier ($W_* \gg k_B T$). (ii) Near-critical bubbles are spherically symmetric; we neglect inhomogeneities of pressure and temperature in them. (iii) The phase surrounding a bubble is a viscous, volatile, and incompressible liquid. (iv) The bubble itself consists of a gas which is described as an ideal gas.

As the thermodynamic variables for the description of a nucleus, we chose its volume, V, and vapor pressure, p'', of the gas in the bubble. These variables are the most convenient parameters as in them (or, more precisely, in the deviations of these variables from their values for a critical nucleus) the second differential of the work of formation of a near-critical bubble has the form of the sum of squares (see Eq. (3.7)), i.e., it does not contain any cross terms. Variables such as the number of molecules in a bubble and the deviation of its pressure from the value corresponding to the conditions of mechanical equilibrium [109] are also capable of bringing the work of bubble formation to the form of the sum of squares.

The growth of a bubble leads to a pressure increase in the vicinity of its wall. This additional pressure is connected with the forces of inertia, and the work performed by the vapor against this pressure is spent for supplying kinetic energy to the liquid. Taking into account inertia effects in the kinetics of boiling-up of a superheated liquid requires inclusion among the determining variables of a nucleus a dynamic parameter, the rate of change of the bubble volume, $\dot{V} = (dV/d\tau)$.

In accordance with the general approach to multiparameter nucleation (see Section 3.8), the problem of determining the stationary rate of formation of aggregates of a new phase consists of two tasks: finding a potential hypersurface in the phase space of the parameters of a vapor bubble and determining the generalized diffusion tensor of the bubble in this space. In order to proceed in this direction, we introduce the dimensionless variables

$$x = \frac{V - V_*}{V_*}, \quad y = \frac{p - p''_*}{p''_*}, \quad v = \frac{1}{V_*}\frac{dV}{dt}, \quad t = \frac{\tau}{\tau_r}, \tag{3.17}$$

where the relaxation time is chosen as the characteristic time, τ_r. It is the time required to establish the equilibrium bubble growth-rate distribution. The dimensionless potential relief in the space of the variables (x, y, v) will be written as [83]

$$\Psi(x, y, v) = \frac{\Delta\Phi(x, y)}{k_B T} + \frac{M_{\text{eff}}}{2k_B T \tau_r^2} v^2. \tag{3.18}$$

Here, $\Delta\Phi(x,y)/k_B T$ is the dimensionless potential of the external field, Eq. (3.1), corresponding to the minimum work of formation of a bubble of volume, x, with the vapor pressure, y, at a temperature, T, of the liquid, the second term is the kinetic energy of a growing bubble, M_{eff} is the effective mass of a bubble in the variables (x, y, v). The potential relief, $\Psi(x, y, v)$, has the form of Eq. (3.18) at times $\tau \gg \tau_r$ when the equilibrium bubble growth-rate distribution is established.

Expanding $\Psi(x, y, v)$ into a series in powers of (x, y, v) in the vicinity of the origin (0, 0, 0), which corresponds to the state of an unstable equilibrium in a "bubble + liquid" system, and restricting ourselves to terms up to second-order, we have

$$\Psi(x, y, v) = \Psi_* + \frac{1}{2}(\vec{r}\widehat{\Psi}\vec{r}) = G_*\left(1 - \frac{1}{3}x^2 + \frac{1}{b}y^2 + \frac{1}{3\chi}v^2\right), \quad (3.19)$$

where

$$\chi = \frac{2k_B T G_* \tau_r^2}{3 M_{\text{eff}}} \quad (3.20)$$

is a dimensionless parameter characterizing the contribution of the internal forces, G_* is the dimensionless work of formation of a critical nucleus (Gibbs number), $\vec{r}(x, y, v)$ is the radius vector of an arbitrary point drawn from the origin of the system of coordinates (0, 0, 0), and

$$\widehat{\Psi} = 2G_* \begin{pmatrix} -\frac{1}{3} & 0 & 0 \\ 0 & \frac{1}{b} & 0 \\ 0 & 0 & \frac{1}{3\chi} \end{pmatrix} \quad (3.21)$$

is the matrix of second-order derivatives of $\Psi(x, y, v)$ at the saddle point of the potential surface. The first derivatives of $\Psi(x, y, v)$ are equal to zero due to the conditions of mechanical and diffusion equilibrium of a critical bubble, with a zero rate of its growth at the saddle point.

The problem of determining the generalized diffusion tensor is reduced to finding the forces acting on a nucleus and the rates of change of its parameters under the action of these forces. The bubble radius (the bubble center is at rest) is determined at any moment of time by

$$\rho_l R \ddot{R} + \frac{3}{2}\rho_l \dot{R}^2 = p'' - p' - \frac{2\sigma}{R} - 4\eta \frac{\dot{R}}{R}, \quad (3.22)$$

where ρ_l is the density of the liquid mass and η is the shear viscosity. The rate of change of the number of molecules in a bubble is given by

$$\dot{N}'' = \frac{\pi \alpha v_t R^2}{k_B T}(p''_R - p''), \quad (3.23)$$

where $v_t = (8k_B T / \pi m)^{1/2}$ is the average velocity of thermal motion of vapor molecules, α is the condensation coefficient, and p''_R is the pressure of saturated vapor in a bubble of radius R.

With regard to bubble growth, a significant role is played by the absorption of heat caused by the evaporation of molecules into a cavity and the resulting temperature decrease at the interface (the opposite situation occurs with respect to bubble collapse). Accounting for this process requires the inclusion of temperature into the set of variables determining the state of a nucleus, and the system of equations, Eqs. (3.22) and (3.23), has to be supplemented by the nonstationary heat conduction equation. The characteristic time of bubble growth is proportional to $(R/\dot{R})$. The time of heat relaxation has a value of the order of (R^2/D_T), where $D_T = \Lambda/\rho c_p$ is the thermal diffusivity. In the vicinity of the saddle point of the potential barrier, the size of the bubble, R, is a small quantity, therefore $(R^2/D_T) \ll (R/\dot{R})$ holds, and in the heat conduction equation it is possible to omit the derivative of temperature with respect to time, and also the term containing the temperature gradient. Thus, the heat conduction equation is reduced to

$$\frac{d^2 T}{dr^2} = 0. \tag{3.24}$$

The solution of this equation with the appropriate boundary conditions is

$$\Delta T = \frac{1}{4\pi\Lambda} \frac{\dot{N}''}{R}, \tag{3.25}$$

where ΔT is the temperature decrease at the bubble boundary.

With respect to the quantities appearing in Eqs. (3.22) and (3.23), only the value of p_R'' will be affected by a relatively small change of temperature. This feature makes it possible not to include the temperature into the variables determining the state of a nucleus, but to account for the thermal effect through $p_R''(T)$ as has been done by Kagan [108], i.e., by assuming the temperature to be constant in all parameters of the problem except with respect to pressure $p_R''(T)$, for which

$$p_R''(T - \Delta T) = p_R''(T) - d\Delta T \tag{3.26}$$

holds. The value of d in Eq. (3.26) may be determined in a first approximation by the well-known temperature dependence of the pressure of saturated vapors over a planar liquid–vapor interface. Using Eqs. (3.25) and (3.26) in Eq. (3.23), we have

$$\dot{N}'' = \frac{\pi \alpha v_t R^2}{k_B T(1+\delta)} \left[p_R''(T) - p'' \right], \tag{3.27}$$

where

$$\delta = \frac{\alpha v_t l R d}{4 \Lambda k_B T}. \tag{3.28}$$

A linerization of Eqs. (3.22) and (3.27) and transformation of the variable from N'' to p'' yield

$$\ddot{R} = \frac{2v}{\rho_l R_*^3}\Delta R + \frac{1}{\rho_l R_*}\Delta p'' - \frac{4\eta}{\rho_l R_*^2}\dot{R}, \tag{3.29}$$

$$\dot{p}'' = \frac{3\alpha v_t}{4 R_*(1+\delta)}\Delta p'' - \frac{p''}{R_*}\dot{R}. \tag{3.30}$$

Let us go over further in Eq. (3.29) to dimensionless quantities (defined via Eq. (3.17)). We then obtain

$$\dot{v} = \frac{2\sigma \tau_r^2}{\rho_l R_*^3}x + \frac{3 p_*'' \tau_r^2}{\rho_l R_*^2}y - \frac{4\eta \tau_r}{\rho_l R_*^2}v. \tag{3.31}$$

To find the effective mass of a bubble, M_{eff}, and the value of τ_r, we bring Eq. (3.31) into the form of the equation of motion in a medium with a friction coefficient, ζ,

$$\frac{M_{\text{eff}}}{\tau_r^2}\dot{v} = -\frac{\partial \Delta \Phi}{\partial x} - \frac{\partial \Delta \Phi}{\partial y} - \frac{M_{\text{eff}}}{\tau_r}gv \tag{3.32}$$

in the field of forces

$$-\frac{\partial \Delta \Phi}{\partial x} = \frac{2}{3}W_* x, \qquad -\frac{\partial \Delta \Phi}{\partial y} = -p_*'' V_* y = -\frac{2}{b}W_* y. \tag{3.33}$$

Substitution of Eq. (3.33) into Eq. (3.32) gives

$$\dot{v} = \frac{2}{3}\frac{G_* k_B T \tau_r^2}{M_{\text{eff}}}x - \frac{2}{b}\frac{G_* k_B T \tau_r^2}{M_{\text{eff}}}y - g\tau_r v. \tag{3.34}$$

In the absence of external forces, the first integral of Eq. (3.34) is

$$v = v_0 \exp(-g\tau_r t), \tag{3.35}$$

where v_0 is the initial rate of change of the dimensionless volume of a bubble, and t is the dimensionless time, the reference parameter which is the time of relaxation. Thus, from Eq. (3.35) taking into account Eq. (3.31) it follows that

$$\tau_r = \frac{1}{g} = \frac{\rho_l R_*^2}{4\eta}. \tag{3.36}$$

Comparing Eqs. (3.31) and (3.34), for M_{eff} we arrive at the result

$$M_{\text{eff}} = \frac{4}{9}\pi \rho_l R_*^5, \tag{3.37}$$

and substitution of Eqs. (3.36) and (3.37) into Eq. (3.20) yields

$$\chi = \frac{\rho_l \sigma R_*}{8\eta^2}. \tag{3.38}$$

We can rewrite Eq. (3.32) as

$$\frac{M_{\text{eff}}}{\tau_r}\left(\frac{1}{\tau_r}\frac{dv}{dx}+g\right)v = -\frac{\partial\Delta\Phi}{\partial x} - \frac{\partial\Delta\Phi}{\partial y}. \tag{3.39}$$

The inertia term in Eq. (3.39) may be neglected if

$$\frac{1}{\tau_r}\frac{dv}{dx} \ll g \tag{3.40}$$

holds or in the notation employed,

$$\chi \ll 1. \tag{3.41}$$

After going over to dimensionless variables in Eqs. (3.29) and (3.30), the system of equations of bubble motion in the vicinity of the saddle point of the pass takes the form

$$\dot{x} = v, \tag{3.42}$$

$$\dot{y} = -\chi\omega y - v, \tag{3.43}$$

$$\dot{v} = \chi x + \frac{3\chi}{b}y - v, \tag{3.44}$$

where

$$\omega = \frac{3}{2}\frac{\alpha v_t \eta}{\sigma(1+\delta)}. \tag{3.45}$$

The matrix of the generalized diffusion tensor can be obtained from the system of equations (2.102), which in this case can be written as

$$D_{xx}F_x + D_{xy}F_y + D_{xv}F_v = \dot{x}, \tag{3.46}$$

$$D_{yx}F_x + D_{yy}F_y + D_{yv}F_v = \dot{y}, \tag{3.47}$$

$$D_{vx}F_x + D_{vy}F_y + D_{vv}F_v = \dot{v}, \tag{3.48}$$

where (F_x, F_y, F_v) are the forces directed parallel to the axes (x, y, v),

$$F_x = -\frac{\partial\Psi}{\partial x} = \frac{2}{3}G_*x, \quad F_y = -\frac{\partial\Psi}{\partial y} = -\frac{2}{b}G_*y, \quad F_v = -\frac{\partial\Psi}{\partial v} = -\frac{2}{3}\frac{G_*}{\chi}v \tag{3.49}$$

and the rates of change of the variables (x, y, v) are determined by the equations of motion, Eqs. (3.42)–(3.44). Setting equal the coefficients at (x, y, v) in Eqs. (3.42)–(3.44) and Eqs. (3.46)–(3.48), we get

$$\hat{D} = \frac{3\chi}{2G_*}\begin{pmatrix} 0 & 0 & -1 \\ 0 & \frac{1}{3}b\omega & 1 \\ 1 & -1 & 1 \end{pmatrix}. \tag{3.50}$$

The absence of symmetry of the matrix of the generalized diffusion coefficients shows that the system under investigation is not dissipative [110]. This property is connected with the inclusion of dynamic variables into the set of parameters describing the state of the nucleus.

The increment of the increase of the nucleus volume, $\tilde{\lambda}_0 = \dot{x}/dx$, is determined by the solutions of the characteristic equation, Eq. (2.101). Substitution of Eqs. (3.21) and (3.50) into Eq. (2.101) yields the cubic equation

$$\tilde{\lambda}^3 + (1 + \omega\chi)\tilde{\lambda}^2 + \chi\left(\omega + \frac{3-b}{b}\right)\tilde{\lambda} - \chi^2\omega = 0 \tag{3.51}$$

with one positive root equal to $\tilde{\lambda}_0 = \lambda_0 \tau_r$.

The kinetic factor in the expression for the stationary nucleation rate is related to the increment λ_0 by Eq. (2.112), where $z_1 = dN''/dV|_* = \rho''_*$ is the density of the vapor phase in a critical bubble. After a transition to dimensionless quantities we have

$$B = \rho''_* \lambda_0 R_*^2 \left(\frac{k_B T}{\sigma}\right)^{1/2}. \tag{3.52}$$

Equation (3.52), when substituting into it the solution of Eq. (3.51) ($B \equiv B_2$), determines the stationary flow of nuclei in a superheated or stretched liquid in the whole range of values of volatility, viscosity, inertia effects taking into account the effect of heat relaxation at the boundary of a growing bubble.

Next we examine different limiting cases of nucleation. According to Eq. (3.41) the effect of inertial forces may be neglected if the viscosity of the liquid and the value of supersaturation are high, and the value of the surface tension is low. In this case, the cubic equation (3.51) is reduced to the quadratic equation

$$\tilde{\lambda}^2 + \chi\left(\omega + \frac{3-b}{b}\right)\tilde{\lambda} - \chi^2\omega = 0. \tag{3.53}$$

Substituting the solution of Eq. (3.53) into Eq. (3.52) yields $B = B_{20}$ or

$$B = \frac{\rho''_* \sigma R_*}{4\eta}\left(\frac{k_B T}{\sigma}\right)^{1/2}\left\{-\left(\omega + \frac{3-b}{b}\right)\right.$$

$$\left. + \left[\left(\omega + \frac{3-b}{b}\right)^2 + 4\omega\right]^{1/2}\right\}. \tag{3.54}$$

At $\omega \gg 1$ and $\omega \gg 3/b$, the limiting factor of bubble growth in the absence of inertia effects is the liquid viscosity, and Eq. (3.54) is transformed into the limiting expression

$$B \equiv B_{21} = \frac{\sigma}{b\eta}\left(\frac{k_B T}{\sigma}\right)^{-1/2}. \tag{3.55}$$

3.2 Boiling-Up Kinetics of Superheated Liquids

The use of Eq. (3.55) in Eq. (2.112) and the choice of the normalization factor in the form of $z_0 = 1$ yield an expression for the nucleation rate which coincides with Eq. (34) derived by Kagan [108]. The corresponding expression to (34) proposed by Deryagin, Prokhorov, and Tunitsky [79] differs from that presented here by the factor $z_0 = \rho'/\rho''_*$.

At $\eta \to 0$, we obtain the case of boiling-up of a nonviscous volatile liquid. We shall restrict ourselves to the region of states where the viscosity is low ($\omega \ll 3/b - 1$), but not to an extent that inertia effects in the liquid become important, i.e., $\chi \ll 1$. If $b \gg 3$, which corresponds to a stretching pressure $-p' \gg 2p''_*$, Eq. (3.54) is reduced then to

$$B \equiv B_{22} = \frac{\rho''_* \sigma R_*(b-3)}{2\eta b} \left(\frac{k_B T}{\sigma}\right)^{1/2}, \qquad (3.56)$$

which, in the case of neglecting the term $3/b$, is transformed into Eq. (3.55). At $b \ll 3$, which corresponds to positive and limited negative values of p', Eq. (3.54) leads to

$$B \equiv B_{23} = \frac{\rho''_* R_* \alpha v_t b}{4(1+\delta)} \left(\frac{k_B T}{\sigma}\right)^{1/2}. \qquad (3.57)$$

The nucleation rate calculated with $B = B_{23}$ and the factor $z_0 = 1$, correcting the equilibrium distribution function, coincides with Kagan's formula (36) [108] and neglecting temperature effects at the bubble boundary ($\delta = 0$) differs from equation (39) proposed by Deryagin, Prokhorov, and Tunitsky [79] by the factor $z_* = \rho'/\rho''_*$.

If we assume that $\alpha = 1$ and $\delta = 0$, Eq. (3.57) yields

$$B \equiv B_{10} = \left(\frac{2\sigma}{\pi m}\right)^{1/2}. \qquad (3.58)$$

This equation coincides with the formula of Döring–Volmer, Eq. (3.15) ($b \ll 3$). At $(3-b)/b \gg 1$ and $\delta \gg 1$, the determining factor of bubble growth is heat supply and

$$B \equiv B_{24} = \frac{2\Lambda k_B T}{l d R_*} \left(\frac{\sigma}{k_B T}\right)^{1/2} \qquad (3.59)$$

holds.

Let us now examine the boiling-up of a nonvolatile liquid in the presence of inertial forces. In the case under consideration, we have $b \to \infty$, $\omega \to 0$ and Eq. (3.51) is written as follows:

$$\tilde{\lambda}^2 - \tilde{\lambda} - \chi = 0. \qquad (3.60)$$

The positive root of this equation is given by

$$\tilde{\lambda}_0 = \frac{1}{2} + \sqrt{\frac{1}{4} + \chi}. \tag{3.61}$$

At high viscosities ($\chi \ll 1$), it follows from Eq. (3.61) that $\tilde{\lambda}_0 = \chi$. The expression for the kinetic factor $B \equiv B_{31}$ in this case coincides with that obtained earlier (Eq. (3.55)). At low viscosities, when $\chi \gg 1$, we have from Eq. (3.61)

$$\tilde{\lambda}_0 = \chi^{1/2}. \tag{3.62}$$

For the kinetic factor, Eq. (3.52), we then obtain

$$B \equiv B_{32} = \rho_*'' \left(\frac{2k_B T R_*}{\rho_l} \right)^{1/2}. \tag{3.63}$$

Equation (3.63) corresponds to the theory of absolute reaction rates developed by Eyring [111, 112], in which the inertial motion of a nucleus in the absence of friction results in an independence of the flow from the form of the upper part of the potential barrier. In the limit $\chi \to \infty$ ($\eta \to 0$), the result of Eq. (3.63) is "nonphysical." At $\eta = 0$, the kinetic factor has to be equal to zero. The finite value of the kinetic factor at a low friction is caused by postulating an equilibrium distribution of precritical new-phase nuclei. However, such distribution is absent in the case under consideration. At $\chi \to \infty$, the nonequilibrium region encompasses not only the critical region of the saddle point, but also the potential well where nuclei of a new phase are formed. In this case, Eqs. (2.112) and (3.52) are no longer valid. Now the energy is the unstable variable and not the characteristic size of a new-phase nucleus. In this case, nucleation theory should be developed based on a model differing from the classical model [86–88].

3.3
Elements of the Stochastic Theory of Nucleation

Fluctuation theory treats the formation of a critical nucleus as a random process proceeding with time in a homogeneous system. A random process is considered to be determined if its distribution, mean, and dispersion functions are known. The problem of finding the distribution function of expectation times for the occurrence of the first nucleus may be solved within the same probabilistic scheme of evolution of precritical bubbles that was used by Volmer [105], Zeldovich [5], and Frenkel [6].

Following the theory of homogeneous random processes [72], we introduce the transition probability

$$p(R, R_0, r) = \langle R - R(\tau) \rangle. \tag{3.64}$$

Here, $R(\tau)$ is the solution of the stochastic equation of motion, Eq. (2.64) ($a = R, i = 1$), where averaging is performed over all realizations of the random force, $\zeta(\tau)$. The function $p(R, R_0, \tau)$ determines the probability that at the moment, τ, the bubble radius will be equal to R if, at $\tau = 0$, the value of R is equal to $R = R_0 < R_*$. In the region $R < R_*$, the transition probability, $p(R, R_0, \tau)$, as well as the bubble size distribution function, $P(R, \tau)$, satisfies the one-dimensional Fokker–Planck equation,

$$\frac{dp}{dR} = \frac{\partial}{\partial R}\left[D\left(\frac{1}{k_B T}\frac{\partial E}{\partial R} + \frac{\partial}{\partial R}\right)\right]p. \tag{3.65}$$

The solution of this equation may be presented as

$$p(R, R_0, \tau) = \sum_{i=1}^{\infty} A_i \pi_i(R) W_i(\tau), \tag{3.66}$$

where the coefficients A_i are determined by the initial distribution $p(0, R_0, 0)$. The eigenfunctions $\pi_i(R)$ which correspond to the eigenvalues λ_i satisfy the condition of orthonormalization and the boundary condition $\pi_i(R_*) = 0$. For $W_i(\tau)$, we have

$$\frac{dW_i}{d\tau} = -\lambda_i W_i. \tag{3.67}$$

If in a one-component metastable system the temperature and the pressure are maintained constant, the eigenvalues, λ_i, are time independent. The condition $G_* \gg 1$ ensures the presence in the spectrum of eigenvalues of a characteristic break, i.e., $\lambda_1 \gg \lambda_0$. This property reflects the establishment in the liquid of a stationary bubble size distribution in a time λ_1^{-1}, which is much shorter than the time of expectation of its subsequent destruction, λ_0^{-1}. The time λ_1^{-1} is the time lag, τ_l. For superheated cryogenic liquids far from the critical point, we have $\tau_l \simeq (10^{-9}\text{–}10^{-8})$ s [113]. Characteristic expectation times of formation of the first critical nucleus during homogeneous nucleation in the region of moderate metastability equal seconds to hours. It means that in Eq. (3.66) the term with $i = 0$ is dominant. According to the theory of stochastic processes [114], the moment of initiation of the first event at sufficiently large τ does not depend on the initial distribution $p(0, R_0, 0)$ if a stationary regime is established. Solving Eq. (3.67) while taking into account the considerations above and the condition of normalization of the transition probability, we have

$$p(R, R_0, \tau) \simeq \exp(-\lambda \tau)\pi_0(R). \tag{3.68}$$

Here and in the subsequent discussion, $\lambda = \lambda_0$ holds.

The probability that within the time, τ, a bubble of radius, R_0, will not reach the critical size $R = R_*$ is equal to

$$W_0(\tau) = \int_0^{R_*} p(R, R_0, \tau)dR. \tag{3.69}$$

In the region $R < R_*$, the eigenfunction $\pi_0(R)$ is close to the stationary distribution function, $P(R)$, and differs from the equilibrium function, $P_{eq}(R)$, only in the critical region of the activation barrier. Substitution of Eq. (3.68) into Eq. (3.69) gives an asymptotic formula for the probability of the absence in the liquid of a critical nucleus within the time τ in the form

$$W_0(\tau) = \exp(-\lambda\tau) \int_0^{R_*} \pi_0(R)dR \simeq \exp(-\lambda\tau). \tag{3.70}$$

The probability that within the time τ the bubble will reach the critical size is $(1 - W_0(\tau))$, and the density of the probability of initiation of the first critical nucleus is

$$w_1(\tau) = -\frac{\partial W_0}{\partial \tau} = \lambda \exp(-\lambda\tau). \tag{3.71}$$

The first moment of the distribution given by Eq. (3.71) is equal to the average expectation time of a critical nucleus,

$$\tau_1 = \bar{\tau} = \int_0^\infty \lambda\tau' \exp(-\lambda\tau')d\tau' = \lambda^{-1}. \tag{3.72}$$

According to the definition of the nucleation rate and Eq. (3.72), between λ, $\bar{\tau}$, J and the volume occupied by the metastable phase, V, the evident relation

$$\bar{\tau} = (JV)^{-1} = \lambda^{-1} \tag{3.73}$$

holds.

Equations (3.71) and (3.72) describe a Poissonian homogeneous stationary process. The dispersion of expectation times of occurrence of the first critical nucleus in the Poisson process

$$(\tau_2 - \tau_1^2)^{1/2} = \left[\int_0^\infty \lambda\tau'^2 \exp(-\lambda\tau')d\tau' - \bar{\tau}^2 \right]^{-1/2} = \lambda^{-1} \tag{3.74}$$

is equal to the average value of τ itself.

If a superheated liquid changes its thermodynamic state steadily in time, the parameter λ is a function of time. In this case, nucleation is described by a nonstationary Poissonian process, and the solution of Eq. (3.67) is written as

$$W_0(\tau) = \exp\left[-\int_0^\tau \lambda(\tau')d\tau'\right]. \tag{3.75}$$

The density of the probability of occurrence of the first critical nucleus is then

$$w_1(\tau) = -\frac{dW_0}{d\tau} = \left[\lambda(\tau) - VP(R_*)\frac{dR_*}{d\tau}\right]W_0(\tau). \tag{3.76}$$

The additional term in Eq. (3.76), as compared with Eq. (3.71), is caused by the fact that at varying supersaturation viable centers of a new phase form not only by the mechanism of diffusive growth, but also as a result of transformation of prenuclei into nuclei with decreasing heights of the activation barrier and critical size (athermal nucleation) [115]. Regarding the ratio of the rates of athermal and ordinary nucleation, according to Eqs. (2.112), (3.4), (3.8), and (3.76), one can write

$$\frac{J_a}{J} = \frac{3n_*\sigma\dot{p}}{BR_*(p_s - p')^2(1 - \rho_s''/\rho_s')^2}. \tag{3.77}$$

In a region of intensive nucleation of cryogenic liquids ($p' = 0.1$ MPa): $B \simeq 10^{-10}$ s^{-1}, $\sigma \simeq 3$ mN m^{-1}, $n_* \simeq 200$, $R_* \simeq 4$ nm [113] and the athermal mechanism will be competitive with the ordinary one at rates of the decrease of pressure of the order $\sim 5 \times 10^8$ MPa s^{-1} or superheating rates $\sim 10^{10}$ K s^{-1}. Such huge rates of change of the state variables are not realized in experiments on nucleation kinetics. The characteristic time of transferring a liquid into the metastable state corresponding to these rates ($\sim 5 \times 10^{-9}$ s) coincides at the order of magnitude with the time of establishment of a stationary distribution for precritical bubbles τ_l, and therefore the contribution of athermal nucleation to Eq. (3.76) may be neglected.

Let us now examine processes of isothermal expansion and isobaric heating of liquids. We shall denote the parameter that changes when entering a metastable region (p or T) by x and assume that $\dot{x} = dx/d\tau = $ const. Going over in Eq. (3.75) from integration with respect to time to integration with respect to x, we have

$$w_1(t) = \frac{J(x)V}{\dot{x}}\exp\left[-\int_{x_0}^x J(y)V\frac{dy}{\dot{x}}\right], \tag{3.78}$$

where $x_0 = x(\tau = 0)$ is the value of the parameter x at the moment of the beginning of observations. In a region of intensive homogeneous nucleation,

$G_x = d \ln J / dx \simeq \text{const}$ [113] holds. In this case, for the density of the probability of formation of a critical bubble we have

$$w_1(x) = \frac{J(x)V}{\dot{x}} \exp\left\{-\frac{[J(x) - J(x_0)]V}{\dot{x} G_x}\right\}. \tag{3.79}$$

The function $w_1(x)$ has a characteristic maximum. Differentiating Eq. (3.79) with respect to x, we obtain for the most probable value of x_n

$$J(x_n)V = \dot{x} G_x(x_n). \tag{3.80}$$

According to Eqs. (3.73) and (3.80), the characteristic time scale in the regime of changing state variables with constant rates may be determined as

$$\tau_1 = \dot{x} G_x(x_n). \tag{3.81}$$

The halfwidth of the distribution of Eq. (3.79) does not depend on the rate of change of the state variables and is determined by

$$\delta x_{1/2} = \frac{2.44}{|G_x|_{x_n}}. \tag{3.82}$$

Equations (3.79) and (3.82) have been obtained in Ref. [116] in examining nucleation as a Markovian discrete process. They were previously used for processing experimental data on the kinetics of spontaneous crystallization of supercooled liquids [117].

3.4
Experimental Procedures in the Analysis of Boiling in Superheated Liquids

The kinetics of nucleation in superheated liquids is studied both under fixed conditions with respect to the state of the ambient phase and in the regime of changing state variables, where the change proceeds with a constant rate. A liquid may be brought to predetermined values of p and T in a metastable region in different ways (Fig. 3.2). In isobaric heating, the degree of metastability of the liquid is characterized by the degree of superheating, $\Delta T = T - T_s(p)$. The *temperature of attainable (limiting) superheating*, T_n, is interpreted as the temperature where boiling occurs with a rate corresponding to the value of the nucleation rate registered in the experiment. According to Eq. (3.80), the value of J depends on the volume, V, of the superheated liquid and the heating rate, $\dot{T}$. By varying the values of V and $\dot{T}$ in the experiment, the dependence $J(T)$ can be determined.

Among all possible pathes of transferring a liquid into the metastable state, those with the lowest sluggishness are preferred. In this respect, a decrease

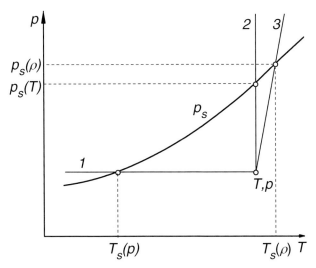

Fig. 3.2 Different ways of transferring liquids to a predetermined metastable state. 1: isobaric; 2: isothermal; 3: isochoric.

in the pressure on the liquid is favored as compared with its isobaric heating. The depth of penetration into a metastable region at $T = $ const is determined by the value of the tensile stretch, $\Delta p = p_s(T) - p$. In an isochoric process, penetration into a metastable region is connected with the cooling of a liquid. The degree of metastability is determined here as $\delta T = T_s(\rho) - T$ or $\delta p = p_s(\rho) - p$. Owing to the large value of the derivative $(\partial p/\partial T)_p$, the inequality $\Delta T > \delta T$ holds. The isochoric process of transferring a liquid into the metastable state has some advantages over the isobaric one. In the vicinity of the critical point, δT determines the scaling value of supersaturation (see Chapter 5).

Experimental investigations of the kinetics of stationary nucleation presuppose the determination of the nucleation rate, J, as a function of temperature and pressure. Information on the nucleation rate may be obtained from data on the distribution functions for the moments of formation of the first critical nucleus. It requires repeated experiments on one sample or measurements with a system of equivalent samples. As in the Poissonian process, the mean-square error of determining the average time of expectation, $\bar{\tau}$, and, consequently, J, in a series of N measurements is

$$\sigma_{\bar{\tau}} = \frac{\bar{\tau}}{N^{1/2}}. \tag{3.83}$$

In order to get values $\sigma_{\bar{\tau}} = 0.1\bar{\tau}$, it is necessary to have $N \simeq 100$ [9].

In an experiment, it is commonly difficult to detect the formation of the first critical nucleus. By this reason, not the moment of formation of a critical nucleus is usually registered but one or another manifestation of the result of formation in the system of a viable center of a new phase (liquid boiling-up, the amount of the vapor phase at a certain moment, etc.). The phase-transition process includes the formation of vapor-phase centers and their subsequent growth. High growth rates of vapor bubbles in highly superheated liquids make it possible to realize conditions in which the average expectation time of occurrence of a viable center, $\bar{\tau}$, will considerably exceed the characteristic time of decay of the metastable phase, τ_g. Thus the characteristic time of decay of one cubic centimeter of liquid argon superheated at atmospheric pressure by $\Delta T_n \simeq 44$ K after the formation of a critical bubble does not exceed 10^{-4} s, which makes it possible to determine the time of formation of a critical nucleus by the moment of boiling-up of the sample under investigation.

As the critical point is approached, owing to the inhibition of heat-transfer processes, the growth rate of the bubbles decreases considerably, and at a certain distance from the critical point the inequality $\bar{\tau} \gg \tau_g$ may be violated. Liquid boiling-up will lose its explosive character, and the characteristic signal of a phase transition will then be the volume fraction of the vapor phase, η, at the moment, τ. Under conditions of constant pressure, p, and temperature, T, the dependence $\eta(\tau)$ is determined by Kolmogorov's formula [118]

$$\eta(\tau) = 1 - \exp\left[-\int_0^\tau J(\tau')V(\tau-\tau')d\tau'\right], \qquad (3.84)$$

where $V(\tau - \tau')$ is the volume of the bubble formed at the moment τ' at time τ. Possessing experimental data on $\eta(\tau)$, information on the nucleation rate can be obtained from Eq. (3.84).

Experimental methods of studying the kinetics of spontaneous boiling-up of superheated liquids are subdivided into quasistatic and dynamic methods. Quasistatic experiments are conducted at relatively low rates of change of state variables. The characteristic feature of such experiments is the fulfillment of conditions of homogeneous nucleation in preparing an experiment, but not in its course. Practice shows [113, 119] that superheated liquids are not critical to the presence of small quantities of dissolved gases and solid particles. A more considerable effect on the liquid limiting superheating may be exerted by the conditions on the walls of a measuring cell. Good wettability and low roughness are the main requirements in choosing a material for a measuring cell. Glasses fulfill these requirements to a large degree. Indeed, employing glass capillaries in their experiments, Wismer [119], and Kenrick, Gilbert, and Wismer [120] obtained the first reproducible data on the attainable superheating for a number of organic liquids. The setup of Wismer et al. was a prototype for further developments of more informative quasistatic methods of studying the kinetics of nucleation [9, 10, 121]. In a number of cases a good wettability of glass and other solid materials by cryogenic liquids is close to complete; a low content of dissolved gases and solid particles in them distinguish these systems as the most convenient objects for studying the kinetics of spontaneous nucleation.

Dynamic methods do not require "purity" of a system. A liquid is superheated to the temperature of homogeneous nucleation as a result of organizing a powerful heat release which considerably exceeds the heat sink into the evaporation centers (shock boiling-up regime) [9, 11]. The required heating rates are evaluated from the inequality [10]

$$\left(\frac{1}{\pi\tilde{\Omega}_d\langle\varphi^d\rangle}\right)^{1/d\alpha}\frac{\dot{T}}{T_* - T_s} \gg 1, \tag{3.85}$$

where $\tilde{\Omega}_d$ is the effective bulk ($d = 3$) or surface ($d = 2$) density of heterogeneous nucleation centers, and T_* is the temperature of intensive homogeneous nucleation. If bubble growth is limited by heat supply ($R = \varphi\tau^\alpha$, $\alpha = 1/2$), the value of φ^d averaged on the interval $T_s, T(\tau)$ is equal to

$$\langle\varphi^d\rangle = \frac{3}{2\tau^{d/2}}\int_0^\tau \varphi^d \tau'^{(d/2-1)} d\tau'. \tag{3.86}$$

The quantity $\tilde{\Omega}_d$ presupposes a replacement of the action of centers of all sizes by the action of a certain effective number of centers where bubble growth begins immediately after passing the saturation temperature, T_s. Information on the bulk and the surface density of centers is extremely limited. With $\tilde{\Omega}_{d=3} = \tilde{\Omega}_V \simeq 10^{10}$ m^{-3} in high-temperature organic liquids, the shock boiling-up regime is realized at heating rates $\dot{T} > 10^6$ K s^{-1} [10]. Owing to lower superheating values, in cryogenic liquids the conditions of intensive fluctuation nucleation will be achieved at lower values of $\dot{T}$. At the same effective density of boiling centers, heating rates for liquid argon are to be about $\sim 10^4$ K s^{-1}, for liquid helium not less than 10 K s^{-1}. Taking into account the purity of cryogenic liquids and the good wettability of solid materials by them, the expected values of $\tilde{\Omega}_d$ in liquefied gases will be smaller than those in high-temperature liquids. This feature is also bound to decrease the characteristic values of the rate of realization of the shock boiling-up regime.

In studying the kinetics of nucleation in superheated cryogenic and low-boiling liquids, methods of continuous isobaric heating [121–124], floating-up droplets [125–127], determination of the average lifetime [113, 128, 129] (quasistatic methods) and the method of pulsed heat transfer [130–132] (dynamic method) were employed.

3.5
Quasistatic Methods of Investigating Limiting Superheatings of Liquids

The method of continuous isobaric heating was first used for studying the nucleation kinetics in condensed inert gases [121–123]. In this method, a liquid is superheated in the lower part of a glass capillary where a constant pressure is maintained. The distribution of events of liquid boiling-up in dependence on temperature is found, and the temperature of attainable superheating and the nucleation rate are determined at given values of pressure and heating rate.

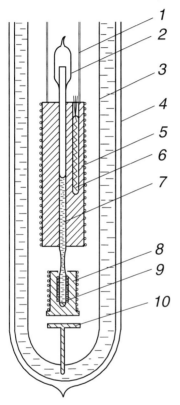

Fig. 3.3 Schematic diagram of a setup for determination of the temperature of attainable superheating for hydrogen by the method of continuous heating [133].

The schema of the measuring device for determining the attainable superheating temperature for liquid hydrogen is given in Fig. 3.3 [133]. The capillary (7) is filled with hydrogen on a special stand at the temperature of liquid helium and soldered. The inner diameter of the capillary is of the order of ~ 1 mm, and the outer one is of the order of ~ 2 mm. The compensation reservoir (2) prevents the capillary from rupture when hydrogen contained in it is heated to room temperatures. The pressure in the capillary is created and maintained by thermostating (± 0.005 K) the liquid vapor interface in an aluminium unit (5). The temperature in the unit is measured by a platinum resistance thermometer (6) (± 0.02 K). The part of the liquid ($V \simeq 20$ mm^3) in the unit (8) is superheated. In order to decrease convection and heat supply

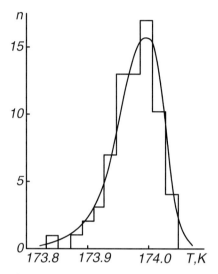

Fig. 3.4 Histogram of experiments on the superheating of methane [113]: $p = 1.83$ MPa; $T_s = 174.0$ K; $N = 80$; $\dot{T} = 1.0$ K s^{-1}. The *smooth curve* shows the results of calculations employing Eq. (3.87).

in the capillary, the superheated volume is separated from the section of pressure creation by a construction on the capillary. The moment of boiling-up is registered by the ejection of the liquid from the superheated volume. For this purpose, unit (5) has a longitudinal through window, and the glass Dewar flask (4) unsilvered vertical strips. The measuring device is suspended on capron threads (1) in a glass vacuum chamber. For fast cooling of unit (8), a thermal switch (10) is used. In experiments on hydrogen, a manganin resistance thermometer (9) was employed as a temperature-sensitive element. The temperature of superheating of the liquid in other variants of the setup was registered with copper–constantan [121–123] and gold–cobalt–copper thermocouples [124, 134, 135]. The error of temperature determination, depending on the substance under investigation and the sensing element employed, was $\pm(0.03$–$0.5)$ K.

In an experiment (with $p = $ const., $\dot{T} = $ const.), 30–50 values of the boiling-up temperature are measured, and a histogram is constructed for the distribution of events of boiling-up, n, per temperature interval $(T, T + \Delta T)$ in a series of N measurements. The characteristic shape of the histograms is shown in Fig. 3.4. On the basis of the histogram, we can determine the temperature of attainable superheating, T_n, as the most probable boiling-up temperature, the derivative G_T from the distribution halfwidth, according to Eq. (3.82), and the nucleation rate, J, by Eq. (3.80). The smooth curve in Fig. 3.4 was calculated by

$$n = \left(\frac{NJV}{\dot{T}}\right) \Delta T \exp\left(-\frac{JV}{G_T \dot{T}}\right). \tag{3.87}$$

In different variants of the device described above, the heating rates varied from $0.01\,\mathrm{K\,s^{-1}}$ to $3\,\mathrm{K\,s^{-1}}$, and the amount of the superheated liquid volume was equal to 0.02–$0.2\,\mathrm{cm^3}$. The values of nucleation rates registered in the experiment are 10^7–$10^{13}\,\mathrm{m^{-3}\,s^{-1}}$. The experimental procedure and the methods of processing experimental data are described in detail in Ref. [123].

The method of floating-up drops was developed by Wakeshima and Takata [136], Moor [137], Skripov et al. [9,10,138], and Blander et al. [125,127]. Liquid drops float up in a vertical column of another, background liquid which does not interact chemically with the liquid under investigation, and has higher values of surface tension, density, and boiling temperature, but a lower freezing temperature than the boiling temperature of the substance under investigation. A stationary temperature gradient is created in the background liquid with the help of a system of heaters. When rising, a drop of the liquid under investigation is superheated and evaporates explosively at a certain height. The temperature of the medium at the site of the drop explosion is taken as the boiling-up temperature.

Figure 3.5 represents the schema of the device used in investigating the attainable superheating of low-boiling liquids [127]. The working section of the setup is a glass tube (4) filled with a background liquid. If the temperature of normal boiling, T_s, of the substance under investigation is not lower than 210 K, water eutectics of ethylene glycol are used as a background liquid, at $T_s \geq 190$ K, water solutions of lithium chloride, and at $T_s \geq 173$ K, water eutectics of ammonia. The temperature gradient (from 0.03 to 10 K cm^{-1}) along the length of the working section is created by pumping a glycol/water solution through the internal jacket (3). The temperature of the solution is predetermined and maintained by an external thermostat. The vacuum jacket (2) serves as a heat insulator. Emulsification of the liquid under study, injected through the tube (5), takes place in the chamber (7) by a mixer. The emulsion temperature is controlled by a thermocouple (9). An emulsifying agent (7) is placed in a Dewar flask (8) with acetone and cooled by pumping nitrogen vapors. Drops 0.1–1 mm in diameter enter the working section through a connecting capillary (6) and float up in the background liquid. The temperature at the site of explosion of drops is measured by two thermocouples (1) located in such a way that one of them is below and the other above the place where up to 95% of all floating-up drops evaporate. The average value of the data supplied by the thermocouples is taken as the temperature of attainable superheating. At characteristic heating rates of 0.03–50 K s^{-1} and drop volumes of 10^{-7}–10^{-4} cm^3, the effective values of nucleation rates are 10^3–10^{14} m^{-3} s^{-1}.

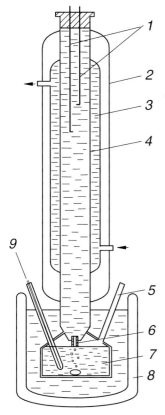

Fig. 3.5 Schematic diagram of a setup determining attainable superheating temperatures of low-temperature boiling liquids by the method of floating-up drops [127].

The advantage of the discussed method is the "purity" of the liquid–liquid interface. However, the absence of appropriate liquid media with freezing temperatures lower than 170 K makes it impossible to use it for studying limiting superheatings of cryogenic liquids. Besides, this method has a comparatively large error of determining the temperature of attainable superheatings, which for the most low-temperature boiling of the investigated liquids, ethane, is ± 1 K. The halfwidth of the distribution of boiling-up events in temperature for this substance is less than 0.5 K. The method of floating-up drops has been used for finding the pressure dependence of T_n, and also for measuring the lifetime of superheated liquids [9, 10, 138].

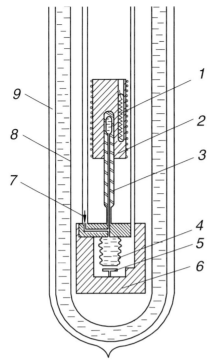

Fig. 3.6 Schematic diagram of a setup for measuring lifetimes of superheated cryogenic liquids [129].

Measuring the lifetimes of a superheated liquid: The first systematic investigations of nucleation kinetics in superheated liquids through the determination of the mean lifetime were made by Sinitsyn and Skripov [139]. Experiments were conducted on liquid hydrocarbons. Penetration into a metastable region was realized by pressure release to a thermostated liquid. The lifetime, τ, of a liquid in the metastable state was determined, and the average value of τ was found at given values of p and T. This average value is connected with the nucleation rate, J, by Eq. (3.73)

Figure 3.6 shows the schema of the measuring device of a setup for studying the attainable superheating of cryogenic liquids [129]. The liquid under investigation is superheated in a glass capillary (3) with an inner diameter of ~ 1 mm and an outer diameter of ~ 6 mm. The liquid has a working volume of $V \simeq 60$ mm^3. The capillary is thermostated in a copper unit (2). The temperature is measured (± 0.02 K) by a platinum resistance thermometer (1). By means of a Kovar–glass junction the capillary is connected to the chamber of the bellows (4). The pressure is created by compressed helium and through the bellows is passed to the liquid under investigation. The bellows (4), the low-temperature valve (7) are thermostated in a massive copper unit (6) at a temperature close to the temperature of normal boiling of the liquid being investigated. The operating chamber of the setup is mounted in a vacuum jacket (8) made of stainless steel. Cooling is realized with liquid nitrogen poured into a metal Dewar flask (9). Pressure release and supply are realized with the help of a system of electromagnetic valves. In order to decrease the effect of adiabatic cooling during pressure release, a two-step transfer of a liquid to the predetermined state in the metastable region is used: release to the pressure p_1 ($p < p_1 < p_s(T)$), delay, the final release to the pressure p, at which τ is measured.

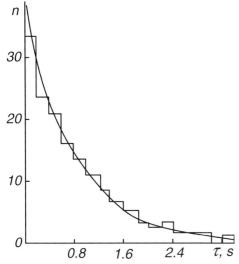

Fig. 3.7 Histogram of experiments on the superheating of argon [128]: $p = 1.4$ MPa; $T = 135.3$ K; $N = 147$; $\bar{\tau} = 0.75$ s. The smooth curve shows results of calculations by Eq. (3.88).

The measurements of the lifetime of a superheated liquid begin from the moment of establishing the lower pressure, p. Boiling-up is registered by a water hammer. Its action results in touching of the bottom of the bellows to the electric contact (5). Under given conditions of temperature and pressure, from 50 to 150 values of the expectation times of boiling-up, τ, are registered in order to realize the required accuracy, and the mean lifetime, $\bar{\tau} = \sum \tau_i / N$, is determined. The density of probability of formation of the first critical nucleus may be found as the ratio of the number of boiling-ups, n, at the time interval $\tau, \tau + \Delta \tau$ to the total number of tests

$$n = N\bar{\tau}^{-1}\Delta\tau \exp\left(-\frac{\tau}{\bar{\tau}}\right). \tag{3.88}$$

In Fig. 3.7, an experimental histogram is compared with the results of calculating $n(\tau)$ by Eq. (3.88). Experiments [128, 129, 140–142] trace the temperature and pressure dependence of the mean lifetime of superheated liquefied gases with τ varying from 0.1 s to 30–40 min, which corresponds to the interval of nucleation rates 10^4–10^8 m^{-3} s^{-1}. The error of determining the temperature of attainable superheating by this method is estimated as $\perp(0.06\ 0.2)$ K.

3.6
Dynamic Methods of Investigating Explosive Boiling-Up of Liquids

The method of pulse heating of a liquid on thin platinum wires in conditions of the shock boiling-up regime has been suggested by Pavlov and Skripov [9–11]. A platinum wire immersed in the liquid under investigation serves simultaneously as a heater and a resistance thermometer. The wire is heated by a current pulse. The moment of the beginning of the explosive boiling-up of the liquid is connected with an abrupt change in the resistance of the wire. This method was used in studying the attainable superheating of high-

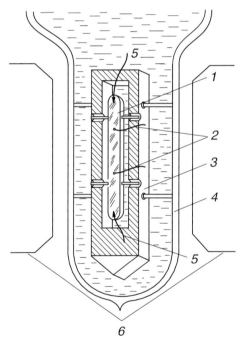

Fig. 3.8 Schematic diagram of the measuring chamber of a setup for pulse helium superheating on a bismuth monocrystal.

temperature organic liquids and water. Since the values of the superheating of cryogenic liquids are some orders of magnitude smaller than those of high-temperature organic liquids, and the electrical resistance of platinum, other metals and alloys become extremely low at the temperatures of liquid helium, platinum wires prove to be inefficient sensors in low-temperature regions.

In Refs. [130–132], it is suggested to use a bismuth monocrystal as a thermometer heater for investigating limiting superheatings of helium isotopes. The electrical resistance of bismuth in magnetic fields increases by several orders making it possible to achieve a sensitivity $S = 1/R_B(dR_B/dT)$ in the range (1.7–5.2 K) from 0.1 to 0.25 K^{-1}. A bismuth thermometer-heater possesses a small time constant (less than 100 µs) and a low resistance (the temperature difference between the crystal surface and liquid helium that is in contact with it does not exceed 20 mK). The total error of measurement of temperatures with a bismuth resistance thermometer, according to the evaluations of the authors of Ref. [130], is equal to ± 10 mK.

The schema of the operating chamber of a setup for investigating the attainable superheating of liquid helium is shown in Fig. 3.8. A bismuth monocrystal (1) in the form of a rectangular parallelepiped which is 6 cm high, 5 mm wide, and 1.5 mm thick with polished surfaces is fixed in a holder (3). Potential leads (2) of a copper wire, 0.05 mm in diameter, are fixed at the center of the crystal by the method of spot welding. Thicker current wires (3) are soldered to the ends of the crystal. The holder with the crystal is mounted in the stem of a glass Dewar flask (4) with liquid helium, which is located between the poles of an electromagnet (6). The direction of the magnetic field with respect to the crystal may be changed by turning the magnet. The magnetic-field induction is equal to 1–1.5 T.

In an experiment, the monocrystal temperature is registered as a function of time with an electronic circuit at several values of the supplied power, q_A. The moment of helium boiling-up manifests itself as a characteristic break on the thermogram (Fig. 3.9). The supplied power, calculated per unit of the monocrystal surface, varies from 5 to 430 mW cm^{-2}, which corresponds to rates of liquid helium heating of 2–1700 K s^{-1}. At $q_A > 30$ mW cm^{-2} ($\dot{T} > 15$ K s^{-1}), the temperature of helium boiling-up does not depend on the value of the supplied power. The dashed lines in Fig. 3.9 show the results of calculating the dependences $T(\tau)$ under the assumption that the whole heat transfer from the surface of the thermometer-heater into liquid helium is realized only by thermal conductivity. The error of determining the temperature of attainable superheating for liquid ^{4}He by data from Refs. [131, 132] is ±0.02 K.

Let us now evaluate the nucleation rates realized in these experiments. In the approximation of free growth for the number of nuclei formed in the time τ in the near-surface layer of a liquid of volume Ax where A is the area of the

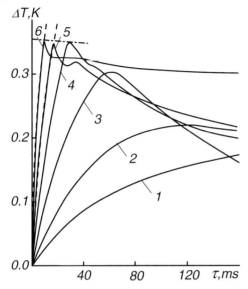

Fig. 3.9 Thermograms of experiments on the superheating of ^{4}He at atmospheric pressure on a bismuth monocrystal. The power supplied per unit surface area of a monocrystal is: 1: 5; 2: 9; 3: 15; 4: 29; 5: 41; 6: 72 mW cm^{-2}.

crystal surface, one can write

$$i = \int_0^\tau \int_0^x JA\,dx\,d\tau'. \tag{3.89}$$

Assuming the heating rate, $\dot{T}$, to be constant and expanding the exponent in the expression for the nucleation rate, Eq. (2.112), into a series in the vicinity of $x = 0$, $\tau' = \tau$ [9], we get

$$G[T(\tau', x)] = G[T(\tau, 0)] + G_T \left(\frac{\partial T}{\partial \tau'}\right)_{x=0,\tau'=\tau} (\tau' - \tau) \tag{3.90}$$

$$+ G_T \left(\frac{\partial T}{\partial x}\right)_{x=0,\tau'=\tau} x + \cdots,$$

and after substitution of Eq. (3.90) into Eq. (3.89) we obtain

$$J \simeq \frac{i}{A} G_T^2 \dot{T} \left(\frac{\partial T}{\partial x}\right)_{x=0,\tau'=\tau}. \tag{3.91}$$

In calculating the integral (3.89), the Laplace asymptotic method [143] was applied. Most of the bubbles are formed in the boundary liquid layer of thickness $\delta \simeq (D_T \tau)^{1/2}$ adjacent to the heating surface where D_T is the thermal diffusivity. Since the bubble radius is about the thickness of the superheated layer, $R(\tau) \simeq \delta$, determining the area of the monocrystal surface occupied by the vapor phase as $A(\tau) = \pi R^2(\tau)$ and using the approximation $\partial T/\partial x \simeq (T_n - T_s)/\delta$, from Eq. (3.91) we have

$$J \simeq \frac{A(\tau)}{A} \frac{G_T^2 \dot{T}^{5/2}}{4\pi D_T^{3/2}(T_n - T_s)^{1/2}}. \tag{3.92}$$

For liquid helium at $p \simeq 0.1$ MPa, $\dot{T} \simeq 35\text{--}200)$ K s^{-1}, $T_n - T_s \simeq 0.35$ K and $G_T \simeq 650$ K^{-1}. Assuming that the boiling-up signal corresponds to $\eta = A(\tau)/A \simeq 0.1$, from Eq. (3.92), we obtain $J = (10^{16}\text{--}10^{21})$ m^{-3} s^{-1}. The authors of Refs. [131, 132] compare their experimental data on the attainable superheating of liquid helium with the theory of homogeneous nucleation at $J = 10^7$ m^{-3} s^{-1}.

3.7
Results of Experiments on Classical Liquids

This section presents the results of experimental investigations of the attainable superheating of cryogenic and low-boiling liquids, with respect to which

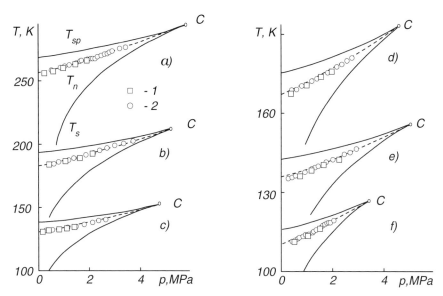

Fig. 3.10 The line of attainable superheating T_n, of the spinodal, T_{sp}, and the binodal, T_s, of xenon (a), krypton (b), argon (c), methane (d), oxygen (e), nitrogen (f). 1: method of measuring the lifetime ($J = 10^7$ m^{-3} s^{-1}); 2: method of continuous heating ($J = 10^{11}$ m^{-3} s^{-1}) [121, 122, 128, 140–142, 145, 146].

the influence of quantum effects on the thermodynamic properties may be neglected. The first experiments in the cryogenic range of temperatures were made on liquid argon [121,128]. Subsequent papers [122,123,140–142,144–146] investigated the limiting superheats of liquid nitrogen, oxygen, methane, xenon, and krypton. Experimental data have been obtained in a wide pressure range by the methods of continuous isobaric heating and measuring the lifetime. At atmospheric pressure, the attainable superheating of liquid nitrogen and oxygen was also investigated by Nishigaki and Saji [124]. The results of this work agree within 0.3 K with the results of earlier investigations [140,142].

Tab. 3.1: Temperature of attainable superheating, T_n, of pure liquids at atmospheric pressure (except for ^{3}He) and nucleation rate, J, in m^{-3} s^{-1}. PM: pulse method; CH: method of continuous isobaric heating; LT: method of measuring the lifetime; FD: method of floating-up drops; (+): data refer to the pressure $p = 0.0135$ MPa; (*): calculation by Eqs. (2.112), (3.52) and $z_0 = 1$ with the use of surface tension values from Refs. [153–161] obtained on samples of the liquid from the lots on which the attainable superheating was investigated. All the remaining theoretical values of T_n have been calculated by Eqs. (2.112), (3.15) with $z_0 = 1$.

Substance	T_s (K)	lg J	T_n (K) Experiment	T_n (K) Theory	Method	Reference
Helium (^{3}He)	1.81	20	2.46	2.50$^+$	PM	[149]
Helium (^{4}He)	4.21	20	4.55	4.56	PM	[132]
	4.21	7	4.45	4.50	CH	[135]
	4.21	6	4.58	4.50	LT	[150]
Neon (Ne)	27.09	11	38.0	38.64*	CH	[100]
	27.20	7	38.0	38.40	CH	[134]
Argon (Ar)	87.29	8	130.5	131.0*	LT	[128]
	87.29	11	130.8	131.5*	CH	[121]
Krypton (Kr)	119.78	7	181.0	182.0	LT	[145]
	119.78	11	181.5	182.7*	CH	[122]
Xenon (Xe)	165.03	7	250.6	252.0*	LT	[145]
	165.03	11	251.9	253.3*	CH	[122]
Hydrogen	20.38	11	27.8	28.05*	CH	[133]
(nH$_2$)	20.4	7	28.1	28.06*	CH	[134]
Nitrogen	77.35	7	109.7	110.3*	LT	[142]
(N$_2$)	77.35	11	109.9	110.7*	CH	[146]
	77.3	7	110.0	109.8	CH	[124]
	77.35	17	110.3	111.1	PM	[147]
Oxygen	90.19	7	134.0	134.8*	LT	[140]
(O$_2$)	90.19	11	134.2	135.3*	CH	[146]
	90.1	7	134.1	134.2	CH	[124]
Chlorine (Cl$_2$)		8	366.5	367.8	LT	[151]
Methane	111.66	7	165.1	166.0*	LT	[141]
(CH$_4$)	111.66	11	166.0	166.6*	CH	[146]
Ethane	184.95	7	267.4		LT	[152]
(C$_2$H$_6$)	185.0	12	269.2	269.7	FD	[127]

Tab. 3.1: (continued)

Substance	T_s (K)	lg J	T_n (K) Experiment	T_n (K) Theory	Method	Reference
Propane	231.1	7	327.2	327.4*	LT	[148]
(C$_3$H$_8$)	231.1	12	326.2	328.5	FD	[127]
	231.1	12	329.2		FD	[126]
Isobutane	261.3	7	361.2	361.7*	LT	[148]
(iC$_4$H$_{10}$)	261.4	12	361.0	360.9	FD	[127]
	261.5	12	359.2		FD	[126]
n-Butane	272.7	12	378.4	378.6	FD	[125]
(nC$_4$H$_{10}$)	272.7	12	376.9	378.3	FD	[127]
Sulphur dioxide (CS$_2$)	263.2	12	323.2		CH	[120]
Propadiene (C$_3$H$_4$)	238.7	12	346.2		FD	[125]
Propene (C$_3$H$_4$)	249.9	12	356.8		FD	[125]
Propylene (C$_3$H$_6$)	225.5	12	325.6	323.5	FD	[125]
Cyclopropane (C$_3$H$_6$)	240.3	12	350.7		FD	[125]
2-Methilpropane (C$_4$H$_6$)	266.3	12	369.6	372.5	FD	[125]
1-Butene (C$_4$H$_6$)	266.9	12	371.0	373.4	FD	[125]
1,3-Butadiene (C$_4$H$_6$)	268.8	12	377.3		FD	[125]
Chloromethane (CH$_3$Cl)	249.0	12	366.2		FD	[125]
Cloroethilene (C$_2$H$_3$Cl)	259.3	12	374.1		FD	[125]
Fluoroethilene (C$_2$H$_3$F)	201.0	12	290.1		FD	[125]
1,1-Difluoro-ethane (C$_2$H$_4$F$_2$)	248.5	12	343.6		FD	[125]
Difluorochloro-methane (CHF$_2$Cl)	232.4	12	232.4		FD	[125]

Sinha, Brodie, and Semura [147] measured superheating temperatures of liquid nitrogen by the method of pulse heating on metal wires. It is noted that the degree of superheating did not depend (± 0.3 K) on the length of the wire, its positioning (vertical, horizontal), and the material employed (platinum, constantan, manganin). The main data array has been obtained on platinum wires which were 30.9 mm in length and 0.1 mm in diameter. The authors of Ref. [147] estimate the effective value of the nucleation rate as 10^{12} m^{-3} s^{-1}.

Calculations with Eq. (3.92) give $(10^{18}-10^{20})$ m^{-3} s^{-1}. When reduced to the same value of the nucleation rate, the results of Refs. [142] and [147] agree within ± 0.2 K.

In contrast to ordinary fluids, cryogenic liquids have a considerably smaller value of the attainable superheating, $\Delta T_n = T_n - T_s$. If at atmospheric pres-

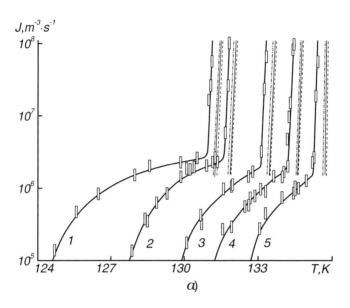

a)

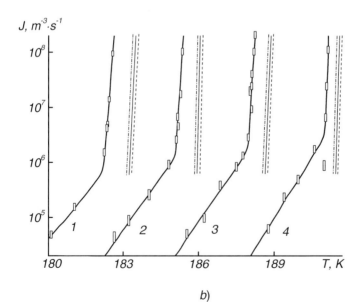

b)

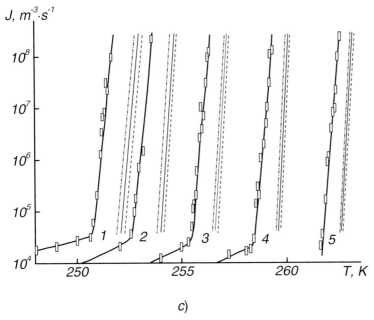

Fig. 3.11 Nucleation rate in superheated liquid argon (a) [128] (1: p = 0.19 MPa; 2: 0.36; 3: 0.81; 4: 1.10; 5: 1.40), krypton (b) [145] (1: p = 0.40 MPa; 2: 1.00; 3: 1.60; 4: 2.20), and xenon (c) [145] (1: p = 0.24 MPa; 2: 0.55; 3: 0.99; 4: 1.48; 5: 1.98). *Dashed lines* show results of calculation by Eqs. (2.112), (3.15) and $z_0 = 1$; *thin solid lines* by Eqs. (2.112), (3.52) and $z_0 = 1$; *dashed-dotted lines* by Eqs. (2.112), (3.52) and $z_0 = \rho'/\rho''_*$. In all cases, the work W_* is determined employing the capillarity approximation, $\sigma = \sigma_\infty$.

sure the limiting superheating of water is estimated as 205 K [10], for propane we have $\Delta T_n \simeq 96$ K [148], and for liquid nitrogen $\Delta T_n \simeq 23$ K [142]. The differences are caused by a relatively weak molecular interaction in cryogenic liquids. This feature also manifests itself in a low value of the surface tension, low values of the critical temperature and the heat of evaporation.

The results of experiments on the determination of the temperatures of attainable superheat for cryogenic liquids and liquids with normal boiling temperatures lower than 273 K (at atmospheric pressure) are presented in Table 3.1. There, the effective values of the nucleation rate are also given. Figure 3.10 shows temperatures of attainable superheating for liquefied gases obtained by the methods of continuous isobaric heating and measuring the lifetime. The results of these experiments refer to different values of the nucleation rate. The value of the attainable superheating decreases with increasing pressure vanishing at the critical point. A similar pressure dependence is found for the spinodal. At atmospheric pressure, the spinodal point of liquid argon is 5.8 K higher than the maximum superheating experimentally registered. The limits of essential instability for one-component liquids presented in Fig. 3.10 have been approximated by the results of (p, ρ, T)-measurements in the region of metastable states [52].

Tab. 3.2 Experimental and theoretical values of the attainable superheating temperature for one-component liquids at different pressures and a fixed nucleation rate. For superheated liquid ^{3}He, ^{4}He, and Cl_2, the theoretical values have been calculated by Eqs. (2.112), (3.15) and $z_0 = 1$. The temperatures of attainable superheating for the remaining liquids have been determined from Eqs. (2.112), (3.52) and $z_0 = 1$ by the data on σ_∞ from Refs. [153–161].

p, MPa	T_n, K		p, MPa	T_n, K	
	Exp.	Theory		Exp.	Theory
	Helium-3 [149]		0.510	39.1	39.44
	$J = 10^{14}$ m^{-3} s^{-1}		0.687	39.5	39.80
			0.833	39.9	40.09
0.00507	2.387	2.439	0.983	40.3	40.41
0.0135	2.460	2.504	1.099	40.5	40.65
0.0199	2.536	2.553	1.203	40.7	40.88
0.0280	2.605	2.616	1.285	41.0	41.06
0.0380	2.683	2.695	1.444	41.4	41.40
0.0501	2.767	2.792	1.451	41.3	41.42
0.0645	2.900	2.908	1.592	41.7	41.74
	Helium-4 [132]			Argon [128]	
	$J = 10^{20}$ m^{-3} s^{-1}			$J = 10^5$ m^{-3} s^{-1}	
0.0374	4.22	4.23	0.19	130.9	131.46
0.0540	4.31	4.31	0.36	131.5	132.06
0.0667	4.37	4.38	0.81	133.3	133.67
0.0801	4.45	4.46	1.10	134.4	134.73
0.1008	4.55	4.56	1.40	135.4	135.84
	Helium-4 [64]			Argon [121]	
	$J = 10^7$ m^{-3} s^{-1}			$J = 10^{11}$ m^{-3} s^{-1}	
0.0374	4.22	4.23	0.10	130.8	131.5
0.0540	4.31	4.31	0.26	131.5	132.1
0.0667	4.37	4.38	0.41	131.9	132.6
0.0801	4.45	4.46	0.60	132.8	133.2
0.1008	4.55	4.56	0.81	133.7	134.0
			1.15	135.1	135.2
	Neon [100]		1.42	136.0	136.2
	$J = 10^{11}$ m^{-3} s^{-1}		1.72	137.1	137.3
			2.14	138.6	138.9
0.131	37.9	38.70	2.45	139.5	140.2
0.351	38.7	39.13	2.71	141.3	141.1

Continued on the next page

p, MPa	T_n, K		p, MPa	T_n, K	
	Exp.	Theory		Exp.	Theory
Krypton [145] $J = 10^7$ m^{-3} s^{-1}			0.83	256.3	256.9
			0.99	256.2	257.8
			1.07	257.2	258.2
0.40	182.40	183.35	1.26	258.2	259.3
1.00	185.24	186.00	1.47	259.6	260.6
1.60	188.10	188.73	1.55	260.3	261.0
2.20	191.11	191.56	1.68	261.0	261.8
			1.75	261.6	261.2
Krypton [122] $J = 10^{11}$ m^{-3} s^{-1}			1.86	261.9	262.9
			1.97	262.8	263.5
			2.07	263.4	264.2
0.40	182.5	184.0	2.17	263.8	264.7
0.82	184.3	185.8	2.37	265.2	265.9
1.20	187.0	187.5	2.48	266.1	266.6
1.41	187.6	188.4	2.63	266.9	267.6
1.63	189.1	189.4	2.75	267.5	268.3
1.90	189.9	190.5	2.85	267.8	268.8
2.20	192.1	191.9	2.97	269.1	269.6
2.43	192.9	193.0	3.05	269.7	270.2
2.80	194.8	194.9	3.13	270	270.6
3.14	196.6	196.5	3.45	272	272.7
3.46	198.0	198.1	3.63	273	273.9
3.80	199.4	199.8			
Xenon [145] $J = 10^7$ m^{-3} s^{-1}			Hydrogen [133] $J = 10^{11}$ m^{-3} s^{-1}		
			0.143	28.1	28.21
0.24	251.41	252.78	0.195	28.2	28.41
0.55	253.34	254.53	0.200	28.5	28.43
0.99	256.06	257.06	0.270	28.6	28.69
1.48	259.03	259.95	0.325	28.7	28.90
1.98	262.18	262.98	0.412	29.1	29.24
			0.480	29.4	29.51
Xenon [122] $J = 10^{11}$ m^{-3} s^{-1}			0.568	29.9	29.86
			0.655	30.1	30.22
			0.660	30.2	30.24
0.50	254.1	255.0	0.760	30.8	30.66

Continued on the next page

3 Attainable Superheating of One-Component Liquids

p, MPa	T_n, K		p, MPa	T_n, K	
	Exp.	Theory		Exp.	Theory
0.898	31.4	31.36	1.667	140.0	140.39
			2.252	142.4	142.61
	Nitrogen [142]				
	$J = 10^7$ m^{-3} s^{-1}			Oxygen [146]	
				$J = 10^{11}$ m^{-3} s^{-1}	
0.5	111.55	112.04			
1.0	113.85	114.18	0.26	134.9	135.8
1.5	116.25	116.43	0.40	135.4	136.3
			0.55	136.0	136.8
	Nitrogen [146]		0.68	136.5	137.2
	$J = 10^{11}$ m^{-3} s^{-1}		0.92	137.4	138.0
			1.18	138.3	138.9
0.41	111.4	112.0	1.35	139.0	139.5
0.52	112.0	112.4	1.48	139.4	140.0
0.61	112.1	112.8	1.74	140.7	140.9
0.70	112.7	113.2	2.03	141.9	142.0
0.82	113.2	113.7	2.26	142.8	142.8
0.94	113.8	114.2	2.50	143.6	143.7
1.06	114.2	117.7	2.70	144.5	144.5
1.21	114.8	115.3	2.97	145.9	145.5
1.33	115.5	115.9			
1.36	115.6	116.0		Chlorine [151]	
1.46	116.2	116.4		$J = 10^{11}$ m^{-3} s^{-1}	
1.59	116.8	117.0			
1.62	117.0	117.1	0.88	373.0	371.4
1.73	117.6	117.6	1.20	372.2	373.0
1.77	117.7	117.8	1.60	374.3	375.0
1.87	118.3	118.3	2.06	377.6	377.4
1.92	118.4	118.5			
2.07	119.1	119.2		Methane [141]	
				$J = 10^7$ m^{-3} s^{-1}	
	Oxygen [140]				
	$J = 10^7$ m^{-3} s^{-1}		0.4	166.58	167.41
			1.0	169.64	170.30
0.686	136.2	136.85	1.6	172.76	173.31
1.171	138.2	138.57	2.0	174.86	175.38

Continued on the next page

p, MPa	T_n, K		p, MPa	T_n, K	
	Exp.	Theory		Exp.	Theory
	Ethane [152]		1.1	373.04	373.17
	$J = 10^7$ m^{-3} s^{-1}		1.9	383.28	383.15
0.335	269.00			Methane [146]	
0.913	273.15			$J = 10^{11}$ m^{-3} s^{-1}	
1.447	277.00				
1.973	281.00		0.40	167.6	167.9
			0.62	168.3	168.9
	Propane [148]		0.82	169.3	169.8
	$J = 10^7$ m^{-3} s^{-1}		1.03	170.5	170.8
			1.23	171.4	171.8
0.7	332.43	332.73	1.43	172.1	172.7
1.3	338.17	338.31	1.63	173.1	173.7
1.9	344.37	344.15	1.83	174.0	174.7
			2.03	175.2	175.8
	Isobutane [148]		2.22	176.4	176.8
	$J = 10^7$ m^{-3} s^{-1}		2.43	177.6	177.8
			2.63	178.6	178.9
			2.63	178.6	178.9
0.4	364.65	365.02	2.82	180.0	179.9

The temperature and pressure dependences of the nucleation rate have been experimentally determined by measuring the lifetime of superheated liquids. The results of experiments on condensed inert gases are presented in Fig. 3.11. Sections with abruptly increasing T_n (the boundary of spontaneous boiling-up of the liquid) correspond to the maximum superheatings. Cryogenic liquids are characterized by a very strong temperature and pressure dependence of the nucleation rate. If in superheated liquid n-pentane at $p = 0.1 p_c$ and $J = 10^7$ m^{-3} s^{-1} the derivative $d \lg J / dT \simeq 2$ K^{-1}, in liquid nitrogen $d \lg J / dT \simeq 12$ K^{-1} is found. At low superheatings, experimental curves have characteristic bends caused by the action of the background radiation initiating the nucleation process [9, 10] and the effect of weak spots in the liquid-measuring-cell system [113] (see Sections 3.12 and 3.13).

Most organic liquids display the initiating effect at rates $J < J_{i*} = 6 \times 10^6$ m^{-3} s^{-1} [9, 10] in cryogenic liquids, condensed inert gases, and benzene at $J < J_{i*} = (1.5-2.5) \times 10^6$ m^{-3} s^{-1} [10, 128, 140–142, 145]. The highest resistance to ionizing radiation is shown by xenon. Here, the threshold value of the rate of initiated nucleation J_{i*} is only 2.2×10^4 m^{-3} s^{-1} [145] (Fig. 3.11c). The low radiation sensitivity of xenon, first discovered by Glaser et al. [162], is caused by its considerable scintillation properties. As a result, the energy of an ionizing particle is de-excited rather than transformed into heat. The scintillation properties of condensed inert gases increase with increasing atomic number of the element [163]. Experimental investigations of the kinetics of fluctuational boiling-up of liquid Ar, Kr, Xe [128,145] confirm such a tendency. The results of investigating the pressure dependence of the attainable superheating temperature for liquefied gases are presented in Table 3.2.

In addition to the experimental data, Tables 3.1, 3.2 and Figs. 3.10–3.12 present the values of the temperatures of attainable superheating obtained from the homogeneous nucleation theory. Calculations of T_n have been made in a macroscopic approximation, i.e., without taking into account the dependence of the properties of critical nuclei on the curvature of the dividing surface. Comparing theory and experiment, besides the error of experimental data, it is necessary to take into account both the errors caused by the approximate character of the theory itself and those caused by the uncertainty of the knowledge of the thermodynamic parameters of the substances under consideration used in the calculations. Different variants of the homogeneous nucleation theory discussed in Section 3.2 differ in the value of the kinetic prefactor B and the factor z_0, which corrects the equilibrium distribution function. Table 3.3 gives the results of computation of the kinetic factor by the theories of Döring–Volmer (Eq. (3.15)), Kagan (Eq. (3.52)), Deryagin, Prokhorov, and Tunitsky (Eq. (3.52)) with $z_0 = \rho'/\rho''_*$. Discrepancies in the values of B do not exceed two or three orders of magnitude. This uncertainty results in differences in the values of T_n from several hundreds to several tenths of kelvin. For all approximations the value of B shows only a weak dependence on temperature and pressure as compared with the exponential factor in Eq. (2.112).

Numerical values of the factor B according to Kagan's theory are approximately 15–20 times lower than those derived by Deryagin, Prokhorov, and Tunitsky. The Döring–Volmer formula (3.15), as compared with Kagan's formula, Eqs. (3.52), (3.51), and $z_0 = 1$, underestimates the value of B by approximately one order of magnitude. If bubble growth in a superheated liquid is limited by viscous forces (Eq. (3.55)), the value of the factor B_{21} differs from the general solution of B_2 by 1.5–2 times. Somewhat larger discrepancies are observed if the approximation B_{24} (Eq. (3.59)) is employed.

The largest contribution to the value of the nucleation rate, Eq. (2.112), is supplied by the exponential factor, which contains in the exponent the work of formation of a critical bubble. The accuracy in the knowledge of the data on the thermodynamic properties of the substance, employed in the computations of W_*, is of great importance for the correct evaluation of its value. Data on the surface tension and the pressure of the saturated vapor in the nucleus are particularly critical for a correct determination of the nucleation rate, J. One percent error in the determination of the surface tension, σ, leads to an error in the value of J of about one order of magnitude. For liquid Ar, Kr, Xe, N_2, O_2, CH_4, C_2H_6, C_3H_8, and C_4H_{10}, in Refs. [153–161] results of mea-

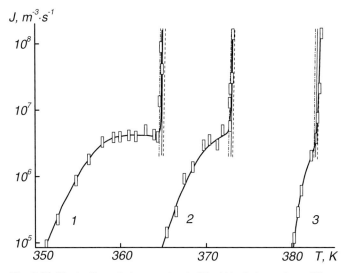

Fig. 3.12 Nucleation rate in superheated liquid isobutane along different isobars. 1: $p = 0.4$ MPa; 2: 1.1; 3: 1.9 [148]. For the meaning of the *dashed*, *thin full*, and *dashed-dotted* lines, see Fig. 3.11.

Tab. 3.3 The kinetic factor, B (in 10^{10} s^{-1}), and the temperature of attainable superheating, T_n (in K), of propane are given according to different versions of the homogeneous nucleation theory. B_1: Eq. (3.15), B_2: Eq. (3.52) and $z_0 = 1$, B_3: Eq. (3.52) and $z_0 = \rho'/\rho''_*$. The experimental values of the temperature of attainable superheating are specified by the superscript *.

lg J	p (MPa)	B_1	B_2	B_3	$T_n(B_1)$	$T_n(B_2)$	$T_n(B_3)$
4	0.7	0.292	3.01	69.5	332.34	332.15	331.90
	1.3	0.347	3.11	67.0	338.00	337.86	337.66
	1.9	0.423	3.02	62.7	343.91	343.82	343.68
7	0.7	0.299	2.98	67.3	332.94	332.73	332.46
						332.43*	
	1.3	0.354	3.09	65.0	338.47	338.31	338.11
						338.17*	
	1.9	0.429	3.01	60.9	344.25	344.15	344.00
						344.37*	
9	0.7	0.304	2.96	65.7	333.37	333.15	322.87
	1.3	0.359	3.07	63.5	338.81	338.65	338.43
	1.9	0.434	3.00	59.6	344.50	344.39	344.24

surements of the surface tension are collected performed on samples of the liquid taken from the same lots of substances for which experiments on the nucleation kinetics have been conducted. This approach makes it possible to compare the results of theory and experiment with higher confidence.

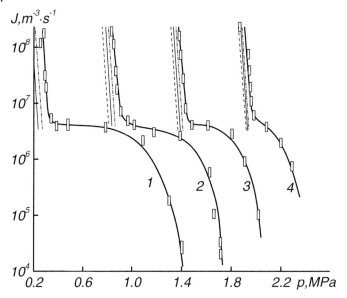

Fig. 3.13 Nucleation rate in superheated liquid ethane for different isotherms. 1: T = 269 K; 2: 273.15; 3: 277; 4: 281 [152]. For the meaning of the *dashed*, *thin full*, and *dashed-dotted* lines, see Fig. 3.11.

Tables 3.1, 3.2 and Figs. 3.10–3.12 show a satisfactory agreement between the homogeneous nucleation theory and experiment. At the same time, a more detailed comparison reveals two principal tendencies. First, for simple liquids (such as Ar, Kr, Xe, N_2, O_2, CH_4) the values of T_n registered experimentally for the boundary of spontaneous boiling-up are systematically lower than the calculated ones. This effect is not observed for liquids with a more complex molecular structure (such as C_3H_8, C_4H_{10}) [148]: the measured values of T_n and those calculated by the homogeneous nucleation theory in a macroscopic approximation agree within the limits of the total error of experiment and calculation. At a pressure of $0.1 p_c$, the value of $\Delta = (T_n^{\text{theor}} - T_n^{\text{exp}})/(T_n^{\text{theor}} - T_s)$ is 2.4% for simple liquids. For liquid propane and isobutane, $\Delta \simeq 0.6\%$. In the methane series, ethane has an intermediate position between a simple liquid (CH_4) and liquids with a complicated molecular structure (C_3H_8, etc.). For this substance, we have $\Delta \simeq 1\%$ (Fig. 3.13).

Discrepancies in T_n can be eliminated assuming that the homogeneous nucleation theory underestimates the value of the pre-exponential factor in Eq. (2.112) by approximately six orders of magnitude or that the surface tension of critical bubbles is about 5–7% smaller than that of a planar interface. Different versions of the homogeneous nucleation theory lead to an uncertainty in the kinetic factor B_2, which does not exceed three orders of magnitude. As will be shown in Section 3.9, abandoning the macroscopic approximation in calculating the work of formation of critical bubbles makes it possible to eliminate the discrepancy between theory and experiment with respect to the values of T_n.

Second, for all liquids investigated, experiments give a weaker curvature for the temperature dependence of the nucleation rate. It should be noted, however, that in experiments the value of the derivative $d \ln J/dT$ is determined with a large error which reaches 20–35%. Owing to a very weak temperature and pressure dependence of the pre-exponential factor in Eq. (2.112), we have

$$\frac{d \ln J}{dT} \simeq -\frac{d(W_*/k_B T)}{dT} = -G_T, \tag{3.93}$$

i.e., the temperature dependence of the nucleation rate is mainly determined by the temperature dependence of the Gibbs number.

The theoretical value of J should be compared with the value of $J_{exp} - J_i$, where J_i is the initiated nucleation rate which is usually unknown. For liquid krypton and xenon in the region of initiated nucleation, $\ln J_{exp}$ is a linear function of temperature (see Figs. 3.11b and c). If we assume that here $J_{exp} \simeq J_i$ and introduce in this approximation a correction to the superheats measured at the boundary of spontaneous boiling-up, at $p = 1.0$ MPa we get $d \ln J/dT \simeq 14.3$ K^{-1} (Kr). According to the homogeneous nucleation theory, $d \ln J/dT \simeq 19$ K^{-1}, and the experimental value is $d \ln J_{exp}/dT \simeq 12$ K^{-1}. The experimental data in Figs. 3.11–3.13 and Tables 3.1 and 3.2 are presented without any correction accounting for initiated nucleation.

In experiments on continuous isobaric heating, the slope of the temperature dependence of J is determined by the halfwidth of the distribution of liquid boiling-up events as a function of temperature. A good "inscription" of the experimental histogram into the theoretical curve (3.87) (see Fig. 3.4) is an indication of the agreement between theory and experiment not only in the value of T_s, but also in the temperature dependence of J. Differentiating the logarithm of the nucleation rate, Eq. (2.112), with respect to pressure (at $T =$ const), we have

$$\frac{d \ln J}{dp} \simeq -\frac{d(W_*/k_B T)}{dp} = G_p. \tag{3.94}$$

Equations (3.4) and (3.7) yield

$$G_p = -\frac{32\pi}{3k_B T} \frac{\sigma^3}{(p''-p)^3} = -\frac{V_*}{k_B T}, \tag{3.95}$$

i.e., the slope of the pressure dependence of J determines the characteristic size of a critical bubble.

The boundary of spontaneous boiling-up of superheated condensed inert gases satisfies the law of corresponding states [122], and in reduced thermodynamic variables ($\widetilde{T} = T/T_c$, $\widetilde{p} = p/p_c$) may be approximated by the equation [100]

$$\widetilde{T}_n = 1 - 0.139(1 - \widetilde{p})^{0.955}(1 + 0.075\widetilde{p}). \tag{3.96}$$

The analysis of the thermodynamic similarity with respect to the conditions of spontaneous nucleation for a wider class of radii requires the intro-

Tab. 3.4 Parameters of the critical point (T_c, p_c), the parameter of thermodynamic similarity (A), and the quantum-mechanical parameter (Λ) of liquefied gases.

Substance	T_c (K)	p_c (MPa)	A	Λ
^{3}He	3.324	0.1149	19.93	1.889
^{4}He	5.189	0.2274	15.07	1.411
Ne	44.40	2.653	4.637	0.238
Ar	150.66	4.860	4.075	0.0749
Kr	209.39	5.510	4.066	0.0410
Xe	289.77	5.842	4.011	0.0255
Rn	377.6	6.32	4.00	
pH_2	32.98	1.293	8.66	0.7600
nH_2	33.24	1.297	8.61	0.7560
HD	35.90	1.484	7.50	0.605
HT	37.13	1.570	6.80	0.520
nD_2	38.35	1.665	6.63	0.516
oD_2	38.26	1.650	6.65	0.515
DT	39.42	1.773	6.17	0.461
nT_2	40.44	1.850	5.87	0.418
N_2	126.20	3.400	3.553	0.0920
O_2	154.58	5.043	3.751	0.0829
F_2	144.50	5.250	3.617	
Cl_2	417.16	7.709	3.137	
CO	132.92	3.498	3.457	
CO_2	304.20	7.382		
CF_4	227.7	3.74	2.23	
CD_4	192.5	4.65		
SF_6	318.69	3.761		
CH_4	190.54	4.598	3.927	0.0953
C_2H_6	305.33	4.871	2.886	
C_3H_8	370.4	4.252	2.396	
iC_4H_{10}	407.85	3.631	2.060	
nC_4H_{10}	425.16	3.796	2.040	

duction into the law of corresponding states of additional criteria [164] reflecting special features of the particular substances. In accordance with the one-parameter version of thermodynamic similarity we have

$$\widetilde{T}_n = \widetilde{T}_n(\widetilde{p}, A), \tag{3.97}$$

where A is an individual parameter which has the meaning of a characteristic similarity criterion. Following Filippov [164], we shall assume

$$A = 100\widetilde{p}_{0,625}. \tag{3.98}$$

Here, $\widetilde{p}_{0,625}$ is the reduced saturation pressure at the reduced temperature, $\widetilde{T} = 0.625$. The values of the similarity parameter A, and also the critical parameters for the substances studied in this book, are given in Table 3.4.

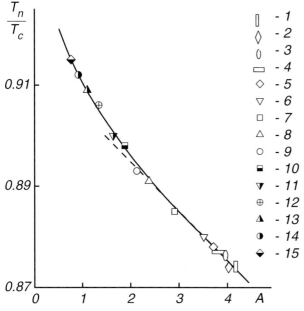

Fig. 3.14 Reduced temperature of attainable superheating ($J = 10^7$ m^{-3} s^{-1}) as a function of the parameter A at $p/p_c = 0.1$. 1: argon [128]; 2: krypton [145]; 3: xenon [145]; 4: methane [141]; 5: oxygen [140]; 6: nitrogen [142]; 7: ethane [152]; 8: propane [148]; 9: isobutane [148]; 10: n-pentane [10]; 11: n-hexane [10]; 12: n-heptane [10]; 13: n-octane [125]; 14: n-nonane [125]; 15: n-decane [125].

In Fig. 3.14, the temperatures of attainable superheating are presented as functions of the similarity criterion A [164] of liquids belonging to different classes of chemical compounds. The smoothness of the A-dependence of T_n confirms the universal character of Eq. (3.97) for normal liquids. For liquefied gases (with the exception of quantum liquids), the function $\widetilde{T}_n(\widetilde{p}, A)$ is linear

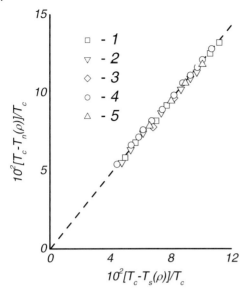

Fig. 3.15 Reduced temperature of limiting supercooling in an isochoric process as a function of the reduced saturation temperature of argon (1), nitrogen (2, 3), and oxygen (4, 5) at nucleation rates: 1, 2, 4: $J = 10^{11}$ m^{-3} s^{-1}; 3, 5: 10^7 m^{-3} s^{-1}.

in A and approximated in Ref. [100] by

$$\widetilde{T}_n = 1 - b(1-\widetilde{p})^q(1+c\widetilde{p}), \tag{3.99}$$

where

$$b = 1 - \widetilde{T}_n(\widetilde{p}=0) = 0.0955 + 0.01045A; \qquad q = 0.955; \qquad c = 0.075. \tag{3.100}$$

Equations (3.96) and (3.99) are not applicable in the near vicinity of the critical point as they do not satisfy the condition $d\widetilde{T}_n/d\widetilde{p}|_c = d\widetilde{T}_s/d\widetilde{p}|_c$.

The temperature dependence of the limiting stretching pressure may be represented for a wide class of normal liquids at different values of nucleation rate as [165]

$$\widetilde{p}_n = \widetilde{p}_s - a\frac{\widetilde{\sigma}^{3/2}}{\widetilde{T}^{1/2}}\left(1 - \frac{\widetilde{\rho}''_s}{\widetilde{\rho}'_s}\right)^{-1}. \tag{3.101}$$

Here, $\widetilde{p}_s$ is the reduced saturation pressure, $\widetilde{\rho}'_s = \rho'_s/\rho_c$ and $\widetilde{\rho}''_s = \rho''_s/\rho_c$ are the reduced liquid and vapor orthobaric densities, and a is a parameter which is a function of the nucleation rate,

$$a = \left[\frac{7.28}{(36.8 - \lg J)}\right]^{1/2}. \tag{3.102}$$

The factor $(1 - \widetilde{\rho}'_s/\widetilde{\rho}_s)^{-1}$ has an appreciable effect on the value of $\widetilde{p}_n$ only at temperatures close to the critical point and may be omitted at $\widetilde{T} < 0.99$. Equation (3.101) makes it possible to calculate the value of the stretching limit for normal liquids employing the known values of (p_c, T_c, A), making use of the dependences $\widetilde{\sigma}(\widetilde{T}, A)$ and $\widetilde{p}_s(\widetilde{T}, A)$ presented in Refs. [10, 57, 164].

Some other correlations for the line of attainable superheating are also discussed in the literature. At a fixed density of the liquid, the following simple relation holds (Fig. 3.15) [38]:

$$\frac{[T_c - T_n(\rho)]}{T_c} = \gamma \frac{[T_c - T_s(\rho)]}{T_c}, \tag{3.103}$$

where $\gamma = 1.182$ if $J = 10^7$ m^{-3} s^{-1} and $\gamma = 1.191$ if $J = 10^{11}$ m^{-3} s^{-1}. According to Ref. [135], at atmospheric pressure the reduced superheating $(T_n - T_s)/(T_c - T_s)$ is for all normal liquids approximately constant and equal to 0.68.

3.8
Superheating of Quantum Liquids

At sufficiently low temperatures the physical properties of macroscopic systems are largely determined by quantum effects. Besides helium, quantum effects show up noticeably in the properties of isotopes of hydrogen and, to a lesser degree, of neon. When the de Broglie wavelength, $\lambda_T = (2\pi\hbar^2/mk_BT)^{1/2}$, exceeds the typical dimension of a critical nucleus, R_*, the thermo-activation mechanism of nucleation is replaced by quantum tunneling of heterophase fluctuations through a potential barrier [8] (see Section 3.1). According to the estimates made in Ref. [101], the temperature of the change of the nucleation regimes, T_*, is equal to 0.3 K for liquid helium and 2–3 K for hydrogen. In liquid neon, we have $T_* \simeq 1.5$–2 K [100]. At $T > T_*$, quantum effects may have an indirect influence on the boiling-up kinetics of a superheated liquid through thermophysical parameters and peculiar nucleation centers not typical for classical systems, vortex lines and rings in superfluid helium [101, 166], thermal spikes of ortho–para conversion in normal hydrogen [133], electronic and positronic bubbles in liquid neon, hydrogen, and helium [100, 101, 133].

Owing to the repulsive exchange forces between the electron and atoms of some cryogenic liquids, in the latter there may occur a localization of the electron inside a spherical cavity. The existence of such electronic bound state (electronic bubble) has been confirmed by experiments on the mobility of charged particles in quantum liquids [101, 167, 168]. Free electrons may develop when ionizing particles of cosmic radiation are passing through a liquid. As a result of a large number of collisions, the electron energy decreases to the thermal energy. This effect makes its effective localization in a number of liquids possible.

The change of the liquid free energy during the formation of an electronic bubble may be presented as [169]

$$\Delta F = F_e + F_R + F_p - E_0, \tag{3.104}$$

where F_e is the energy of the ground state of the electron in a spherically symmetric potential well, F_R is the work of formation of a vapor bubble of radius R in the liquid, F_p is the energy of polarization interaction of the electron with the surrounding particles, and E_0 is the value of the energy barrier between the electron and the atoms of the liquid. In writing the excess thermodynamic potential in the form given by Eq. (3.104), we assume that the lifetime of the electron on a density fluctuation considerably exceeds the characteristic time of thermal relaxation in the medium. We assume furthermore that there is enough time for the establishment of a local equilibrium in the liquid.

The minimum value of the work of bubble formation is given by Eq. (3.11). In a polar liquid, owing to the interaction with induced dipoles of molecules, a localized electron gains some additional energy

$$F_p = -\frac{\varepsilon - 1}{2\varepsilon} \frac{e^2}{R}, \tag{3.105}$$

where ε is the dielectric constant of the liquid and e is the electron charge. If in helium the polarization contribution is small ($F_p \simeq 3.2 \times 10^{-21}$ J) and may be neglected, in liquid hydrogen it is comparable in magnitude with the other terms in Eq. (3.104) ($F_p \simeq 1.6 \times 10^{-20}$ J) and has to be taken into account when one is calculating the energy of localization of the electron.

In the Wigner–Seitz approximation [170], the energy of an electron in the field of the atoms surrounding it is

$$E_0 = \frac{\hbar^2 q_0^2}{2m_e}, \tag{3.106}$$

$$q_0 R_s = \mathrm{tg}\left[q_0 \left(R_s - l\right)\right]. \tag{3.107}$$

Here, m_e is the electron mass, $R_s = ((4/3)\pi\rho)^{-1/3}$ is the radius of the Wigner–Seitz sphere, and l is the electron scattering length on a liquid atom. The value of F_e is found by solving the problem of description of the motion of an electron in a spherically symmetric potential well, the depth of which taking into account the polarization interaction is equal to $E_0' = E_0 - F_p$ [59].

The dependence $\Delta F(R)$ for an electronic bubble in normal liquid hydrogen is shown in Fig. 3.16. The electron scattering length was assumed to be equal to $1.51 a_0$ [171], where a_0 is the Bohr radius. Electron localization is possible if $\Delta F < 0$ holds. The equilibrium radius of an electronic bubble, R_0, is determined by the condition $d\Delta F = 0$. Penetrating deeper into a metastable region, an increase in the volume work against the pressure forces can be observed determined by the second component in Eq. (3.104). At a certain degree of superheating, on the curve describing the dependence of ΔF on R the minimum disappears. In such a case, the existence of a stable equilibrium of an electronic bubble becomes impossible. At the boundary of stability $\partial^2 \Delta F / \partial R^2 = 0$, and $R_0 = R_+$.

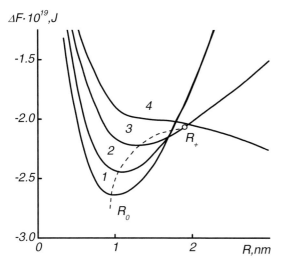

Fig. 3.16 Hydrogen excess free energy as a function of the electronic bubble radius at pressures $p = 0.1$ MPa and temperatures T (in K). 1: 14; 2: 20; 3: 25; 4: 29. The *dashed line* corresponds to the equilibrium bubble.

According to the calculations made in Refs. [100, 101, 133], electronic bubbles in superheated liquid helium, hydrogen, and neon are stable in the whole range of state variables from the curve of phase equilibrium to the line of attainable superheating corresponding to the nucleation rate $J \simeq 10^{15}$ m^{-3} s^{-1}. At pressures $0 < p < 0.7 p_c$, the equilibrium radius of an electronic bubble in liquid helium is 2.0–3.2 nm, in hydrogen 1.4–2.5 nm, and in neon 1–2 nm. The size of an electronic bubble increases with temperature and approaches infinity at the critical point.

With a subsystem of electronic bubbles in a superheated liquid, nucleation may take place not only as a result of spontaneous formation of viable centers of a new phase, but also by fluctuational "growing-up" of electronic bubbles up to critical dimensions. If homogeneous nucleation is connected with an overcoming of the potential barrier, W_*, nucleation initiated by electronic bubbles requires lower activation barriers [101, 133], i.e.,

$$W_e = W_* - W_0, \qquad (3.108)$$

where

$$W_0 = 4\pi R_0^2 \sigma - \frac{4}{3}\pi R_0^3(p'' - p). \tag{3.109}$$

In the stationary case, the nucleation rate on the subsystem of electronic bubbles may be presented in the form of Eq. (2.199). The total flux of nuclei taking into account the homogeneous nucleation and that initiated by electronic bubbles is

$$J' = J + J_e. \tag{3.110}$$

Electronic bubbles will have an appreciable effect on the boiling-up process at $J' \geq J$.

Depending on the orientation of nuclear spins, the molecules of hydrogen and its isotopes may be in different energy (ortho or para) states. The equilibrium ortho–para composition is determined mainly by temperature. Normal hydrogen, i.e., hydrogen that is in equilibrium at room temperatures, contains 75% of ortho-hydrogen and 25% of para-hydrogen. When keeping liquefied normal hydrogen, a spontaneous transformation of an ortho-modification into a para-modification can be observed. The ortho–para transition is accompanied by local heat emissions and, consequently, by an increase in the probability of formation of a nucleus in the vicinity of the ortho-molecules that have interacted.

We now examine the possibility of initiating the boiling-up of superheated normal hydrogen by elementary acts of spontaneous ortho–para conversion [133]. In the absence of a catalyst, ortho–para conversion is caused by the magnetic interaction of ortho-molecules. The frequency of the elementary acts of conversion is proportional to the square of concentration of ortho-molecules, x [172],

$$J'' = c_0 \rho' x^2, \tag{3.111}$$

where $c_0 = 3.51 \times 10^{-6}$ s^{-1}, and ρ' is the particle number density in the liquid. The conversion will result in an appreciable contribution to the nucleation process if the amount of heat released is sufficient for the formation of a critical nucleus, and the frequency of the elementary acts of conversion exceeds the rate of homogeneous nucleation and that of nucleation initiated by some other factors. At $T = 28$–30 K, we get $J'' = 10^{22}$ m^{-3} s^{-1}. The energy released in the transformation of two hydrogen molecules from the ortho- into the para-state is

$$\Delta h \simeq 2(u_{oH_2} - u_{pH_2}), \tag{3.112}$$

where

$$u = k_B \frac{\partial \ln Z}{\partial (1/T)}, \tag{3.113}$$

and $Z = Z_o$ and $Z = Z_p$ are the rotational components of the partition function of ortho-particles and para-particles [20]. At sufficiently low temperatures ($T < T_R = \hbar^2/2k_B I_m$, where I_m is the moment of inertia of the molecule) limiting ourselves to the first terms of the expansion of the partition function into a series, we have

$$\Delta h \simeq 4k_B T_R. \tag{3.114}$$

For hydrogen $T_R = 85$ K; then $\Delta h = 5 \times 10^{-21}$ J. The total energy, W''', required for the formation of a critical-size bubble consists of the energy of reversible formation of the bubble and the sum of irreversible energy losses connected with liquid inertia, viscosity, and thermal conductivity [173]. On the line of attainable superheating ($J = 10^{11}$ m^{-3} s^{-1}) at $T = 29$ K we have $W''' = 2.5 \times 10^{-18}$ J, the equilibrium work of formation of a critical bubble equals $W_* = 2.7 \times 10^{-20}$ J. The value of W''' changes only slightly with pressure, decreasing with it. Thus, natural ortho–para conversion in normal hydrogen may not have a considerable effect on the process of spontaneous nucleation. The effect of initiation may be observed if paramagnetic impurities exist in the system.

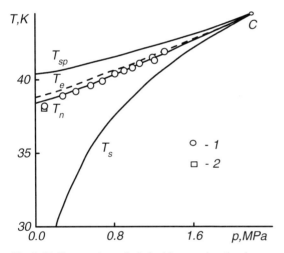

Fig. 3.17 Temperature of attainable superheating for neon. 1: experimental data [100]; 2: [134]; T_{sp}: spinodal; T_e: electronic bubble stability boundary; T_n: boundary of spontaneous boiling-up ($J = 10^{11}$ m^{-3} s^{-1}); T_s: binodal.

Data of measurements of the temperature of attainable superheating for neon [100, 134], hydrogen [133, 134], and helium isotopes [132, 135, 149, 150] are presented in Tables 3.1 and 3.2. They have been obtained by the methods of continuous isobaric heating, lifetime measurement, and by the pulse method. The results of Refs. [100, 134] on the superheating of liquid neon by the method of continuous isobaric heating in glass capillaries agree well with each other and with the homogeneous nucleation theory (Fig. 3.17). Similar to other liquefied inert gases, at low pressures there is a systematic liquid "underheating" to theoretical values of T_n. At a pressure $p = 0.1 p_c$, the attainable superheating is 8.0 K ($J = 10^{11}$ m^{-3} s^{-1}) [100]. On the spinodal, $\Delta T_s \simeq 9.9$ K. The stability boundary of an electronic bubble corresponds to a superheating that exceeds the attainable superheating by only 0.3 K. The closeness of superheats achieved by experiment to those calculated by the homogeneous nucleation theory indicates that the concentration of free electrons in liquid neon capable of forming electronic bubbles does not exceed $(10^{11}$–$10^{16})$ m^{-3}.

The first investigations of the superheating of liquid hydrogen were directed to the development of liquid-hydrogen bubble chambers and cryopumps [174, 175]. The authors of Ref. [174], employing a pure bubble chamber of volume $\simeq 3$ cm^3, determined the mean expectation time of liquid hydrogen boiling-up ($\bar{\tau} = 22$ s) at atmospheric pressure and temperature $T = 24.7$ K. Figure 3.18 shows the lower boundary of the zone of liquid-hydrogen sensitivity to high-energy particles obtained on the bubble chamber of the Joint Institute for Nuclear Research in Dubna, Russia, with 100 cm in diameter. Hord et al. [175] investigated superheated liquid hydrogen in a glass Dewar flask ($V \simeq 1$ l) creating an abrupt pressure decrease in the vapor cavity. The moment of boiling-up was registered by a pressure burst. The rates of the pressure decrease were equal to 0.14–9.0 MPa s^{-1}. The results of the experiments are shown in Fig. 3.18. At pressure-release rates exceeding 7 MPa s^{-1}, limiting stretches of liquid hydrogen are close to their predictions based on the homogeneous nucleation theory.

In experiments on continuous isobaric heating, a considerable spread in boiling-up temperatures of normal liquid hydrogen can be observed [133,134]. Thus, at a heating rate of 0.03 K s^{-1} and a pressure of 0.15 MPa, the boiling-up of nH$_2$ was found to take place in the temperature range from 26.0 to 28.1 K. With increasing heating rate the halfwidth of the temperature distribution of liquid boiling-up events decreased, always remaining larger than the theoretical value of $\delta T_{1/2} \simeq 0.03$ K. In the papers [133,134], the maximum value of the registered boiling-up temperature was taken as the attainable superheating temperature of hydrogen.

The first measurements of the temperature of attainable superheating for liquid helium (^{4}He) were made applying the method of pulse heating on bismuth monocrystals (see Tables 3.1 and 3.2, Fig. 3.19) [131, 132]. The experimental data have been approximated by the equations (note that Eq. (3.116) is not applicable in the nearest vicinity of the critical point)

$$\Delta T_n = 4.322 \left(1 - \frac{T_s}{T_c}\right)^{1.534}, \tag{3.115}$$

$$\widetilde{T}_n = 0.783 + 0.224 \widetilde{p}. \tag{3.116}$$

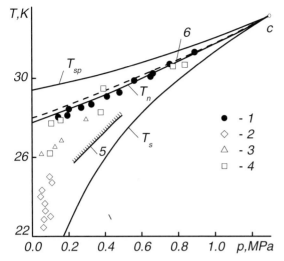

Fig. 3.18 Limiting superheating of liquid hydrogen. 1: experimental data [133]; 2–4: [175] at rates of pressure decrease equal to $\dot p =$ 1.8 MPa s^{-1}; 3.3; 8.0, respectively; 5: boundary of radiation sensitivity [173]; 6: boundary of stability of electronic bubbles; T_n: boundary of spontaneous boiling-up ($J = 10^{11}$ m^{-3} s^{-1}), T_s: binodal, T_{sp}: spinodal.

It was noted that the superheating temperature, T_n, does not depend on the quality of processing a bismuth thermometer–heater surface, the external magnetic field, and the intensity of X-ray radiation. The authors of Ref. [132] refer their data to the nucleation rate $J = 10^7$ m^{-3} s^{-1}. Our evaluations (see Section 3.3) resulted in $J = 10^{18}$–10^{21} m^{-3} s^{-1}.

In Fig. 3.19, experimental data on the temperature of attainable superheating of liquid helium [132, 135, 150] are compared with the results of calculation by the homogeneous nucleation theory (the Döring–Volmer variant, Eqs. (2.112), (3.15), and $z = 1$). The values of the surface tension recommended in Ref. [57] are used.

The results of Nishigaki and Saji [135] obtained by the method of continuous isobaric heating refer to the nucleation rate $J = 10^7$ m^{-3} s^{-1}, and taking this into account agree well with the data of the Brodie group [132]. The authors of Ref. [135] point out that, having achieved a satisfactory agreement between experiment and theory, the values of T_n calculated from the homogeneous nucleation theory are systematically higher than the experimental values.

Experiments on the mean lifetime of superheated ^{4}He [150, 176] have yielded dependences $\bar\tau = \bar\tau(T)$ qualitatively different from those of other liquids: the value of τ was practically temperature independent in the whole range from the saturation line to the boundary of spontaneous boiling-up (Fig. 3.20). The superheating of helium was realized in glass cells of volume $V \simeq 2$ cm^3 with a limiting rate of initiated nucleation $J_{i*} \simeq 5 \times 10^5$ m^{-3} s^{-1}.

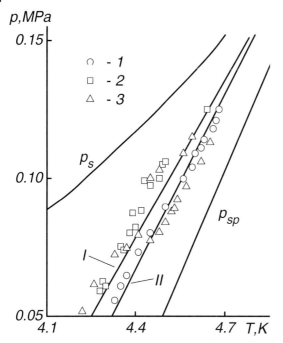

Fig. 3.19 Attainable superheating of liquid ^{4}He. 1: experimental data [132]; 2: [135]; 3: [150]. I, II: calculation by Eqs. (2.112), (3.15), and $z_0 = 1$ for $J = 10^7$ m^{-3} s^{-1} and $J = 10^{20}$ m^{-3} s^{-1}, respectively; p_s: binodal; p_{sp}: spinodal.

This result is less than the respective one for argon and krypton, but larger than that for xenon. The reasons for such a behavior of ^{4}He are yet to be explained. Helium is used as a working substance in bubble chambers, but no information was given on any peculiarities for the boundary of the initiation zone in this substance [173]. Besides, the limiting superheats of liquid helium obtained in Ref. [150] at $p > 0.08$ MPa considerably exceed those registered in Ref. [132], although they refer to much lower nucleation rates (see Fig. 3.19). The phenomenon of superheating is also characteristic of the superfluid phase of ^{4}He (HeII).

In Refs. [177, 178] superfluid helium was superheated in glass cells of volume $V \simeq 0.72$ cm^3. An electric heater and a resistance thermometer were transferred into a liquid. The change in temperature of the liquid was registered at a fixed heating rate. Collapse of the metastable state of HeII resulted in an abrupt temperature decrease or increase in the cell. The first case corresponds to the boiling-up of liquid helium, the second one, to the HeII–HeI phase transition. In the (p, T)-diagram (Fig. 3.21), the points of the HeII–HeI transition form a smooth curve, which is an extension of the λ-line into the region of metastable states.

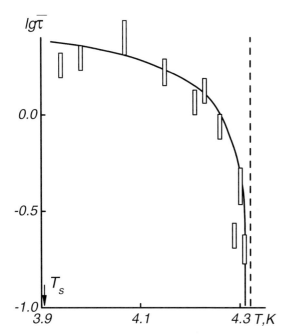

Fig. 3.20 Temperature dependence of the mean lifetime for superheated ^{4}He at a pressure of $p = 0.073$ MPa.

In experiments on the pulse heating of liquid ^{3}He on bismuth monocrystals, it was possible to determine the temperature of attainable superheating only at pressures that were lower than $0.5 p_c$ [149]. The comparison of the data obtained with the homogeneous nucleation theory is complicated due to the lack of reliable information on the surface tension of the light helium isotope. At temperatures above 2.5 K, discrepancies between the results of different authors in σ reach 30% [57]. Table 3.2 presents the theoretical values of T_n obtained by using the values of surface tension from Ref. [179].

Figure 3.22 shows the lines of attainable superheating for liquefied inert gases, hydrogen, and helium isotopes in reduced thermodynamic coordinates [100]. Deviations of the boundaries of spontaneous boiling-up of neon, hydrogen, and helium from the law of corresponding states are caused by quantum effects. The degree of quantization of the system is determined by the value of the de Boer parameter, which may be written in terms of the critical parameters of the respective substances as

$$\Lambda = 2\pi\hbar \frac{p_c^{1/3}}{m^{1/2}(k_B T_c)^{5/6}}. \tag{3.117}$$

In a quantum liquid $\tilde{T}_n = \tilde{T}_n(\tilde{p}, \Lambda)$. With an error that does not exceed 0.5%, the temperature of attainable superheating (for a nucleation rate $J =$

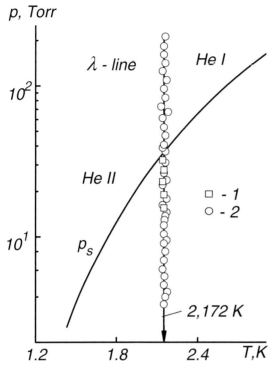

Fig. 3.21 λ-line of helium-4. 1: by data of Ref. [177]; 2: [178]. p_s: line of liquid–vapor phase equilibrium.

10^{11} m^{-3} s^{-1}) for liquids exhibiting quantum effects may be approximated by Eq. (3.99) where

$$b = 0.1288 \exp(0.4397\Lambda), \quad q = 0.948 + 0.0823\Lambda,$$
$$c = 0.0866 - 0.155\Lambda. \tag{3.118}$$

In a wide range of nucleation rates, the cavitation strength of quantum (as well as classical) liquids is calculated by Eq. (3.101). The explicit form of the dependence $\tilde{\sigma}(\tilde{T}, \Lambda)$ is given in Ref. [57]. The reduced orthobaric densities $\tilde{\rho}_s$ and $\tilde{\rho}'_s$ also depend on the parameter Λ, but the ratio $\tilde{\rho}''_s / \tilde{\rho}'_s$ is a much weaker function of Λ than $\tilde{\sigma}(\tilde{T}, \Lambda)$. The factor $(1 - \tilde{\rho}''_s / \tilde{\rho}'_s)^{-1}$ should be taken into account only in the nearest vicinity of the critical point.

3.9
Surface Tension of Vapor Nuclei

Spontaneous boiling-up of a superheated liquid is preceded by the formation of a small nucleus, the critical bubble. Critical-bubble radii range from 3.5 to

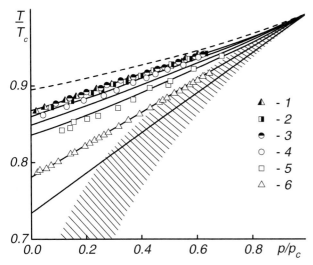

Fig. 3.22 Temperature of attainable superheating for cryogenic liquids according to experimental data. 1: Ar [122]; 2: Kr [122]; 3: Xe [122]; 4: Ne [100]; 5: nH_2 [133]; 6: ^{4}He [132]. *Solid lines* show the results of calculations by Eqs. (3.96) (Ar, Kr, Xe) and (3.99), (3.118) (Ne, nH_2, ^{4}He, ^{3}He). The *dashed line* shows the spinodal curve of neon. The region of states between the saturation lines of Ar and ^{3}He is shaded.

5.0 nm; the number of molecules contained in them is 300–1000 (for $p < 0.7 p_c$). In the classical homogeneous nucleation theory it is usually assumed that the properties of such small objects do not differ from those of macroscopic phases (macroscopic or capillary approximation). If in calculating the work of formation of a critical bubble Gibbs' method of separating surfaces is used, the capillarity approximation implies to use for bubbles of all sizes the value of the surface tension for equilibrium coexistence of the respective liquid and gas at a planar interface, i.e., $\sigma = \sigma_\infty$.

Tab. 3.5 Radius and number of molecules in a critical bubble and value of the surface tension of the liquid–vapor interface in xenon.

T (K)	p (MPa)	R_* (nm)	N_*''	σ (mN m^{-1})	
				$\sigma(R_*)$	σ_∞ [156]
251.41	0.24	3.9	245	3.97	4.25
253.34	0.55	4.0	285	3.73	3.98
256.06	0.99	4.2	360	3.40	3.61
259.03	1.48	4.5	480	3.02	3.22
262.18	1.98	4.9	660	2.63	2.81

Experimental data on the kinetics of nucleation in simple liquids show that, with satisfactory agreement between the homogeneous nucleation theory and experiment, the measured values of temperatures of attainable superheating are systematically lower than the calculated values. Neglecting the dependence of the properties of the critical nuclei on the radius of the separating surface can be considered as a possible reason for such a discrepancy [180]. If the homogeneous nucleation mechanism is experimentally achieved, then from data on T_n and J and from Eqs. (2.112) and (3.8) we can evaluate the excess free energy of a nucleus, $\sigma(R_*)$ [9, 10]. In Table 3.5, values of the surface tension for critical bubbles of xenon, obtained in such a way, are compared with data referring to a planar interface [145]. The parameters T and p correspond to the nucleation rate $J = 10^7$ m^{-3} s^{-1}. The radii of the critical bubbles have been calculated by Eq. (3.4); the number of molecules in a bubble has been computed from the well-known vapor density. The values of $\sigma(R_*)$ are systematically lower than those of σ_∞; the discrepancy is of the order of 5–7%. Similar discrepancies are found for all other simple liquids [180]. In contrast, if systems are considered where the molecular structure of the liquid is more complicated, then the differences in $\sigma(R_*)$ and σ_∞ decrease. Thus, for propane and isobutane the value of $\Delta\sigma = \sigma_\infty - \sigma(R_*)$ is already comparable with the total error of calculation and experiment [148].

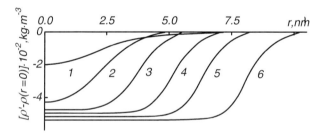

Fig. 3.23 Density distribution in critical bubbles of nitrogen at a pressure of $p = 1.0$ MPa and different temperatures [142]. 1: $T = 117$ K; 2: 116; 3: 114; 4: 113; 5: 112; 6: 111.

The effect of the nucleus curvature can be taken into account when, in calculating the work of formation of critical bubbles, the Cahn–Hilliard approach [58] is used which is based on the theory of capillarity developed by van der Waals [7]. Besides the nucleation work, this method makes it possible to obtain information on the structure of the critical heterophase fluctuation. According to Eq. (2.36), the functional of the excess free energy of a system containing a heterophase fluctuation will be given if the free energy of a homogeneous system, f, which is to be determined in the metastable and unstable states, and the coefficient, κ, are known. In Refs. [142, 181], the function $f(T, \rho)$ was computed based on the experimental data on thermodynamic properties

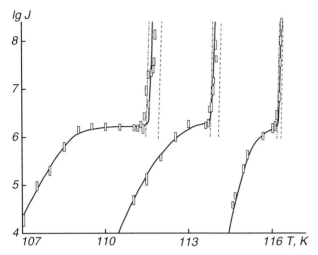

Fig. 3.24 Nucleation rate in superheated liquid nitrogen along different isobars [142]. 1: $p = 0.5$ MPa; 2: 1.0; 3: 1.5. *Dashed lines* show calculations performed by Eqs. (2.112), (3.8), (3.52), and $z_0 = 1$ with $\sigma = \sigma_\infty$; *dashed-dotted lines*: Eqs. (2.112), (3.11), (3.52), and $z_0 = 1$.

of superheated liquids [52,53]. The coefficient κ was calculated from the equation

$$\sigma_\infty = 2 \int_{\rho'_s}^{\rho''_s} (\kappa \Delta f)^{1/2} d\rho, \qquad (3.119)$$

employing data on the surface tension at a planar interface [57]. It was assumed that the value of κ does not depend on density. The numerical solution of the Euler equation (2.39) for the search for an extremum of the functional gives the density distribution in bubbles of critical sizes. The results of such a calculation for liquid nitrogen are presented in Fig. 3.23 [142]. At low superheatings, in a nucleus one can distinguish between a homogeneous core and a transition layer, the thickness of which is smaller than the bubble size. Penetrating deeper into the metastable region, the homogeneous core decreases, and at a certain degree of superheating, the density distribution function takes the form of a dome. The density at the center of a vapor bubble begins to differ, at very high superheatings, from the value determined by the equilibrium condition, Eq. (3.5). A coefficient of metastability, $\beta = (p_s - p)/(p_s - p_{sp}) \simeq 0.45$, corresponds to such degrees of superheating. The work of formation of a critical nucleus obtained in the framework of the Cahn–Hilliard approach (Eq. (3.10)) is smaller than that calculated by the Gibbs formula (3.8) employing in its evaluation the capillarity approximation, $\sigma(R_*) = \sigma_\infty$. At the boundary of spontaneous boiling-up

($J = 10^7$–10^{11} m^{-3} s^{-1}) the discrepancies are of the order of 15%. In this case, the equilibrium density differs from the density at the center of a vapor bubble only by 0.05%. Figure 3.24 presents, in a semilogarithmic scale, the results of experiments on the measurements of the nucleation rate in liquid nitrogen [142]. The agreement between theory and experiment improves if in the calculations of J one uses the expression for the work of formation of a critical nucleus, Eq. (3.10).

When solving the system of equations (3.4), (3.5), (3.8), and (3.10), the dependence of the surface tension of the nucleus on the curvature of the separating surface can be determined [142, 181]. The results of such a calculation for bubbles of nitrogen, oxygen, and methane at $T = 0.95T_c$ in the region of intensive homogeneous nucleation are shown in Fig. 3.25. The values for the surface tension of bubbles of radius $R_* \simeq 5$ nm are by 4%, and of radius $R_* \simeq 2.5$ nm by 15% smaller than that at a planar interface. The stratification of the curves $\tilde{\sigma}(\tilde{R}_*)$ is caused by the absence of a complete thermodynamic similarity of the investigated liquids.

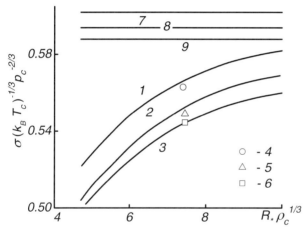

Fig. 3.25 Surface tension of critical bubbles versus the radius of curvature of the surface of tension (1–3) [181], the values of σ by data on the attainable superheating (4–6), and σ_∞ (7–9) of nitrogen (1, 4, 7), oxygen (2, 5, 8), and methane (3, 6, 9).

Figure 3.26 presents temperature dependences of the surface tension for critical bubbles of liquid nitrogen along isobars. The arrows show the temperatures of attainable superheating [142]. Different approaches to the determination of the dependence of the surface tension of new-phase nuclei on the curvature of the separating surface and the influence of the size effect on the work of nucleus formation are discussed in Refs. [182–185].

In addition to the surface of tension we consider the equimolecular dividing surface. For the radius of the equimolecular dividing surface of a vapor bubble

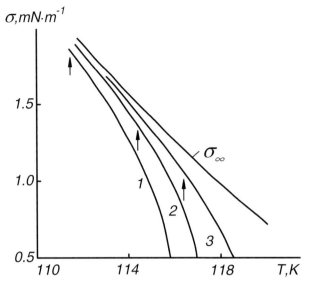

Fig. 3.26 Temperature dependence of the surface tension of critical nitrogen bubbles at different pressures. 1: $p = 0.5$ MPa; 2: 1.0; 3: 1.5. σ_∞ is the surface tension at a planar interface.

we have

$$R_e = \left\{ \frac{3 \int_0^\infty [\rho - \rho(r)] r^2 dr}{\rho - \rho''(0)} \right\}^{1/3} = \left[\frac{n_+}{\rho - \rho''(0)} \right]^{1/3}, \quad (3.120)$$

where $\rho''(0)$ is the density at the center of the nucleus calculated from the equilibrium condition, Eq. (3.5). The quantity n_+ denoted the number of particles transferred to infinity during the formation of a bubble. The surface tension of the equimolecular dividing surface and that of the surface of tension are related by the following equation:

$$\sigma_e = \sigma \left[\frac{1}{3} \left(\frac{R_*}{R_e} \right)^2 + \frac{2}{3} \left(\frac{R_e}{R_*} \right) \right]. \quad (3.121)$$

Some characteristic parameters for critical bubbles of nitrogen, oxygen, and methane at $T = 0.9 T_c$ and $J = 10^{11}$ m^{-3} s^{-1} are given in Table 3.6.

At the boundary of spontaneous boiling-up of a superheated liquid, σ_e is larger than σ by approximately 0.3%. For $R_* = 2.5$–5.0 nm, the value of $\delta(R_*) = R_e - R_*$ is positive and essentially depends on the curvature of the

Tab. 3.6 Parameters of critical bubbles of nitrogen, oxygen and methane.

Parameter	N_2	O_2	CH_4
p (MPa)	0.9	1.39	1.31
R_* (nm)	3.9	3.7	4.1
N_*''	400	380	410
$\sigma(R_*)$ (mN m^{-1})	1.54	2.09	2.09
$\delta(R_*)$ (nm)	0.19	0.21	0.20
σ_e (mN m^{-1})	1.64	2.25	2.28
δ_∞ (nm)	0.44	0.42	0.45

dividing surface. The positive value of $\delta(R_*)$ means that the separating surface is located closer to the liquid than the surface of tension. In the limit of infinitely large radii of curvature of the surface of tension, the values of $\sigma(R_*)$ and $\delta(R_*)$ tend to their values at a planar interface, σ_∞ and δ_∞. The following asymptotic formula for $\delta(R_*)$ [186–188] holds:

$$\delta(R_*) = \delta_\infty + \frac{\alpha}{R_*}, \qquad (3.122)$$

where the parameter $\alpha = d\delta/d(1/R_*)|_{R_* \to \infty}$ is a function of temperature only. According to the Tolman equation [76]

$$\sigma = \sigma_\infty \left(1 - 2\frac{\delta_\infty}{R_*}\right), \qquad (3.123)$$

the surface tension of vapor bubbles ($\delta_\infty > 0$) at $R_* \to \infty$ is always smaller than σ_∞.

The question of the dependence of the surface tension of new-phase nuclei on the curvature of the dividing surface is a topic of intensive discussions. At present there are several papers available devoted to the experimental investigation of $\sigma(R)$ for vapor bubbles. Wingrave et al. [189] measured times of saturation with liquid for glass mesoporous spheres. The rate of adsorption of the liquid is related to the driving capillary pressure and, consequently, to the surface tension. The experiments were performed on water and five alkanes (pentane, heptane, etc.). The effective radius of curvature of the liquid meniscus in mesopores is $R_{\text{eff}} \simeq 4.2$ nm. Values of the surface tension exceeding the value of σ_∞ by 5–50% were obtained at $T = (0.45\text{–}0.62)T_c$ (smaller deviations refer to water).

The opposite to the analysis of the result of Ref. [189] was registered in experiments performed by Fisher and Israelachvili [190, 191]. These authors [190] measured the adhesion force $\Im$ between two crossed cylinders containing the liquid under investigation in the contact zone. The value of $\Im$ does not depend on the curvature of the liquid menisci and is equal to $\Im = 4\pi R_0 \sigma \cos\theta$, where R_0 is the radius of the cylinders, and θ is the wetting angle. The effective radii of curvature of the liquid menisci ($R_{\text{eff}} \simeq 0.3$–50 nm) were determined in a separate experiment [191] or calculated from the Kelvin equation using the well-known degree of supersaturation, $S = p/p_s$. With respect to liquid hydrocarbons (cyclohexane, benzene, n-hexane), the surface tension at a curved interface ($R_{\text{eff}} > 0.5$ nm) coincides within 10% with its macroscopic value, σ_∞. A stronger dependence $\sigma(R_{\text{eff}})$ was revealed in experiments on water. A deviation of $\sigma(R_{\text{eff}})$ from σ_∞ was observed at $R_{\text{eff}} \leq 5$ nm. All measurements were performed at temperatures $T = 0.45$–$0.58T_c$. The data from Refs. [190, 191] are in qualitative agreement with the results of determination of the surface tension of vapor bubbles in experiments on nucleation kinetics and calculations of the dependence $\sigma(R_*)$ in the framework of the van der Waals, Cahn–Hilliard model. According to this model, at a temperature $T = 0.6T_c$ and $R_* = 1$ nm the surface tension of a vapor bubble of a simple liquid differs from the macroscopic value of σ by approximately 10%.

3.10
Cavitation Strength of Cryogenic Liquids

In the form of metastable phases, liquids can exist also at negative pressures. There are no fundamental differences between a superheated ($p \geq 0$) and a stretched ($p < 0$) liquid. Boiling-up and cavitation at temperatures sufficiently far from absolute zero can be described by the classical theory of thermofluctuation nucleation.

Up to now, extensive experimental data have been accumulated concerning the value of the cavitation (or tensile) strength of liquids of different chemical nature [9, 10, 192–199]. Limiting degrees of stretching have been registered in experiments, but the data of most of the authors [192–198] are usually by tens and even hundreds of times smaller than the theoretical values. And only in recent years limiting stretches close to theoretical predictions have been achieved for a number of liquids [10, 199]. The reason for the disagreement between theory and experiment is usually connected with the imperfection of the contact of the liquid with the interior walls of the measuring cell, and the presence of dissolved gases and solid weighted particles in the liquid. In this respect, cryogenic liquids are sufficiently pure. A lot of foreign gases contained in them is frozen out and can be easily separated by filtration. The good close to complete wettability of solid bodies with cryogenic liquids does not allow vapor cavities to exist on the interior walls of a vessel for a long time. This statement especially concerns liquid HeII, which possesses the property of superfluidity.

Tab. 3.7 Rupture tensile stress in cryogenic liquids.

Liquid	T (K)	p (MPa) Experiment	p (MPa) Theory	References
Helium	2	−0.03		[194]
	1.85	−0.014	−0.04	[195]
	1.9	−0.016	−0.044	[196]
	1.2	−0.003		[198]
	2.09	−0.12		[200]
Argon	85	−1.2	−19.0	[196]
Nitrogen	71	−0.35	−6.0	[193]
	75	−1.0	−14.0	[196]
Oxygen	75	−1.5	−35.0	[196]

In studying the cavitation strength of cryogenic liquids, both quasistatic [193–198] and dynamic [199–201] methods have been employed. Misener et al. [193, 194] investigated the cavitation strength of nitrogen and superfluid helium in a device which was a system of two metal bellows connected by a rigid link. The liquid that filled one of the bellows was transferred into a metastable state when a compressed gas was filled into the inner space of the other. The moment of boiling-up was observed visually. The results of measurements are compared with calculations employing the homogeneous nucleation theory in Table 3.7. In experiments on HeII no abrupt cavitation effect was discovered. The detachment of a liquid from the bellow walls took place practically at the moment of application of a tensile stress. The authors of Ref. [194] conclude that if in HeII a negative pressure was realized, its value did not exceed 0.03 MPa.

Experiments on cryogenic liquids were continued by Beams [195, 196]. In these experiments [195], use was made of the centrifugal method. The limiting values of the tensile strengths of liquid nitrogen, oxygen, argon, and helium were investigated by applying inertial forces to them. A U-shaped glass tube was immersed into a Dewar flask with the liquid to be investigated. The forces that were stretching the liquid developed during tube braking. The moment of cavitation was registered visually or with a carbon resistor (HeII). The results of measurements are given in Table 3.7. The reproducibility of the data obtained was ±0.1 MPa for nitrogen, oxygen, argon and ±0.003 MPa for helium. Changing from a glass to a metal (stainless steel) tube did not affect the value of the cavitation strength of HeII. All attempts of beams to determine where—at the liquid–glass interface or in the volume of a stretched liquid—its initial break takes place, did not give an unambiguous answer.

The cavitation strength of liquid helium was also investigated using the methods of spouting [197] and osmotic pressure [198]. In Ref. [197], the tensile stress dependence of the mean lifetime of HeII is determined at a temperature of 0.85 K ($\bar{\tau} = 100$–600 s, $V \simeq 1$ cm^3). In the method of osmotic pressure, the measuring device was a system of two cells, one of which contained ^{4}He to be investigated, the other, a solution of ^{4}He–^{3}He [198]. The cells were connected by a capillary, practically impermeable to the light helium isotope. The stretch was created at the expense of directed diffusion of ^{4}He into the solution of ^{4}He–^{3}He. In analyzing the results of their experiments, the authors of Ref. [198] came to the conclusion that the main reason for the discrepancy between the homogeneous nucleation theory and experiment is the radiation background and cosmic radiation. Ref. [202] mentions the high sensitivity of stretched HeII to vibrations of the measuring device.

In the 1960s-1970s, much work was done on investigating the cavitation strength of superfluid helium applying acoustic methods [200, 203–209]. The first experiments [203–205] already revealed the dependence of the value of the registered breaking load on the methods employed to register the initial moment of the cavitation process. The pressure in an expansion wave corresponding to the appearance of a cavitation noise in a number of experiments was an order of magnitude lower than that required for the generation of visible bubbles [204, 209]. Some authors [203, 207] noted a cavitation noise even at positive pressures. In all the first papers on acoustic cavitation in liquid helium, the obtained stretches were much smaller than both the theoretical values and the values achieved in quasistatic experiments [195–198].

Experimental data from Refs. [194–198,203–209] most probably support heterogeneous cavitation as the mechanism of nucleation in cryogenic liquids. All the attempts to remove from the system under investigation possible completed and easily activated boiling-up centers did not lead to positive results. The idea of realizing homogeneous cavitation in HeII based on the ideal wettability of practically all solid materials with this substance did not prove to work either. The strategy adopted in subsequent papers [199, 210–213] was focused on an achievement of homogeneous nucleation in a stretched liquid not at the expense of removing possible cavitation centers from a liquid, but by means of their neutralization in the phase-transition shock regime. For this purpose, focused acoustic fields were used making it possible to separate the cavitation zone from the converter walls and reduce its volume to $(\lambda/2)^3$, where λ is the wavelength of an acoustic wave.

Acoustic vibrations in a liquid were generated and focused by a hemispherical piezoelectric radiator, which operated at frequencies from 500 kHz to 1 MHz [199,213] (Fig. 3.27). The piezoradiator (1) was excited by a radio-wave pulse with a duration of about 1 ms. The initiation of cavitation was registered with the help of a helium–neon laser. A laser beam was focused by an optical lens (2) on the cavitation zone. The intensity of transmitted or scattered light was measured by a photodiode (3). The piezoradiator was mounted in

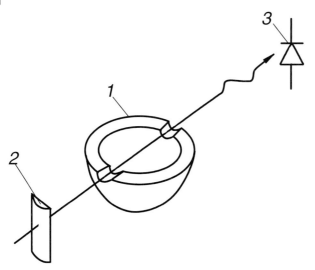

Fig. 3.27 Schematic diagram of a piezoradiator with a system of recording the initiation of cavitation.

a helium bath, the temperature of which was changed by the pumping out of vapors [199,210,211]. In Refs. [212,213] a cell with ^{4}He was used cooled by ^{3}He vapors and capable of withstanding pressures up to 2 MPa.

The main !difficulty in investigating cavitation with the help of focusing acoustic fields consists in the determination of the pressure in the liquid at the moment preceding the cavitation initiation. The authors of Refs. [199,210,211] used two approaches for its evaluation. First, the pressure was calculated by data on the emitted power, the ultrasound absorption factor, and the piezoradiator geometry. The error of determination of the pressure amplitude in this case was $\pm 10\%$. Second, the amplitude of the stretching pressure in the piezoradiator focus was found by light diffraction on the ultrasound lattice created by acoustic vibrations. Up to temperatures of about 3 K, both methods gave sufficiently consistent results. At higher temperatures, subboiling of the liquid at the inner surface of the piezoradiator could be observed. It resulted in a screening of the radiating surface by the vapor phase and disagreement in pressure evaluations.

At rates of ultrasound vibrations in the range $f = 500–1000$ kHz, characteristic times of liquid stretch are $\bar{\tau} \simeq 0.2$–$0.5\,\mu$s. The volume of the cavitation zone is $V \simeq (\lambda/2)^3 = 10^{-5}\,\text{cm}^{-3}$. Hence for the nucleation rate, we have $J = (\bar{\tau}V)^{-1} \simeq 10^{18}\,\text{m}^{-3}\,\text{s}^{-1}$. This value should be regarded as a lower estimate. An upper estimate may be obtained from data on the acoustic energy absorbed by vapor bubbles in the time from the initiation of nucleation till the moment when the bubble size becomes comparable with the length of a transmitted light wave. Calculations made in Ref. [210] for the nucleation rate in this case resulted in a value of $J \simeq 10^{25}\,\text{m}^{-3}\,\text{s}^{-1}$.

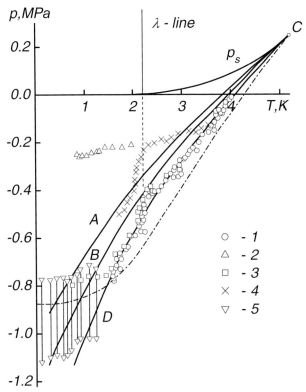

Fig. 3.28 Limiting stretches of ^{4}He. 1: experimental data [211]; 2: [212]; 3: [213]; 4: [214]; 5: [215]. A, B, D: classical nucleation theory employing the macroscopic approximation, for $J = 10^{-9}\,\mathrm{m}^{-3}\,\mathrm{s}^{-1}$; 10^8; 10^{21}, respectively. The *dashed-dotted* curve shows the spinodal; p_s is the binodal.

By Brodie's group [199, 210, 211], acoustic cavitation in liquid helium-4 was investigated in the temperature range 1.6–4.2 K (Fig. 3.28). At the lower boundary of the temperature interval, the maximum value of the stretching pressure was −0.85 MPa. The data from Refs. [199, 210, 211] are in good agreement with the classical homogeneous nucleation theory when the macroscopic surface tension is used in the computations. Xiong and Maris [212] repeated the experiments of Brodie's group using similar equipment and bringing the lower boundary of the temperature range under investigation down to 0.8 K. According to these authors [212], the maximum stretches achieved for superfluid helium did not exceed 0.3 MPa (see Fig. 3.28).

Petterson, Balibar, and Maris [213] are inclined to see the reason for such great disagreement between two identical experiments in the difficulties of evaluating the pressure at the moment when cavitation starts. In their experiments on acoustic cavitation in ^{4}He, they gave up evaluations of breaking loads and investigated the statistic laws of cavitation depending on the temperature ($\Delta T = 0.8$–1.5 K) and the value of the voltage supplied to the piezoradiator.

The probability, W_1, of a break-up of the liquid was determined in a series of one hundred attempts to excite cavitation at given values of temperature, the amplitude of voltage, U_m, on the piezoradiator, and the duration of the exciting radio-wave pulse. The temperature dependence of W_1 for four values of U_m is given in Fig. 3.29. On the basis of Eqs. (3.73) and (3.75) for the

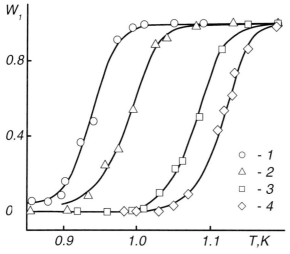

Fig. 3.29 Probability of cavitation in liquid ^{4}He as a function of the temperature at several voltage amplitudes on the piezoradiator [213]. 1: $U_m = 23.20$ V; 2: 22.76; 3: 22.27; 4: 21.85.

probability of initiation of cavitation in the volume V in the time τ, one can write

$$W_1 = 1 - \exp\left[-V\tau J_0 \exp\left(-\frac{W_*}{k_B T}\right)\right], \tag{3.124}$$

where J_0 is the pre-exponential factor in the expression for the stationary nucleation rate,

$$J = J_0 \exp\left(-\frac{W_*}{k_B T}\right), \tag{3.125}$$

and W_* is the height of the energy barrier corresponding to the pressure of the beginning of cavitation in the volume, V.

Within the cavitation zone, the pressure is a complex function of the spatial, r, and time, τ, coordinates. There is a certain minimum pressure, $p_{\min}$, achieved in each half-period of an acoustic wave. Assuming that it happens at moment $\tau = 0$ at point $r = 0$, then in the vicinity of p_{min}, where the nucleation work, W, differs from W_* by no more than $k_B T$, one can write [213]

$$p(r,\tau) \simeq p_{min}\left(1 - \frac{2\pi^2 r^2}{3\lambda^2}\right)\left(1 - \frac{2\pi^2 \tau^2}{T_\sim^2}\right), \qquad (3.126)$$

where $T_\sim$ is the period of an ultrasound wave in a liquid.

Taking in a first approximation the pre-exponential factor, J_0, independent of pressure and the activation energy, W_*, independent of temperature, the latter justified by a very weak temperature dependence of the surface tension of superfluid helium (in the interval from 0.95 to 1.13 K variations of σ do not exceed 1%), then we have for the distribution function of expectation times of appearance of a cavitation center in the piezoradiator focus

$$W_1 = 1 - \exp\left(-N_\sim \int dV \int J dt\right), \qquad (3.127)$$

where $N_\sim$ is the number of half-periods of stretch in a radio-wave pulse, and after substitution of Eqs. (3.125) and (3.126) into Eq. (3.127) and integration we obtain

$$W_1 = 1 - \exp\left[-\frac{3^{3/2} N_\sim \lambda^3 T_\sim (k_B T)^2 J_0}{4\pi^2 (d\ln W_*/d\ln \Delta p)^2 W_{min}^2} \exp\left(-\frac{W_{min}}{k_B T}\right)\right]. \qquad (3.128)$$

Equation (3.128) makes it possible to estimate the values of J_0 and W_{min} by experimental data on W_1. In the case of spontaneous cavitation in HeII the derivative $d\ln W/d\ln \Delta p \simeq 7$. According to the theory of cavitation for classical liquids developed by Fisher [216], $J_0 = J_0' T$. The authors of Ref. [213] extended this result to superfluid liquids, and in the temperature range 0.95–1.13 K they obtained the following values: $J_0 \simeq 10^{31}$–10^{33} m^{-3} s^{-1}, $W_{min} \simeq (27$–$33)k_B T$. For superfluid helium ($T < T_\lambda$), the thermofluctuation theory gives [101]

$$J_0 = \rho'\left(\frac{4k_B T \sigma \rho'}{m\Delta p_n}\right)^{1/2}, \qquad (3.129)$$

where m is the mass of the helium atom. According to Eq. (3.129), $J_0 \simeq 6.5 \times 10^{39}$ m^{-3} s^{-1}. If we assume that thermal phonons are responsible for pressure fluctuations in HeII [213], then

$$J_0 = \frac{\nu_R}{\tau_R^3} = \frac{(3k_B T/\hbar)^4}{C^3}. \qquad (3.130)$$

Here, ν_R is the effective frequency, $\lambda_R = \hbar c/3k_B T$ is the effective wavelength of a thermal phonon, and C is the sound velocity. At $T = 1$ K, from Eq. (3.130) we have $J_0 = 2 \times 10^{36}\,\text{m}^{-3}\,\text{s}^{-1}$. Thus, the values of J_0 obtained on the basis of experimental data are underestimated with respect to their theoretical evaluations by five-seven orders of magnitude, a fact that in the opinion of the author of Ref. [213] considerably exceeds the error of determination of the pre-exponential factor (one or two orders) from Eq. (3.128). The height of the activation barrier, $W_{\min}$, also turns out to be underestimated by approximately two or three times.

An attempt to obtain absolute values of the limiting tensile strength for ^{4}He in acoustic experiments was made by Caupin and Balibar [215]. The authors [215] measured threshold cavitation stresses $U_{*1/2}$ (stresses corresponding to $W_1 = 1/2$) depending on the static helium pressure in a cell. The upper boundary of the cavitation threshold has been evaluated employing the assumption that the amplitude of a sound wave in the cavitation zone is proportional to the voltage supplied to the radiator, the sound absorption in helium is absent, and the piezoradiator characteristics are independent of the voltage supplied. In this case, the voltage of the cavitation threshold is a linear function of the static pressure. Extrapolation of this dependence into the region of negative pressures to the value of $U_{*1/2} = 0$ gives the upper boundary of the cavitation threshold. The lower boundary of the limiting tensile strength for ^{4}He was calculated through the characteristics of the acoustic piezoradiator and the properties of liquid helium. The results are presented in Fig. 3.28. The discrepancies between the two means of evaluating p_n are approximately 0.25 MPa.

Despite the apparently good agreement of experimental data on the cavitation strength of liquid helium with the homogeneous nucleation theory (see Fig. 3.28), a number of factors (differences in the values of the pre-exponential factor, the work of nucleus formation, etc.) make it impossible to draw an unambiguous conclusion about the nucleation mechanism. The authors of Ref. [217] estimated the value of the cavitation strength of superfluid helium in the vicinity of a glass surface, i.e., at a place where one can observe heterogeneous nucleation. Despite the fact that helium wets glass excellently, the focus of acoustic vibrations at points of contact of liquid and glass decreased the cavitation threshold from $\sim -(0.8$–$1.0)$ MPa to -0.3 MPa, and in a number of experiments even to zero. Another possible reason for a decrease in the cavitation strength of a superfluid liquid may be nucleation on quantum filaments or rings, which are capable of producing high-energy particles and acoustic fields in a quantum liquid. The liquid circulation around a vortex filament leads to a pressure decrease in the latter and, as a consequence, an increase in the probability of formation of a cavitation cavity [218]. Correlation between the vorticity and cavitation strength of HeII was experimentally observed by Finch et al. [219–221].

If in an incompressible liquid of density ρ' an infinitely long vortex filament with a hollow ($\rho' = 0$) core of radius, R, is formed, then the energy of the filament per unit length can be written as [222]

$$E_{\text{vor}} = \pi R^2 \Delta p + 2\pi R \sigma + \frac{\pi \hbar \rho'}{m^2} \ln\left(\frac{R_{\max}}{R}\right), \qquad (3.131)$$

where $\Delta p = p_s - p \simeq -p$ and $R_{\max}$ is the external radius of the vortex, $R_{\max} \simeq$ 3 nm. The last term on the right-hand side of Eq. (3.131) determines the kinetic energy of a rotating liquid. At equilibrium $dE_{\text{vor}} = 0$, and from Eq. (3.131) we have

$$\frac{R}{R_0} = \frac{-2|p_{\text{vor}}^*|}{p}\left[1 \pm \left(\frac{R}{|p_{\text{vor}}^*|}\right)^{1/2}\right]. \qquad (3.132)$$

Here, the notations $R_0 = \hbar^2 \rho'/2\sigma m^2$ and $p_{\text{vor}}^* = -\sigma^2 m^2 / 2\hbar \rho'$ are introduced.

At $p > 0$, Eq. (3.132) has only one stable solution (Fig. 3.30). The radius of the vortex core for very high pressures is approximately equal to $(\hbar^2 \rho'/2pm^2)^{1/2}$. If $p < 0$, then along with a solution that is stable with respect to small perturbations, there also exists a solution which reflects the overcoming of a certain activation barrier by the vortex. When $p \leq 0$, the radius of the vortex core is $R \simeq 2\sigma/|p|$, and this solution may by interpreted as the formation of a bubble on the vortex. The stable and the unstable solution merge at $p = p_{\text{vor}}^*$, the vortex filament becoming unstable along the whole length with respect to a radial stretch. The discussed simplified model of the vortex filament reproduces in a qualitatively correct way the results of more rigorous approaches based on the van der Waals theory of inhomogeneous systems [222] and density functional methods [223]. At the absolute zero of temperature, according to data of Ref. [222], $p_{\text{vor}} = -(0.6–0.7)$ MPa holds. According to Ref. [223], $p_{\text{vor}}^* = -0.8$ MPa. Equation (3.132) gives $p_{\text{vor}}^* = -1.95$ MPa. The pressure on the spinodal of superfluid helium is $p_{\text{sp}} \simeq -0.95$ MPa [224].

The value of the energy barrier, $\Delta E_{\text{vor}} = E_{\max} - E_{\text{vor}}$, determining the work of formation of a cavitation cavity on a quantum vortex has been calculated by Maris in the framework of the van der Waals square gradient approximation [222]. At all pressures, ΔE_{vor} is smaller than the work of formation of a critical nucleus in a homogeneous liquid and tends to zero when $p \to p_{\text{vor}}^*$. If at small stretches the shape of a vapor bubble forming on a vortex is close to a spherical, then with increasing $|p|$ the bubble size along the vortex axis begins to exceed its diameter, and at $p \to p_{\text{vor}}^*$ the whole vortex becomes a cavitation cavity.

If it is granted that in experiments of Petterson et al. [213] cavitation proceeds on quantum vortices, limiting stretches $\Delta p \simeq 0.55$ MPa correspond to the cavitation energy $W_{\min} = 30k_BT$. According to Gibbs' classical formula, the work of formation of a critical nucleus at a given pressure is $W_* \simeq 100k_BT$.

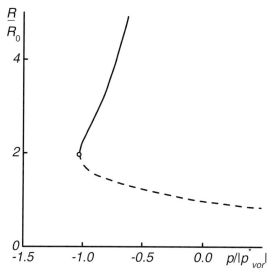

Fig. 3.30 Radius of a quantum filament in superfluid helium-4 as a function of pressure [222]. The *dashed* line corresponds to stable vortices, the *full* curve to vortices on which the vapor phase originates.

The number of nuclei forming in a unit time on a unit length of a vortex filament is given by

$$J_{vor} = J_{0,vor} \exp\left(-\frac{\Delta E_{vor}}{k_B T}\right). \tag{3.133}$$

In bulk boiling-up $J_0 = J_{0,vor} L_{vor}$, where L_{vor} is the length of the vortex filament per unit volume of a liquid. Using the values of J_0 calculated by experimental data [213] and setting

$$J_{0,vor} = \frac{\nu_R}{\lambda_R} = \frac{(3k_B T/\hbar)^2}{c}, \tag{3.134}$$

we have $L_{vor} \simeq (10^{12}-10^{14})$ m^{-2}. The authors of Ref. [213] suppose that such high densities of quantum vortices can be realized in experiments on acoustic cavitation. Thus, densities of vortex filaments of the order of 10^8 m^{-2} were registered in experiments of Schwarz and Smith [225] at intensities of ultrasound fields of several milliwatts per square centimeter. The intensities of acoustic fields in the investigation of cavitation in HeII in Refs. [199, 210, 211, 213] were several Watt per square centimeter.

With respect to the behavior in the transition through the λ-point, in a number of papers [199, 213] the existence of a break point was reported in the temperature dependence of the line of the limiting stretch for liquid helium. If cavitation in superfluid helium proceeds on quantum vortices, the formation

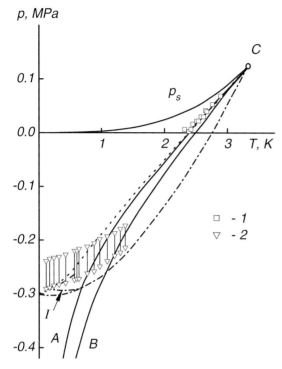

Fig. 3.31 Limiting stretches and spinodal of ^{3}He. 1: experimental data [149]; 2: evaluations of the upper and the lower boundary of cavitation strength by data of Ref. [231]. A and B: classical nucleation theory in macroscopic approximation for $J = 2.5 \times 10^{10}$ m^{-3} s^{-1}. *Dotted line*: homogeneous nucleation theory and density functional method for $J = 10^{10}$ m^{-3} s^{-1} [233, 234]. *Dashed-dotted line*: spinodal [233]. At $T < 0.5$ K, the branch on the spinodal is the spinodal (l) approximation by data of Ref. [235], the curve with a minimum at $T = 0.4$ K. p_s: binodal.

of a normal phase is accompanied by the disintegration of vortices with an abrupt increase in the cavitation strength of helium. This effect was not observed in experiment. The break point observed is more likely to be connected with the change of helium properties during the transition across the λ-line, which has an extension in the range of negative pressures. The character of this extension was discussed by Skripov [226]. The latest data on this problem are presented in Refs. [227, 228].

Helium-3, a lighter isotope than helium-4, is subject to more considerable quantum fluctuations. Its molar volume in the liquid phase is large, and the cohesive forces of the molecules are smaller. The cavitation strength of normal liquid ^{3}He was investigated in Refs. [229–231]. In experiments, equipment was used similar to that employed in studying acoustic cavitation in Ref. [232]. A piezoceramic radiator with an internal radius of 8 mm generated an acoustic pulse with a filling rate of 1 MHz which was focused in a region with a linear dimension of the order of 100 μm. Experiments were performed in the temperature range 40 mK–1 K at a rate of arrival of pulses exciting cavitation of 0.2 Hz. In this case, the dissipation connected with the mechanical vibrations of the radiator and the absorption of the laser beam in the liquid did not exceed 2 μW. In Refs. [229, 230], only the stochastic laws of cavitation are investigated. The authors did not reveal any qualitative differences for ^{3}He and ^{4}He in the dependences of W_1 on the value of the voltage amplitude on the piezoradiator and temperature.

In experiments on ^{3}He, $V\tau = 1.2 \times 10^{-22}$ m^3 s holds. The uncertainty in theoretical evaluations of J_0 results in differences in the values of the complex $J_0 V\tau$ reaching almost two orders (6×10^{14}–1.8×10^{16}) and small changes in the value of the nucleation work, W_*. At $W_1 = 0.5$, from Eq. (3.124) for W_* we have $(34 \pm 3)k_B T$. Caupin and Balibar [231] evaluated the upper, $p_{n,\max}$, and the lower, $p_{n,\min}$, boundary of the cavitation threshold in helium-3 (Fig. 3.31). The error of determination of $p_{n,\max}$ was ± 0.005 MPa, and that of $p_{n,\min}$ was ± 0.015 MPa. The differences between the values of $p_{n,\max}$ and $p_{n,\min}$ in ^{3}He are much smaller than those in ^{4}He which is connected with less expressed nonlinear effects in liquid helium-3. Extrapolation of the values of $p_{n,\max}$ and $p_{n,\min}$ to $T = 0$ gives -0.305 MPa and -0.24 MPa, respectively.

Owing to the fact that experiments on acoustic cavitation in normal ^{3}He were conducted in the nearest vicinity of the spinodal, in calculating the work of formation of a critical nucleus in the framework of density functional methods, the accuracy of determination of the position of the boundary of essential instability of the liquid phase is of great significance. The first evaluations of the pressure on the spinodal of liquid helium-3 at $T = 0$ were made in Refs. [214, 231, 233, 236, 237]. According to data of Maris [214], $p_{sp}(0) = -0.309$ MPa holds. The author of Ref. [231] obtained $p_{sp}(0) = -0.314$ MPa.

Caupin, Balibar, and Maris [235] paid attention to the fact that the coefficient of thermal expansion of the Fermi liquid changes its sign with decreasing temperature and becomes negative in the vicinity of the absolute zero of temperature. Using the experimental data for the sound velocity of Roach et al. [238] and the methods of approximating the spinodal described by Maris [214], they recalculated data for the boundary of essential instability of helium-3 in the temperature range 0–0.6 K and obtained a curve with a minimum at $T_{\min} = 0.4$ K and pressure $p_{\min} = -0.29$ MPa. The line of points of density maxima of the liquid phase extrapolated from the stable into the metastable region in (p, T)-coordinates is directed into the point of the spinodal extremum.

The presence of a minimum on the spinodal should also affect the temperature dependence of the limiting stretches of ^{3}He. In the vicinity of the spinodal and $T \simeq 0$, Maris [214] presented the pressure dependence of the work of for-

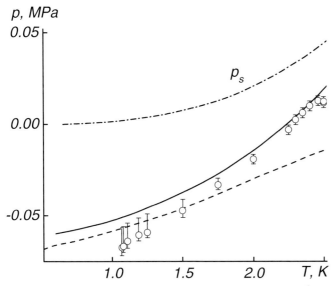

Fig. 3.32 Temperature dependence of the cavitation strength of ^{3}He. *Dots*: experimental data [235], *dashed line*: calculation by Eq. (3.136) employing data from Ref. [233], *full curve*: the same by data for p_{sp} from Ref. [235], p_s: binodal.

mation of a critical nucleus as

$$\frac{W_*}{k} = \beta[p - p_{sp}(T)]^\delta, \tag{3.135}$$

where β is an individual constant of the substance equal to $\delta = 3/4$. A substitution of Eq. (3.135) into Eq. (3.128) gives the following expression for the temperature dependence of the cavitation pressure

$$p_n(T) = p_{sp}(T) + \left[\frac{T}{\beta}\ln\left(\frac{J_0 V \tau}{\ln 2}\right)\right]^{1/\delta}. \tag{3.136}$$

In Fig. 3.32, the results of calculating the cavitation strength of ^{3}He are compared with experimental data [239, 240]. It was assumed that $J_0 V \tau = 6 \times 10^{14}$, $\beta = 47.13$. Applying in calculations of $p_n(T)$ the values of $p_{sp}(T)$ obtained by Guilleumas et al. [233], one can observe a considerable disagreement of experimental and calculated data at elevated temperatures. This disagreement is explained in Ref. [240] as a result of an incorrect approximation of the spinodal of ^{3}He performed in Refs. [214, 231, 233, 236, 237, 239]. Application in Eq. (3.136) of the data for $p_{sp}(T)$ of Caupin et al. [235] results in a good agreement between theory and experiment in the whole investigated temperature range (see Fig. 3.32).

As has already been mentioned (Section 3.8), electrons may form in liquid ^{4}He and ^{3}He bound states in the form of electronic bubbles. Su et al. [241]

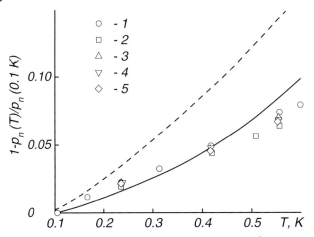

Fig. 3.33 Stability limit for electronic bubbles in liquid ^{3}He. *Dots*: experimental data [241], *dashed line*: calculation without regard for the gas pressure in an electronic bubble, *full curve*: results of calculation taking into account the presence of gas in electronic bubbles.

investigated cavitation on electronic bubbles in liquid helium-3. The design of the measuring cell and the experimental procedure were similar to those described in Refs. [229, 232]. Electrons injected into liquid helium lost their kinetic energy and formed electronic bubbles. Using the probability of appearance of cavitation W_1, which depends on the electrons in the system, the amplitude of voltage on the piezoradiator was determined for each given temperature, when $W_1 = 0$. This voltage was related to the boundary of stability of an electronic bubble in the liquid and assumed to determine the value of the pressure at the boundary of its region of stability, p_e.

Since for ^{3}He, the polarization contribution to the excess free energy of an electronic bubble is too small (see Eq. (3.105)), and the value of the potential barrier keeping the electron in the cavity considerably exceeds the energy of the electron, at low temperatures, with no regard for the gas pressure in the bubble, from Eq. (3.104) we have

$$\Delta F = \frac{\pi^2 \hbar^2}{2 m_e R^2} + 4\pi R^2 \sigma + \frac{4\pi}{3} R^3 p. \qquad (3.137)$$

Hence for the pressure p_e at the boundary of stability of an electronic bubble ($\partial^2 \Delta F / \partial R^2 = 0$) it follows that

$$p_e = -\frac{16}{5} \left(\frac{m_e}{10\pi \hbar^2} \right)^{1/4} \sigma^{5/4}. \qquad (3.138)$$

With respect to ^{3}He at $T = 0$, the value of $p_e = -0.7$ MPa is obtained. At finite temperatures it is necessary to take into account the gas pressure in the

Tab. 3.8 Limiting tensile strength of liquid argon ($J = 10^7$ m^{-3} s^{-1}), surface tension of critical nuclei, their radius and the number of atoms contained in them.

| T | p_s | p_n (MPa) | | σ (mN m^{-1}) | | R_* | N_*'' |
(K)	(MPa)	W_*, σ_∞	$W_*, \sigma(R_*)$	σ_∞	$\sigma(R_*)$	(nm)	
95	0.214	−14.78	−12.23	10.71	9.46	1.5	2
105	0.474	−9.43	−8.05	8.39	7.59	1.8	8
115	0.912	−5.16	−4.44	6.20	5.69	2.2	29
125	1.585	−1.67	−1.34	4.12	3.84	2.8	111
135	2.552	1.23	1.34	2.22	2.10	4.0	555

electronic bubble. For ^{3}He, a simplified model of a gas-filled electronic bubble is suggested in Ref. [241], in the framework of which the dependence $p_e(T)$ has been obtained. The boundary of stability for electronic bubbles in normal liquid ^{3}He has been determined in the temperature range from 1.07 to 2.5 K (Fig. 3.33). Experimental data are in good agreement with the results of theoretical calculations if the latter take into account the presence of the gas phase in an electronic bubble. Equation (3.137) presupposes that the electron in a bubble is in the ground state. A light wave may excite still higher energy levels of the electron in the cavity, which results in a decrease in the threshold value of p_e. These problems are discussed in Refs. [242–244].

At low temperatures, superfluid helium-4 and normal helium-3 are degenerate quantum liquids, and perhaps some peculiarities in the cavitation strength of these liquids have to be expected which are not typical of classical cryogenic systems. However, reproducible experimental data in this respect are not yet available. Experiments on superheated ($p > 0$) cryogenic liquids give an indication of a rise in activity of completed and easily activated evaporation centers with decreasing pressure [113]. This effect hinders both the performance of quasistatic experiments and the realization of the shock phase-transition regime in the region of negative pressures.

Comparing the predictions of the homogeneous nucleation theory and experiment in the region of negative pressures, one should take more care about the approximations used in its construction. The classical nucleation theory is based on the Szilard–Farkas schema, according to which the formation of a viable nucleus is a chain of random acts of evaporation and condensation of molecules. Since at negative external pressures a critical bubble is practically empty, such an approach causes difficulties in consideration of cavitation in a highly stretched liquid. This drawback is absent in a model which considers the formation of a nucleus as the result of a single fluctuation [245].

A decrease in temperature leads to a decrease in the critical bubble radius (J = const). Table 3.8 presents characteristic sizes of critical bubbles in stretched liquid argon. The cavitation strength of argon has been calculated by the classical nucleation theory under the assumption that the bubble growth is limited by the viscosity forces for two approaches of determining the work of nucleus formation, employing either the Gibbs formula in a macroscopic approximation ($\sigma = \sigma_\infty$) or the Cahn–Hilliard formula, which takes into account the dependence $\sigma = \sigma(R_*)$. At the temperature of the triple point, the discrepancies between σ_∞ and $\sigma(R_*)$ reach 15%. This deviation results in differences in the values of the cavitation strength exceeding 20% [113]. It should be mentioned, however, that at low temperatures the thickness of the nucleus interfacial transition layer is comparable with the radius of molecular interaction. This coincidence leads to restrictions in the applicability of the Cahn–Hilliard approach based on the square-gradient approximation of the van der Waals capillarity theory and requires to take into account higher order terms in the expansion with respect to the gradients in calculations of the work of nucleus formation [184, 246].

3.11
Attainable Superheating of Liquid Argon at Negative Pressures

As already mentioned in Sections 3.7 and 3.10, at temperatures below $\sim 0.9 T_c$ homogeneous nucleation proceeds at an appreciable rate only when a liquid is stretched ($p < 0$). Numerous attempts [193, 196] to achieve stretches predicted by the homogeneous nucleation theory in simple classical liquids were not successful. A considerable breakthrough was achieved in experiments on helium isotopes [199, 210–213], for which tensile strengths close to theoretical predictions were obtained. However, the realization of the method of bringing acoustic vibrations into focus used in Refs. [199, 210–213] cannot be extended to classical cryogenic liquids without a considerable elaboration as their limiting stretches at $0.5 T_c$ are approximately 20 times larger than those of helium-4.

Considerable tensile stresses in liquids are realized when a pressure wave is reflected from the free surface of the liquid [248]. Rupture tensile stresses close to theoretical predictions were achieved in experiments on the cavitation of organic liquids in an expansion wave at nucleation rates $J > 10^{10}\,\mathrm{m}^{-3}\,\mathrm{s}^{-1}$ [249]. To advance into a region of higher values of nucleation rate, a combination of two dynamic methods was suggested [250], i.e., pulse stretching in an expansion wave and pulse liquid superheating on thin wire heaters. Later on this method was used for investigating limiting superheats of liquid argon at negative pressures [251]. The schematic diagram of the setup of the measuring chamber is given in Fig. 3.34. The liquid under investigation is filled into a cylindrical chamber (8) with a diameter 40 mm and volume $\sim 80\,\mathrm{cm}^3$. The liquid free surface is located at a distance of 3–5 mm from the chamber cover. A platinum wire (2) of a length 1 cm and a diameter 20 µm is fixed along the

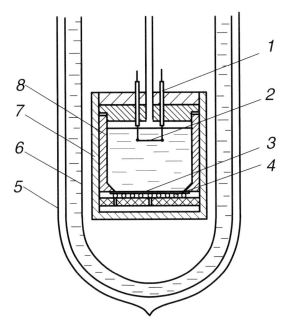

Fig. 3.34 Schema of a setup determining the attainable superheating of argon on a platinum wire in an expansion wave.

chamber axis with the help of lead-in wires (1) and immersed in liquid argon to a length of 5–10 mm. A pressure pulse in the liquid was created with the help of a bronze membrane (3), which is the chamber bottom. The membrane is excited when a low-inductance capacitor is discharged onto a flat coil (4). This setup was calibrated with the help of a pulse pressure transducer which served as a substitute for the platinum wire. During calibration the value of the pressure-pulse amplitude was determined as a function of the capacitor supply voltage. The error of pressure determination in stretching a liquid is ∼ 5%.

The chamber was thermostated in a massive copper block (7) at temperatures close to that of the normal boiling of the liquid under investigation. The block-chamber system was located in a vacuum jacket (6) of stainless steel. Cooling was realized with the use of liquid nitrogen poured into a metal dewar (5). The temperature of the liquid in the chamber was measured by a platinum resistance thermometer and controlled by the pressure of saturated vapors over the liquid under investigation. At a given voltage across the capacitor C ($U \simeq 5$–10 kV, $C \simeq 3$ μF), a signal from the generator (5) is supplied to the unit of control of its discharge (6) (Fig. 3.35). Simultaneously, the oscillograph (12) was started up. The capacitor discharge switched on the prescribing generator (7), the output pulse of which controlled the operation of a square-wave generator (8). This generator formed a square heating pulse on a platinum wire (1) with a duration of 15 to 100 μs. The wire is incorporated in a bridge measuring circuit (9). A high-frequency signal caused by the boiling-up of a liquid on the wire was separated with the help of a differential amplifier (10), a comparator (11) and supplied to the oscillograph (12). A pulse from the generator (7) was synchronized in such a way that the heating-up of the platinum wire took place after the reflection of a pressure wave from the liquid free surface. The amplitude of the heating pulse increased smoothly till the moment of appearance of a boiling-up signal on the oscillogram. The balance of the bridge (9) determined the wire temperature at the moment of appearance of a boiling-up pulse.

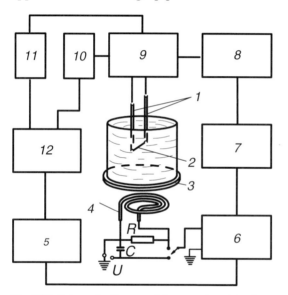

Fig. 3.35 Schematic diagram of the experimental setup.

In the regime of liquid pulse heating, up to the process of explosive boiling-up the rate of increase of the temperature of the heater is constant in a first approximation ($\dot{T}_0$ = const) and is calculated taking into account the heat transfer into the liquid. The thermal effect caused by the heat insulation of parts of the heater and the heat transfer due to evaporation is added to the heat balance at the stage of development of explosive boiling-up. As a result, the rate of heating-up of a cylindrical heater registered at time t, counted from the beginning of the heating-up, is an integral function of a complex of thermophysical and geometrical characteristics at the liquid–vapor–heater interface,

$$T_0 - T(t) = F\left\{q'(t), S(t), L(t,\tau), \bar{q}(\tau), R(\tau), J(t), \ldots\right\}, \tag{3.139}$$

where $q'(t)$ is the density of the heat flow into the liquid, $S(t)$ is the relative heat-insulating area of the heater, $L(t,\tau)$ is the distribution of the wetting-line

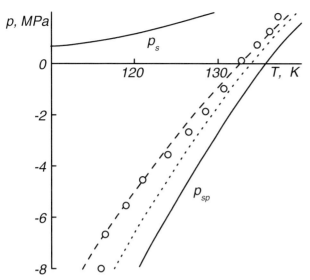

Fig. 3.36 Attainable superheating of liquid argon. Dots show experimental data, the *dashed line* the results of computations performed with Eqs. (2.112), (3.52), and $z_0 = 1$ in a macroscopic approximation ($J = 10^{26}$ m^{-3} s^{-1}), and the *dotted line* shows the same taking into account the size effect; p_s: binodal, p_{sp}: spinodal.

length in the time of its existence, τ, related to the unit of the heater area, $\bar{q}(t)$ is the heat-flow rate related to the unit of the wetting-line length existing for the time τ, $R(\tau)$ is the bubble growth law, and $J(t)$ is the nucleation rate at the heater temperature. The functions $S(t), L(t,\tau), \bar{q}(\tau)$ are calculated in the framework of the general theory of heat removal from a rapidly heated wire into a boiling-up liquid [252]. The form of the function $\bar{q}(\tau)$ at limiting superheatings depends on the regime (critical or presound) of the vapor flow at the surface of an evaporating bubble. It is assumed that the vapor radius varies according to the power law $R(\tau) = \phi \tau^\kappa$, where for negative pressures the values of ϕ and κ are determined by the Rayleigh formula. The solution of Eq. (3.139) makes it possible to obtain an approximate analytical expression for the rate of spontaneous nucleation observed in an experiment and to relate the rate of increase in the heater temperature to the slope of the increase of the nucleation rate, $G_T = d \ln J / dT$. Methods of measuring the wire temperature by a boiling-up signal and determining the nucleation rate are described in Ref. [11].

Figure 3.36 presents the results of measuring the attainable superheating temperature for liquid argon at positive and negative pressures. The data obtained refer to different values of the nucleation rate. The averaged (effective) value of the nucleation rate over all the measurements is $J = (1 \pm 100) \times 10^{26}$ m^{-3} s^{-1}. In experiments at atmospheric pressure on the pulse heating of

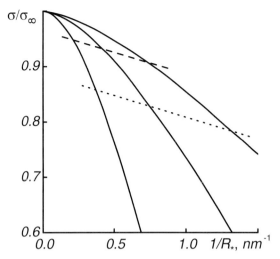

Fig. 3.37 Surface tension (referred to the surface of tension) of argon critical bubbles versus curvature of the surface of tension at different temperatures. 1: $T = 140$ K, 2: 120, 3: 90. The *dashed line* corresponds to the reduced height of the activation barrier (Gibbs number) $G_* = 70$, the *dotted line* to $G_* = 20$.

liquid argon on a platinum wire, the achieved temperature was about 2.5 K higher than that in quasistatic experiments ($J = 10^7 \text{ m}^{-3}\text{ s}^{-1}$) [128]. In this case, a superheating 3 K higher than that recorded in the experiment of the maximum superheating corresponds to the spinodal. A decrease in pressure to $p = -8.0$ MPa increases the discrepancy between the superheating temperature experimentally recorded and the temperature of the spinodal to 4.5 K (Fig. 3.36).

The results of calculating the temperature of attainable superheating ($J = 10^{26} \text{ m}^{-3}\text{ s}^{-1}$) for liquid argon by the classical homogeneous nucleation theory (Eqs. (2.112) and (3.52)) in a macroscopic approximation with the use of the data on the surface tension from Ref. [153] are presented in Fig. 3.36 by a dashed line. In the whole pressure range investigated, a systematic "underheating" of liquid argon with respect to theoretical values of T_n can be observed. The magnitude of the "underheating" increases with decreasing pressure, and at $p = -(7-8)$ MPa reaches 2–3 K. Since the basic quantity determining the value of T_n in Eq. (2.112) is the work of formation of a critical nucleus, which is directly proportional to the cube of the surface tension at the liquid–vapor bubble interface, it can be assumed that the recorded discrepancy between theory and experiment is first of all connected with neglecting theory of the size effect with respect to the surface tension.

In Fig. 3.36, a dotted line shows the results of calculating the temperature of attainable superheating ($J = 10^{26} \text{ m}^{-3}\text{ s}^{-1}$) for argon by the classical nu-

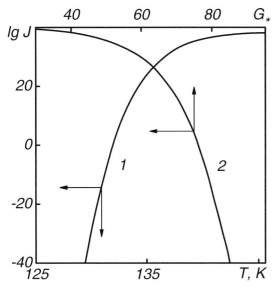

Fig. 3.38 Theoretical dependence of the nucleation rate (1) and Gibbs' number (2) in superheated liquid argon at a pressure $p = 0.5$ MPa ($T_s = 105.9$ K) on temperature.

cleation theory (Eqs. (2.112) and (3.52)) taking into account the size effect of nucleation in the framework of the Cahn–Hilliard approach [182]. As can be seen in Fig. 3.36, the abandonment of the macroscopic approximation improves the agreement between theory and experiment. Another important argument in favor of the presence of size effects during nucleation in a region of negative pressures is the value of the derivative $G_T = d \ln J/dT$, which is recorded in an experiment irrespective of the value of J. According to Eq. (3.93), the value of this derivative is mainly determined by the temperature dependence of the Gibbs number. With increasing superheating temperature and decreasing pressure, the value of G_T decreases. If in liquid argon at pressures close to atmospheric and $J \simeq (10^1–10^{10})\,\text{m}^{-3}\,\text{s}^{-1}$ the derivative $G_T \simeq 20$, which corresponds to an increase in the nucleation rate of about nine orders with an increase in the superheating of 1 K, then at the same pressure and $J \simeq 10^{20}\,\text{m}^{-3}\,\text{s}^{-1}$ the value of G_T is two times lower.

A decrease in temperature is accompanied by an increase in the width of the metastable region in pressure, and consequently in the pressure difference inside and outside a critical bubble. This effect leads to a considerable reduction of the dimensions of critical bubbles at temperatures close to the triple point as compared with the vicinity of the critical point. Thus, if in liquid argon at $T/T_c = 0.95$ ($T \simeq 143$ K) and $J = 10^7\,\text{m}^{-3}\,\text{s}^{-1}$ the radius of a critical bubble is 6.4 nm, and at $J = 10^{26}\,\text{m}^{-3}\,\text{s}^{-1}$ we have $R_* \simeq 3.4$ nm, at $T/T_c = 0.6$ ($T \simeq 90$ K), the radii of critical bubbles for the indicated values of the nucleation rate are 1.4 and 0.9 nm, respectively. At low temperatures the vapor density is low, therefore critical bubbles in a stretched liquid prove to be practically empty. All these properties give rise to a number of difficulties in the theoretical description of the nucleation process at negative pressures.

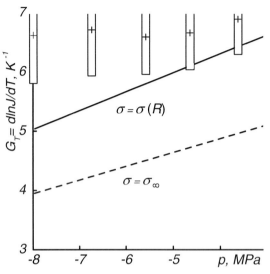

Fig. 3.39 Derivative G_T as a function of pressure at $J = 10^{26}$ m^{-3} s^{-1}. Rectangles present experimental data, and their height defines the error of determination of G_T. A *dashed line* shows calculations by the homogeneous nucleation theory in a macroscopic approximation, and the *full curve* gives the same dependences taking into account $\sigma = \sigma(R_*)$.

Figure 3.37 presents the results of calculating the size dependence of the surface tension of critical bubbles for argon in the framework of the Cahn–Hilliard approach (see Section 3.9). The interface curvature leads to a monotonic decrease in the surface tension. The size effect increases with decreasing temperature. Some quantitative characteristics of the nucleation process for argon are given in Table 3.8. The cavitation strength of argon ($J = 10^7$ m^{-3} s^{-1}) has been calculated by Eqs. (2.112) and (3.52) for the cases when the work of formation of a critical nucleus has been determined by the Gibbs formula, Eq. (3.8), under the assumption that $\sigma = \sigma_\infty$ and by the Cahn–Hilliard formula, Eq. (3.11). At the temperature of the triple point, discrepancies between σ_∞ and $\sigma(R_*)$ reach 15%. This difference leads to deviations in the values of the cavitation strength exceeding 20% [113].

Figure 3.38 presents J versus the temperature and the Gibbs number on a semilogarithmic scale. Data on the nucleation rate have been obtained from calculations via Eqs. (2.112) and (3.52) in a macroscopic approximation. The existence of saturation at high values of $J\,[\lg J \rightarrow \lg(\rho B) \approx 38]$ is connected with the exhaustion of places for the formation of bubbles and the finite rate of molecular exchange processes. In the saturation region, the results of calculation are known to be inaccurate as the theory presupposes that nuclei bubbles have no effect on each other, and the mean temperature and pressure do not change.

The baric dependence of the derivative G_T obtained by experiment and calculated in the framework of the classical nucleation theory with and without taking into account the size effect is shown in Fig. 3.39. It follows from Fig. 3.39 that neglecting the size effect in nucleation theory underestimates the value of G_T as compared with experiment by 60% at $p = -8.0$ MPa and by 20% at $p = -3.0$ MPa. Taking into account the size effect in the Cahn–Hilliard approximation improves agreement between theory and experiment. At $p > -6.5$ MPa, the discrepancies do not exceed the experimental error. At $p < -6.5$ MPa, underestimates in the calculated values of G_T may be connected with neglecting higher order gradient terms in the functional, Eq. (2.36). As is shown in Refs. [246, 247], the Cahn–Hilliard square-gradient approximation is valid at temperatures exceeding $(0.8$–$0.85)T_c$.

3.12
Initiated Nucleation

Phase transitions often proceed in the presence of external factors of different physical nature (radiation, ultrasonic, electromagnetic, and some other fields). Such factors usually initiate the nucleation process, but in principle the effect of its suppression is not excluded either. In the course of initiated boiling-up of a liquid, the potential barrier that separates the stable from the metastable state of the system can be overcome, at least, partly at the expense of the energy introduced into the system, which may considerably exceed the level of thermal fluctuations. Despite the fact that in most cases the initiation mechanism is known, its inclusion into the traditional schema of nucleation kinetics proves to be a rather complicated task. Besides, external fields may have indirect effects on superheated liquids, giving rise to boiling-up centers such as, for instance, electronic and positronic bubbles in some quantum liquids, vortex lines and rings in superfluid helium. Commonly, a superheated liquid is exposed to the influence of numerous factors which are difficult to control, and it may be hard to distinguish the main source of nucleation initiation. One can perform a check of the activity of different factors in the easiest way by acting on a system purposefully with one or another initiating the nucleation factor.

The sensitivity of a superheated liquid to ionizing radiation was first demonstrated by Glaser [253, 254]. This phenomenon has not only found a direct application in nuclear physics (the development of bubble chambers), but has also allowed the extension of the notion of possible nucleation centers in superheated liquids. All bodies on the surface of the Earth are exposed to the action of cosmic radiation and natural radiation background. On average, one ionizing particle of secondary cosmic radiation (μ-mesons, electrons) passes through 1 cm^2 of the Earth's surface every minute [101]. In interacting with the atoms of the substance under consideration, high-energy particles knock electrons out of them. If the energy of an expelled electron is sufficiently high, in subsequent collisions with atoms it may cause their ionization again. Such an electron is commonly called a δ-electron [173]. The energy of δ-electrons is efficiently transformed into heat. Calculations show [173] that in the course of braking of a δ-electron in a time of the order of $\sim 10^{-10}$ s over a distance of $\sim 10^{-6}$ mm an energy of about 100 eV is released, which proves to be sufficient for a critical nucleus to form on such a thermal spike. The sizes of bubbles initiated by high-energy particles depend on the thermodynamic parameters and molecular properties of the liquid and the energy of the δ-electrons. Since under actual conditions one does not generally manage to completely eliminate cosmic radiation and the natural radioactivity of materials, the flux, J_{exp}, measured in experiments on nucleation kinetics is the superposition of spontaneous, J, and initiated, J_i, nucleation fluxes. Owing to a very strong dependence of J on the thermodynamic state variables, the radiation background manifests itself in a strong deviation of experimental and theoretical nucleation rate curves.

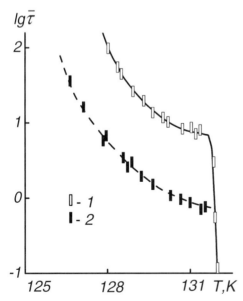

Fig. 3.40 Temperature dependence of the mean lifetime of superheated liquid argon on an isobar $p = 0.36$ MPa. 1: under natural conditions; 2: in the field of γ-radiation [128].

The first systematic investigations of the action of γ-radiation on the boiling-up kinetics of superheated liquids were made by Sinitsyn and Skripov [9, 10, 255]. Figure 3.40 shows the temperature dependence of the mean lifetime of superheated liquid argon under natural conditions and in the field of γ-radiation (^{60}Co of activity 0.2 mg eq. Ra) [128]. An increase in the intensity of ionizing radiation reduces the expectation time of boiling-up. In the vicinity of the boundary of spontaneous boiling-up in a number of liquids, a plateau can be observed (see Figs. 3.12 and 3.13), which is connected with the hundred percent probability of origination of a supercritical bubble on each thermal spike created by an ionizing particle [255]. In phenomenological initiation models [9, 256], they usually postulate the energy distribution law of thermal spikes and the probability of origination of bubbles on them. Experimental data on the mean lifetime of a superheated liquid on one isobar (isotherm) make it possible to determine the free parameters of such a distribution and calculate the rate of initiated nucleation. The coincidence of the character of the temperature (pressure) dependence of J_i with experimental data on other isobars (isotherms) is regarded as a corroboration of the initiation mechanism. At present, there is no full confidence that the action of ionizing radiation is the only reason for discrepancy between theory and experiment at small nucleation rates. The similarity of temperature dependences of $\bar{\tau}$ and J obtained under natural conditions and in the presence of a γ-source indicates only the undeniable relation between these phenomena.

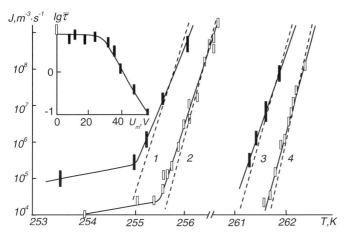

Fig. 3.41 Nucleation rate in stretched liquid xenon under natural conditions (*light marks*) and in an ultrasonic field (*dark marks*) with an amplitude, $p_m = 0.1$ MPa, at different initial pressures. 1, 2: $p_0 = 0.99$ MPa; 3, 4: 1.98 [258]. *Dashed lines* show calculations by the homogeneous nucleation theory taking into account the dependence $\sigma(R)$. In the inset one can see the mean lifetime as a function of the amplitude of voltage across the piezoradiator, $T = 255.8$ K.

If under the action of radiation the boiling-up of a superheated liquid is the result of a local heat release, in acoustic fields the formation of nuclei is initiated by the creation of tensions in a liquid at moments of pressure lowering. Apart from spontaneous nucleation, an acoustic wave may initiate the activity of completed boiling centers on various accidental inclusions. Owing to a rectified heat transfer, vapor bubbles forming in such centers in a periodic pressure field have a possibility of growing [101]. Let a superheated liquid be under pressure p_0. At the moment of time $\tau = 0$, periodic pulsations of pressure, $p_\sim(\tau)$, with an amplitude, $p_m (p_m \ll p_s - p_0)$, and frequency, f, are excited in the liquid. It is assumed that pressure pulsations do not disrupt the stationarity of the bubble size distribution. Far from the critical point, this condition is fulfilled if $f \leq 10^9$ Hz. In an ultrasonic field, the rate J is a function of the pressure $p = p_0 + p_\sim(\tau)$. For the simplest case of a periodic process described by

$$p_\sim(\tau) = (-1)^{i+1}\left(\frac{4\tau}{T_\sim} - 2i - 1\right) p_m, \tag{3.140}$$

where $T_\sim$ is the period of oscillations and $i = E(2\tau/T_\sim)$ is the integral number of half-periods, the density of probability of appearance in a superheated liquid of the first critical nucleus, Eq. (3.76), may be presented as [257]

$$w_1(\tau) = \frac{(J_+ - J_-)V}{2G_p p_m} \exp\left(-\frac{(J_+ - J_-)V}{2G_p p_m}\tau\right). \tag{3.141}$$

Here, J_+ and J_- are the nucleation rates at pressures $p_0 + p_m$ and $p_0 - p_m$, respectively. In an adiabatic process, we get

$$G_p = \frac{2G(p_0)}{p_s - p_0} + \left.\frac{\partial G}{\partial T}\right|_p \left.\frac{\partial T}{\partial p}\right|_s. \tag{3.142}$$

According to Eq. (3.141), in weak fields of variable pressure, cavitation is a Poissonian process. When expanding the expression for the nucleation rate, Eq. (2.112), into a series in terms of $p_\sim$ and limiting ourselves to the first term of the expansion, we have

$$J_+ - J_- \simeq 2J(p_0)\text{sh}(G_p p_m). \tag{3.143}$$

For the mean expectation time of the first nucleus, from Eqs. (3.72), (3.141), and (3.143) we get

$$\bar{\tau} = \bar{\tau}_0 \frac{G_p p_m}{\text{sh}(G_p p_m)}, \tag{3.144}$$

where $\bar{\tau}_0 = [J(p_0)V]^{-1}$ is the mean lifetime of a superheated (stretched) liquid in the absence of a variable pressure field. The change in the form of the field $p_\sim(\tau)$ does not alter the pattern of the phenomenon examined.

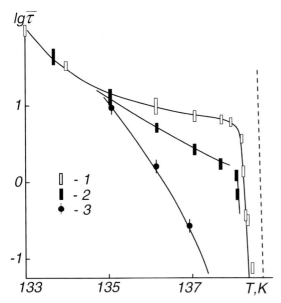

Fig. 3.42 Temperature dependence of the mean lifetime of superheated liquid oxygen at an initial pressure $p_0 = 1.171$ MPa under natural conditions (1) and ultrasonic field (2: $U_m = 10$ V, 3: 15) [258]. The *dashed line* shows calculations by the homogeneous nucleation theory in a macroscopic approximation.

In Refs. [257, 258], the method of measuring the lifetime is used to investigate the boiling-up kinetics of superheated liquid xenon and oxygen in an ultrasonic field of frequency $f = 700$ kHz. A liquid was superheated in glass cells. Ultrasonic oscillations were excited by a piezoceramic radiator at the moment of the transfer of the liquid into the metastable state. The incorporation of an acoustic field increased the probability of nucleation (Fig. 3.41). The temperatures of attainable superheating for liquid xenon decreased by approximately 0.5 K when a variable voltage with an amplitude $U_m = 52$ V was supplied to the piezoradiator. The amplitude dependence of the mean lifetime is shown in the upper-left corner of Fig. 3.41. As in the case of the quasistatic pressure release, histograms of time distribution of boiling-up events at ultrasonic cavitation were of the Poissonian form. The small volume of the superheated liquid ($V \simeq 80$ mm^3) made it impossible to measure the amplitude of the sound pressure. The value of p_m was determined from Eq. (3.144) by experimental data on the mean lifetime of a cavitating liquid. For liquid xenon at $U_m = 52$ V and $J = 10^7$ m^{-3} s^{-1} it gives $p_m \simeq 0.1$ MPa. From the obtained value of p_m and experimental data on $d \ln J(p_0)/dT$, the nucleation rate was determined in the whole temperature range investigated.

The mechanism of acoustic cavitation in the temperature range where the liquid boiling-up may be initiated by high-energy particles is more complicated. The result of the action of acoustic fields depends on the radiation resistance of the liquid. As is seen from Fig. 3.42, the increase of the amplitude of acoustic oscillations in liquid oxygen leads to a stronger decrease of $\bar{\tau}$ in the proximity to the boundary of spontaneous boiling-up of the liquid, where experimental isobars have a characteristic plateau. If we assume, as has been mentioned above, that every ionizing particle on the plateau creates a supercritical-sized vapor bubble, it is difficult to understand the mechanism of the nucleation initiation by ultrasound. It is quite possible that the character of the dependence $\bar{\tau}(T)$ observed is a result of the joint action of the external radiation background and weak spots at the liquid–glass boundary.

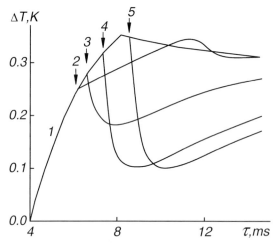

Fig. 3.43 Thermograms of experiments on pulse heating of ^{4}He on a bismuth monocrystal under natural conditions (1) and under the action of a light pulse of duration 24 μs and intensity 24 mW cm^{-2} (2–5) [259].

The effect of a strong initiating action of background radiation was revealed in investigations of the superheating of liquid ^{4}He on bismuth monocrystals [259–261]. If liquid helium was superheated above a certain threshold value T_{LT}, the supply of a light pulse of duration from 2 μs to 1 ms and intensity 1–200 mW cm^{-2} resulted in an abrupt temperature decrease at the monocrystal surface. Within 3 μs from the moment of supply of a light signal, the monocrystal superheating with respect to the temperature of normal boiling of liquid helium ($\simeq 4.2$ K) decreased by 75%. Oscillograms of several experiments, in which a light pulse of fixed duration and power but at different moments of time (shown with arrows) was supplied to a heated monocrystal, are collected in Fig. 3.43. The initiating action of light was observed only after the superheating of liquid helium had reached $\Delta T_{LT} = 0.29$ K (the attainable superheating of liquid helium at atmospheric pressure, ΔT_n, is equal to

0.35 K [132]). The value of T_{LT} does not depend on the size and shape of the heater, the state of its surface, the value of the magnetic field (from 0.4 to 2 T), the duration of the light pulse, the orientation of the light flow with respect to the monocrystal surface, and the induction lines of the magnetic field. The increase of the light intensity to $I = 200$ mW cm^{-2} resulted in an increase of the cooling effect. At $I > 200$ mW cm^{-2}, and also with an increase in the wavelength of the light above 420–450 nm, the monocrystal-cooling effect was not recorded. The authors of Ref. [259] point out that the initiation of boiling-up of superheated helium at the surface of a bismuth heater was observed even at ordinary room light.

A photoelectronic hypothesis has been suggested for explaining the initiating action of light on the boiling-up of superheated liquid helium [261]. According to this hypothesis, the energy of photons is transferred to electrons of the surface layer of a bismuth monocrystal. The absorption of phonons is accompanied by the emission of photoelectrons, which form electronic bubbles in liquid helium. Close to the heater surface the probability of localization of electrons into vapor cavities increases owing to helium superheating and a higher level of density fluctuations. It is assumed that at temperatures $T > T_{LT}$ electronic bubbles lose their stability and are transformed into liquid boiling-up centers. The growth of such evaporation centers is accompanied by the absorption of energy from the thermal boundary layer of a monocrystal and its cooling. A photographic study of the process of formation of a vapor film is made in Ref. [260].

3.13
Heterogeneous Nucleation

Since in daily practice it is impossible to eliminate the effect of accidental inclusions, heterogeneous nucleation is more widespread than homogeneous. Boiling centers may also be present or originate in a system owing to various pollutions of the volume at appropriate conditions on the wall of the system. Like homogeneous nucleation, heterogeneous nucleation is a fluctuation phenomenon. Foreign particles reduce the work of formation of a nucleus, as compared to its evaluations in a pure system, thereby increasing the probability of liquid boiling-up. The impossibility of taking into account the whole variety of incontrollable factors affecting the nucleation process is the main difficulty in constructing a theory of heterogeneous nucleation processes in metastable systems. Sources of nuclei are usually gases dissolved in liquids and adsorbed in the cracks of walls, areas with decreased wettability on solid weighted particles and surfaces that are in contact with liquids. The effect of gas saturation on the superheating of cryogenic liquids is examined in Chapter 4. The main subject of discussion of the present section is heterogeneous nucleation at the surfaces of solids.

The work of formation of nuclei on sufficiently small accidental inclusions depends on their size and physical nature. The simplest case of heterogeneous nucleation is nucleation on a smooth clean wall. All liquid molecules that are in contact with such a wall may be regarded as potential centers of a new phase with the same probability, and for the rate of heterogeneous nucleation, by analogy with homogeneous, we can write

$$J_{het} = \rho_{het} B_{het} \exp\left(-\Psi \frac{W_*}{k_B T}\right), \qquad (3.145)$$

where ρ_{het} is the number of molecules contacting the wall per unit volume; B_{het} is the kinetic factor, which in a first approximation may be considered equal to the kinetic factor at homogeneous nucleation; and Ψ is the factor that takes into account the decrease in the work of nucleus formation on the wall, i.e., $W_{het} = \Psi W_*$. If the formation of a spherical nucleus takes place at a flat surface of a solid, a critical nucleus has the same radius as in the liquid volume, and the work of its formation is smaller than that at a homogeneous process by the value [1]

$$\Psi = \frac{1}{4}(1 + \cos\theta_0)^2 (2 - \cos\theta_0). \qquad (3.146)$$

Here, θ_0 is the equilibrium (macroscopic) wetting angle. Similar to Eq. (3.146), formulae may be written for a convex and a concave surface of a solid [76].

Actual surfaces of solids are not ideally clean and smooth. The origination of vapor-phase centers usually takes place in microcracks, slots, and some other defects of rough surfaces, and is characterized by lower values of the activation energy [262]. The model of a conic cavity [263] is widely used for the analysis of evaporation processes. In this case, $\Psi = \Psi(\theta_0, \beta)$ holds, where β is the angle between the axis and the generator of the cone. The nucleation work is also reduced by impurities in the form of surfactants adsorbed at the surface of a solid [6]. Such impurities may be concentrated on separate isolated areas of the surface. If the wettability of a contaminated island is anomalously low ($\theta_0 \simeq \pi$), then $\Psi \simeq 3(\pi - \theta_0)^2/16$.

The relative purity and the good wettability of solid materials with cryogenic liquids do not ensure a total absence of initiating boiling centers in them, but makes it possible to obtain high superheats in cells with a rough surface, too. An experimental corroboration of this expectation was obtained in Refs. [264, 265] in investigating the lifetime of superheated liquid oxygen and nitrogen in U-shaped tubes of stainless steel and copper with an internal diameter $\simeq 1.1$ mm. The area of the surface contacting a superheated liquid was 4.25 mm, and the volume was 95 mm^3. The treatment of the inner surfaces of the tubes corresponded to the sixth or seventh class of roughness (the

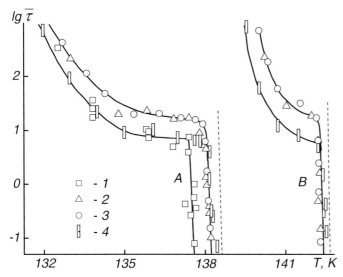

Fig. 3.44 Mean lifetime of superheated liquid oxygen in tubes of stainless steel (1, 2), copper (3) and glass (4) along different isobars [264]. A: $p = 1.171$ MPa; B: 2.252. The *dashed line* shows results of calculations by Eqs. (2.112), (3.8), (3.52), and $\sigma = \sigma_\infty$, $z_0 = 1$.

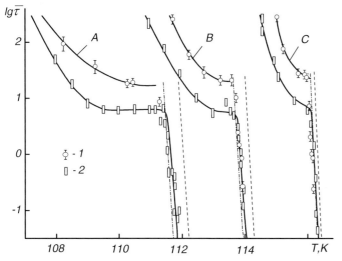

Fig. 3.45 Mean lifetime of superheated liquid nitrogen in tubes of copper (1) and glass (2) along isobars [265]. A: $p = 0.5$ MPa; B: 1.0; C: 1.5. For the origin of the *dashed line*, see Fig. 3.44; the *dashed-dotted curve* shows results of calculations by Eqs. (2.112), (3.11), (3.52), and $z_0 = 1$.

average size of microinhomogeneities was $\simeq 5$ μm). Copper tubes were subjected to an additional polishment. Figures 3.44 and 3.45 show temperature dependences of the mean lifetime for superheated liquid oxygen and nitrogen in metal and glass tubes. The experimental data were reduced to one value of the superheated volume ($V = 100$ mm^3) by the formula $\bar\tau_1/\bar\tau_2 = V_2/V_1$.

Three characteristic regimes of boiling-up of a superheated liquid have been distinguished in experiments on liquid oxygen on tubes of stainless steel (Fig. 3.44) [264]. The initial stage of an experiment (regime I, run-in regime) is characterized by a considerable spread of the values of $\bar\tau$ and the nonreproducibility of data obtained under changes of the thermodynamic state of the liquid. After several boiling-up acts, the mean lifetimes were reproduced on the same tube under changes in (p,T) (regime II). The experimental data form a smooth curve. The reproducibility is retained when liquid portions in a tube are replaced. In the process of long measurements there always came the moment when an abrupt increase in the expectation time of boiling-up (a transition to regime III) was observed. In regime III, the values of $\bar\tau$ were reproduced within the limits of statistical error during all start-ups of the setup, and when the cells were replaced.

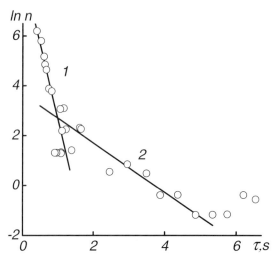

Fig. 3.46 Distribution of expectation times of boiling-up of superheated liquid oxygen in a stainless steel tube: $p = 1.171$ MPa; $T = 137.66$ K, $N = 191$. The lines show results of calculations by Eq. (3.88) with $\bar\tau = 0.15$ s (1) and $\bar\tau = 0.98$ s (2).

The run-in duration depends on the state of the surface contacting the liquid and the pressure. Under the conditions of the same roughness, a longer run-in was required for the surface of tubes of stainless steel than for the surface of copper tubes. In polished copper tubes, reproducible results were already achieved after several tens of boiling-up acts. This result makes it possible to assume that boiling centers were evidently a gas adsorbed in metal pores. In the course of experiments, cases of an abrupt decrease in the lifetime of a metastable liquid with a subsequent restoration of the previous regime could be observed. If the times registered were close to or lower than the setup solubility limit ($\tau_0 \simeq 0.1$ s), they were not taken into account in calculating the mean lifetime of a superheated liquid.

In a number of cases, the distribution of expectation times of boiling-up had a more complicated form in metal tubes than in glass ones. A histogram of one of such experiments is given in Fig. 3.46 on a semilogarithmic scale (regime II). The distribution observed may be presented as the superposition of two Poissonian distributions with different values of $\bar{\tau}$. The presence of two distinct scales of $\bar{\tau}$ is evidently connected with the action of two boiling centers possessing different activation energies.

As follows from Figs. 3.44 and 3.45, temperature dependences of the mean lifetime obtained in metal and glass tubes are qualitatively similar. With decreasing temperature, sections with an abrupt increase of $\bar{\tau}$ (short times) give way to sections with a weaker dependence, $\bar{\tau}(T)$. On the bends of experimental isobars, mean lifetimes of a superheated liquid in metal tubes exceed values of $\bar{\tau}$ obtained in glass tubes from three to five times. This result may be connected with a somewhat higher intensity of the radiation background in glass. However, as already noted, the bends of experimental isobars are, evidently, a result of not only the initiation of volume liquid boiling-up with high-energy particles, but also of the influence of the walls of the measuring cell. For radiation-resistant liquids (Kr, Xe), even the presence of one defect (a crack at the inner surface of a glass cell) disrupts the linear character of the T-dependence of $\bar{\tau}$ in the region of initiated nucleation (Fig. 3.47) [113]. In this case, experimental isobars have a form similar to that observed in liquids with poorly defined scintillation properties (Ar, N_2, O_2, CH_4). It is of fundamental importance that experimental isobar bends, characteristic of low supersaturations, may be observed in the presence of only one easily activated boiling center at the inner surface of the cell.

Close to the boundary of liquid spontaneous boiling-up, the results obtained in metal and in glass tubes agree with each other within the limits of experimental error regarding both the superheating temperature and the value of the derivative $d \ln \bar{\tau}/dT$. Taking into account the dependence of the properties of critical bubbles on the curvature of the separating surface, discrepancies between the homogeneous nucleation theory and experiment here do not exceed 0.1–0.2 K. Experiments with metal tubes have shown [265] that at low pressures sections of isobars with a step T-dependence of $\bar{\tau}$ are not necessarily achieved. A rise in temperature (Fig. 3.45, $p = 0.5$ MPa) resulted in increasing boiling-up acts with times shorter than the limit of the setup solvability. Thus, at $T = 111.26$ K, 54 boiling-up acts of liquid nitrogen led to only two values of $\tau = 7.19$ s and $\tau = 7.58$ s, for the rest, $\tau < 0.1$ s holds. These results allow us to assume that the main reason for disagreement between

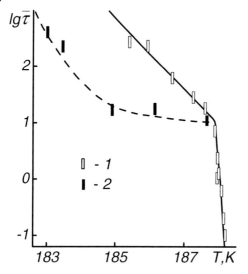

Fig. 3.47 Mean lifetime of superheated liquid krypton at a pressure of $p = 1.60$ MPa in a "clean" cell (1) and in the presence of a completed boiling center (2).

theory and the results of quasistatic experiments on cavitation [193–198] is the action of heterogeneous centers whose activity at negative pressures has to be considerably higher.

In the process of run-in of tubes, gases adsorbed in microheterogeneities of the solid surface are removed and dissolved in a liquid. At high superheatings, critical nuclei are of the order of several or tens of nanometers. For such small new-phase formations many microinhomogeneities may be regarded as a flat smooth wall, but Eq. (3.146) is no longer applicable here as has been obtained by neglecting the dependence of nuclei properties on the curvature of the separating surface and does not take into account the excess energy of the wetted perimeter line tension, γ [265]. The value and the sign of γ are determined by the peculiarities of interaction of contacting phases in the vicinity of the wetted perimeter [266, 267].

Taking into account the effect of line tension, the factor that corrects the nucleation work has the form [268]

$$\Psi = \frac{1}{4}(1+\cos\theta_*)^2(2-\cos\theta_*) + \frac{3}{4}\frac{\kappa_*}{R_* \sin\theta_*}, \qquad (3.147)$$

where θ_* is the critical value of the microscopic wetting angle, $\kappa_* = \gamma/\sigma$. The surface and the line of tension have been chosen as the separating surface and line. It is assumed that the line of tension coincides with the intersection of the surfaces of tension. The R-dependences of σ and γ are not taken into account. At $R \gg \kappa_*$, Eq. (3.147) is reduced to Eq. (3.146).

If it is assumed that in steep sections of isobars nucleation is also initiated by a metal surface, then from Eq. (3.145) we have $\Psi = 0.38$; $\rho_{het} = 2.6 \times 10^{30}$ m^{-3} at $p = 0.5$ MPa and $\Psi = 0.41$; $\rho_{het} = 3.8 \times 10^{30}$ m^{-3} at $p = 1.0$ MPa (nitrogen). According to Eq. (3.146), the mentioned value of Ψ corresponds to an equilibrium wetting angle $\theta_0 \simeq 98°$. Such high values of wetting angles are not characteristic of cryogenic liquids [269]. Thus, for liquid nitrogen θ_0 on copper and stainless steel does not exceed $10°$.

Taking into account the line tension, the equilibrium condition for the critical bubble on a smooth wall is written as [268]

$$\sigma_{SV} - \sigma_{SL} = \sigma \cos\theta_* - \frac{\gamma}{(R_* \sin\theta_*)}, \quad (3.148)$$

where σ_{SV} and σ_{SL} are the surface tensions at solid–vapor and solid–liquid interfaces. Substituting into Eq. (3.148) the equation

$$\sigma_{SV} = \sigma_{SL} + \sigma \cos\theta_0, \quad (3.149)$$

we get

$$\frac{\kappa_*}{R_*} = \sin\theta_*(\cos\theta_* - \cos\theta_0). \quad (3.150)$$

From Eqs. (3.4), (3.147), (3.150), and the data for Ψ given above, at $\theta_0 = 10°$ we have $\theta_* = 50–65°$. A smaller value of θ_* refers to a higher pressure in the liquid. The value of the line tension is negative and equals $-(1.5–4.0) \times 10^{-12}$ J m^{-1}. A statistical calculation [270] for a drop of nitrogen on graphite gives $\gamma = -3.3 \times 10^{-12}$ J m^{-1}, $\theta_* = 10.7°$. Thus, although taking into account the line tension results in decreasing values of the wetting angle as compared with its evaluations from Eqs. (3.145) and (3.146), the values of θ_* continue to considerably exceed the equilibrium value of θ_0. Suggestions that are sometimes made [271] about the deterioration of wettability for highly curved surfaces are not confirmed by direct measurements. When investigating breaking loads under capillary condensation of liquids between mica cylinders, Fisher and Israelachvili [191] did not discover any changes in the wetting angle down to $R \simeq 4$ nm.

The values of ρ_{het} calculated from data on nucleation kinetics are approximately $10^7–10^8$ times lower than the surface density of particles. This result may be interpreted as a consequence of different activities of boiling centers on the cell walls [272]. At the same time, owing to a practically total agreement between data obtained in glass and in metal tubes, it should be assumed that the distribution of "weak spots" is not sensitive to the wall material, i.e., its roughness. The independence of the number of heterogeneous nucleation sites and the resulting reproducibility of the results on nucleation measurements on the methods of preparation of the measuring cell (degassing, washing, oxidation) and the replacement of the liquid by other samples give a strong indication of the weak effect of these factors on the distribution and the activity of surface centers. Thus, experimental data on the superheating of cryogenic liquids in polished metal tubes point to the homogeneous rather than the heterogeneous mechanism of nucleation. At least, these data cannot be interpreted straightforwardly in the framework of traditional models of the heterogeneous nucleation theory without the introduction of physically unjustified assumptions.

4
Nucleation in Solutions of Liquefied Gases

4.1
Critical Nucleus and the Work of its Formation

Liquid solutions are characterized by a variety of phase transition phenomena. Along with ordinary critical points at which two phases become identical, in solutions a coexistence at higher order critical points is possible (tricritical points, critical finite points, etc.). Besides the instability with regard to the formation of a vapor phase, a solution may also be unstable with respect to dissolution of a gaseous phase inside the liquid (decay of a gas-supersaturated solution) and separation into different liquid phases. In the present chapter, the kinetics of boiling-up of liquid mixtures with full solubility of the components and of gas-saturated liquids is considered.

At isothermal formation in a superheated binary ($\nu = 2$) solution of a vapor bubble of volume V with N_1'' molecules of the first component and N_2'' molecules of the second component, the change of the Gibbs thermodynamic potential is written as follows [76]:

$$\Delta \Phi = (p' - p'')V + \sigma A + \sum_{i=1}^{\nu}(\mu_i'' - \mu_i')N_i'', \qquad p'' = \sum_{i=1}^{\nu} p_i'', \tag{4.1}$$

where p_i'' presents the partial pressures of the components of the mixture in a bubble, and the other notations are the same as in Eq. (3.1). Equation (4.1) presumes that the appearance of a bubble in a system does not change the state of the phase that surrounds it.

The equality $(d\Delta\Phi)_* = 0$ corresponds to an equilibrium of a bubble in the otherwise homogeneous solution. In this case, the conditions of mechanical, Eq. (3.4), and diffusion equilibrium

$$\mu_i''(p_*'', T) = \mu_i'(p', T), \qquad i = 1, 2, \ldots, \nu, \tag{4.2}$$

are fulfilled. Substitution of Eqs. (3.4) and (4.2) into Eq. (4.1) gives a relation for the work of formation of an equilibrium (critical) nucleus in the solution which has a form similar to that for a one-component liquid, Eq. (3.8). The type of the extremum is determined by the sign of the second-order differential

$$(d^2\Delta\Phi)_* = \sigma \frac{d^2 A}{dV^2}\bigg|_* (dV)^2 + d\sigma dA - dp''dV + \sum_{i=1}^{v} d\mu_i'' dN_i''. \tag{4.3}$$

Considering the vapor mixture in a bubble as an ideal gas and replacing the chemical potential of the liquid in Eq. (4.1) by chemical potential of the vapor in an equilibrium bubble equal to it, in the vicinity of the extremum of $\Delta\Phi$ we have

$$\Delta\Phi = (p' - p'')V + \sigma A + V \sum_{i=1}^{v} p_i'' \ln \frac{p_i''}{p_{i*}''}, \tag{4.4}$$

$$(d^2\Delta\Phi)_* = -\frac{2\sigma A_*}{9V_*^2}(dV)^2 + \sum_{i=1}^{v} \frac{V_*}{p_{i*}''}(dp_i'')^2. \tag{4.5}$$

The absence of cross-terms in the quadratic form (4.5) implies that fluctuations of the variables V, p_1'', and p_2'' are statistically independent. Owing to the presence of one negative coefficient (at the square of dV) in Eq. (4.5), the surface of the thermodynamic potential in the space (V, p_1'', p_2'') in the vicinity of the extremum point is a hyperbolic paraboloid, and the extremum point itself is a saddle point. The equilibrium of a bubble with the metastable solution, determined by Eqs. (3.4) and (4.2), is unstable with respect to the variable V at constants p_1'' and p_2''. The line of the energetically most favorable trajectory of fluctuational change of the size of a precritical bubble follows the valley of the potential and crosses the pass of the thermodynamic potential, $\Delta\Phi(V, p_1'', p_2'')$. The equilibrium of a critical bubble with respect to pressure is stable. The equalities $V = V_* = \mathrm{const}$, $p_1'' = p_{1*}'' = \mathrm{const}$ and $V = V_* = \mathrm{const}$, $p_2'' = p_{2*}'' = \mathrm{const}$ give the lines of "water sheds" of the potential barrier. The saddle point in this case is the intersection point of the lines of "water sheds" and the valley, as well as of the lines of mechanical, Eq. (3.4), and diffusion, Eq. (4.2), equilibria of a vapor bubble in solution. The minus sign at the variable V in Eq. (4.5) indicates that this variable is unstable, while the variables p_1'' and p_2'' are stable.

We define the molar fraction of the second component in a bubble as $c'' = N_2''/(N_1'' + N_2'') = N_2''/N''$. Employing this definition, in the vicinity of the saddle point of the potential barrier we may then write

$$\Delta\Phi = (p' - p'')V + \sigma A + p''V \ln \frac{p''}{p_*''}$$

$$+ p''V \left[(1 - c'') \ln \frac{1 - c''}{1 - c_*''} + c'' \ln \frac{c''}{c_*''}\right]. \tag{4.6}$$

Accordingly, the second-order differential of Eq. (4.6) is

$$(d^2\Delta\Phi)_* = \left(-\frac{2\sigma A_*}{9V_*} + \frac{N''_* k_B T}{V_*^2}\right)(dv)^2 - 2\frac{k_B T}{V_*}dVdN'' + \frac{k_B T}{N''_*}(dN'')^2$$
$$+ N'' k_B T \left(\frac{1}{1-c''_*} + \frac{1}{c''_*}\right)(dc'')^2$$
$$= -\frac{2\sigma A_*}{9V_*}(dV)^2 + \frac{V_*}{p''_*}(dp'')^2 + p''_* \left(\frac{1}{1-c''_*} + \frac{1}{c''_*}\right)(dc'')^2. \quad (4.7)$$

Equations (4.6) and (4.7) give the surface of the Gibbs thermodynamic potential in variables (V, p'', c''). The fluctuations of these variables are also statistically independent. At the saddle point, the equilibrium of a bubble with its surroundings is stable against variations in composition and the pressure in it but unstable against variations in size.

Considering condensation in a binary mixture, Reiss [273] postulated the incompressibility of liquid droplets in supersaturated vapors. In this case, the number of independent variables of a nucleus decreases so that there are only two of them left. Choosing N'_1 and N'_2 as such variables, we have

$$(d^2\Delta\Phi)_* = \left(-\frac{2\sigma A_*}{9}\frac{1}{N'^2_*} + kT\frac{N'_{2*}}{N'_{1*}N'_*}\right)(dN'_1)^2 + \left(-\frac{2\sigma A_*}{9}\frac{1}{N'^2_*} - kT\frac{1}{N'_*}\right)dN'_1 dN'_2$$
$$+ \left(-\frac{2\sigma A_*}{9}\frac{1}{N'^2_*} + kT\frac{N'_{1*}}{N'_{2*}N'_*}\right)(dN'_2)^2. \quad (4.8)$$

The lines of the "water shed" and the valley of the surface of the hyperbolic paraboloid, $\Delta\Phi(N'_1, N'_2)$, determining the stable and the unstable variable do not coincide with the axes of N'_1 and N'_2, which is a consequence of the statistical dependence of fluctuations in the variables N'_1 and N'_2.

As for one-component liquids, in a region of weak metastability ($G_* \gg 1$) it is possible to obtain approximate relationships relating the work of formation of a critical bubble, its radius and composition to the parameters of the solution at the line of phase equilibrium. At a sufficiently large distance from the line of critical points, a liquid solution may be considered as incompressible, and the vapor in a bubble as ideal. In such cases, Eqs. (3.4) and (4.2) imply [173, 274]

$$p''_* = p_s \left\{c''_s \exp\left[-\frac{(p_s - p')v'_2}{p_s v''}\right] + (1 - c''_s)\exp\left[-\frac{(p_s - p')v'_1}{p_s v''}\right]\right\}, \quad (4.9)$$

$$c''_* = c''_s \left\{c''_s + (1 - c''_s)\exp\left[\frac{(p_s - p')(v'_1 - v'_2)}{p_s v''}\right]\right\}^{-1}. \quad (4.10)$$

Here, v is the specific volume of the solution; the subscript s refers, as before, to a flat interface.

If $v'_i \ll v''$ ($i = 1, 2$), expanding the exponents into a series and restricting ourselves to the first terms of the expansion, after substituting the result into Eq. (3.4), for the critical-bubble radius we obtain

$$R_* = \frac{2\sigma}{(p_s - p')\left[1 - \dfrac{v'_1}{v''} - c''_s \dfrac{(v'_2 - v'_1)}{v''}\right]}, \qquad (4.11)$$

making it possible to present the work of formation of a critical nucleus in a binary mixture, Eq. (3.8), as [274]

$$W_* = \frac{16\pi}{3} \frac{\sigma^3}{(p_s - p')^2 \left[1 - \dfrac{v'_1}{v''} - c''_s \dfrac{(v'_2 - v'_1)}{v''}\right]^2}. \qquad (4.12)$$

Taking into account that $v'_2 - v'_1 \ll v''$ and $v'_1 \simeq v'$, Eq. (4.12) is transformed into the formula for the work of formation of a critical bubble in a one-component system if the approximation, Eq. (3.9), is accepted. In the case of a weak solution of the gas in the liquid, considering the pure solvent as incompressible and a vapor–gas mixture in a bubble as an ideal gas, one can obtain formulae expressing the pressure and the gas concentration in a bubble, and the work of formation of a critical nucleus only in terms of the parameters of a pure solvent [275].

In the framework of the van der Waals square gradient expansion, the change in the Helmholtz free energy of a two-component ($\nu = 2$) system connected with the density distribution of the heterophase fluctuation $\delta\vec{\rho}(\vec{r}) = \vec{\rho}(\vec{r}) - \vec{\rho}$, is written as [276]

$$\Delta F[\vec{\rho}] = \int \left(\Delta f + \sum_{i,j=1}^{\nu} \kappa_{ij} \nabla \rho_i \nabla \rho_j\right) d\vec{r}, \qquad (4.13)$$

where

$$\Delta f = f(\vec{\rho}) - f' - \sum_{i=1}^{\nu} (\rho_i - \rho'_i)\mu'_i. \qquad (4.14)$$

Here, $f(\vec{\rho})$ is the free energy of a unit volume of a homogeneous solution with a local density of the components $\vec{\rho}(\vec{r}) = \{\rho_1(\vec{r}), \ldots, \rho_\nu(\vec{r})\}$, $\nabla\rho_i$ is the density gradient of the ith component, κ_{ij} is the symmetrical matrix of the influence coefficients connected with the interfacial contributions, and μ'_i is the chemical potential of the ith component of the metastable phase.

A homogeneous metastable phase is stable against infinitesimal density perturbations. Therefore, the quadratic form in the second term of the integrand in Eq. (1.13) must be positive. This condition is fulfilled if the inequalities

$$\det \kappa_{ij} = \kappa_{11}\kappa_{22} - \kappa_{12}^2 > 0, \qquad \kappa_{11} > 0, \qquad \kappa_{22} > 0 \qquad (4.15)$$

hold.

The density distribution of the components in a heterophase fluctuation corresponding to the extremum value of $\Delta F[\vec{\rho}]$ is found by solving the Euler system of equations for the functional, Eq. (4.13),

$$2\sum_{j=1}^{\nu}\nabla(\kappa_{ij}\nabla\rho_j) - \sum_{j,\kappa=1}^{\nu}\frac{\partial\kappa_{j,\kappa}}{\partial\rho_i}\nabla\rho_j\nabla\rho_\kappa = \mu_i(\vec{\rho}) - \mu'_i = \frac{\partial\Delta f}{\partial\rho_i},$$

$$i = 1, 2, \ldots, \nu. \qquad (4.16)$$

Neglecting the dependence of the influence coefficients, $\kappa_{i,j}$, on ρ and considering only spherical inhomogeneities, the system of equations, Eq. (4.16), can be transformed to

$$2\sum_{j=1}^{\nu}\kappa_{ij}\left(\frac{d^2\rho_j}{dr^2} + \frac{2}{r}\frac{d\rho_j}{dr}\right) = \mu_i(\vec{\rho}) - \mu'_i = \frac{\partial\Delta f}{\partial\rho_i}, \quad i = 1, 2, \ldots, \nu, \qquad (4.17)$$

with the boundary conditions

$$\rho_i = \rho'_i \quad \text{at} \quad r \to \infty,$$

$$\frac{d\rho_i}{dr} = 0 \quad \text{at} \quad r = 0 \quad \text{and} \quad r \to \infty. \qquad (4.18)$$

As has already been done for a one-component system (see Section 2.5), it can be shown that all solutions of Eq. (4.17) in a metastable region with boundary conditions, given by Eq. (4.18), correspond to saddle points of the functional $\Delta F[\vec{\rho}]$, i.e., they correspond to critical nuclei. The work of formation of a critical nucleus is found by substituting the solutions of Eq. (4.17) into Eq. (4.13)

$$W_* = 4\pi\int_0^\infty\left\{\Delta f - \frac{1}{2}\sum_{i=1}^{\nu}(\rho_i - \rho'_i)[\mu_i(\vec{\rho}) - \mu'_i]\right\}r^2 dr. \qquad (4.19)$$

The equality sign in the first expression in Eq. (4.15) corresponds to the state of an indifferent equilibrium. In this case, the expression

$$\kappa_{12} = (\kappa_{11}\kappa_{22})^{1/2} \qquad (4.20)$$

determines the limiting value of the cross-coefficient, κ_{12}. The use of Eq. (4.20) in Eq. (4.13) for spherical inhomogeneities yields an equation similar to that for a one-component liquid, Eq. (2.36),

$$\Delta F[\vec{\rho}] = 4\pi\int_0^\infty\left[\Delta f + \kappa_{22}\left(\frac{d\rho_\beta}{dr}\right)^2\right]r^2 dr, \qquad (4.21)$$

where $\rho_\beta = \rho_2 + (\kappa_{11}/\kappa_{22})^{1/2}\rho_1$.

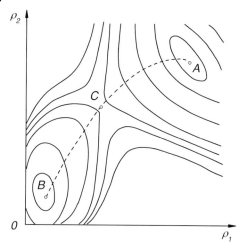

Fig. 4.1 Projection of the surface of the excess density of the free energy of a homogeneous binary system onto the plane (ρ_1, ρ_2).

The density distribution of the components of the mixture, $\rho_i (i = 1, 2)$, in a critical bubble may be obtained by solving the Euler system of equations for the functional (4.21), which includes the one differential equation

$$\frac{d^2 \rho_\beta}{dr^2} + \frac{2}{r} \frac{d\rho_\beta}{dr} = \frac{\mu_2 - \mu_2'}{2\kappa_{22}}, \tag{4.22}$$

with boundary conditions similar to Eq. (4.18) and one algebraic equation

$$\mu_1 - \mu_1' = \left(\frac{\kappa_{11}}{\kappa_{22}}\right)^{1/2} (\mu_2 - \mu_2'). \tag{4.23}$$

In this case, the work of formation of a critical bubble is given by

$$W_* = \min \max \Delta F[\vec{\rho}]. \tag{4.24}$$

Equation (4.24) gives an alternative way as compared with Eq. (3.8) of calculating W_* and makes it possible with Eqs. (3.4), (4.2), and (3.8) to determine the dependence $\sigma(R_*)$.

As distinct from a one-component system, where Eq. (2.39) inevitably leads to a monotonic dependence of the density, $\rho = \rho(r)$, in a binary system, Eqs. (4.17), (4.22) and (4.23) allow the existence of nonmonotonic density profiles. Figure 4.1 shows the lines of a constant value of $\Delta f(\rho_1, \rho_2)$ in coordinates (ρ_1, ρ_2). This figure is similar to Fig. 2.6 for a one-component system. The local minimum Δf (point A), separated from another deeper minimum (point B) by a certain barrier with a saddle point C, corresponds to the metastable state. The trajectory of motion between these minima (cf. Fig. 4.1, dashed line) is determined by the solution of the system of equations, Eq. (4.17). Its position depends on the values of the influence coefficients, κ_{ij}. Assuming that, for instance, the function ρ_1 is monotonic, then, according to Fig. 4.1, one must observe a nonmonotonic behavior in the density profile of the second component, ρ_2. The adsorption of some of the components of the mixture may lead to considerable changes in the surface tension of the nucleus, and therefore in the work of its formation.

4.2
Theory of Nucleation in Binary Solutions

The kinetic nucleation theory in multicomponent systems was mainly developed in application to the problem of condensation in mixtures of supersaturated vapors. The first attempt to calculate the nucleation rate in a binary mixture was made by Döring and Neumann [277]. However, these authors did not manage to obtain quantitative results. Two-component nucleation in supersaturated vapors, as a generalized diffusion in the variables (N_1', N_2') in the field of thermodynamic forces, was first considered by Reiss [273], and Nesis and Frenkel [278]. This approach was further developed by Stauffer [279], Trinkaus [80], Shi and Seinfeld [81], Kuni et al. [280], and Wu [281].

As distinct from a liquid droplet, a bubble in a liquid binary system cannot be regarded as incompressible, and its thermodynamic state is already given by three variables: the volume, V, and the partial pressures of the components of the mixtures, p_1'' and p_2'', or the volume V, the pressure in the bubble, p'', and its composition, c''. The fact that three variables are required for describing the state of a bubble in a binary system in isothermal conditions is explained by the ability of the bubble to have at any given volume, V, and total number of molecules, N'', different compositions, i.e., different values of N_1'' and N_2''. An attempt to develop a theory of boiling-up of superheated binary solutions was made in Ref. [274]. However, these authors mistakenly ignored pressure fluctuations in a bubble and treated the process as a two-dimensional problem in variables, (V, c''). The boiling-up of a gas-filled liquid was examined by Deryagin and Prokhorov [282] and also by Kuni et al. [283,284]. In Refs. [282–285], the solvent is assumed to be nonvolatile, and for this reason, the problem was reduced to a two-dimensional one. A theory of boiling-up for binary solutions taking into account all the factors limiting the bubble growth (volatility of the mixture components, viscous and inertial forces, thermal effects at the bubble boundary, diffusive substance supply to a growing bubble) was developed by Baidakov [286–288].

We examine now the kinetics of spontaneous boiling-up of a binary liquid solution whose initial state is given by temperature, T, pressure, p', and concentration, c'. The initial supersaturation of the solution is considered small, so $G_* \gg 1$ holds, and bubbles forming in the process of decay may be regarded as spherical formations whose properties are only slightly different from those of the bulk phase. Isothermal conditions in the process of phase transition are ensured by a high thermal conductivity and a high heat capacity of the liquid. Temperature effects at the bubble boundary, which we do not examine here, may be taken into account by an appropriate correction, as it is done in our analysis of boiling in a one-component liquid (see Section 3.2). In one-component liquids, only the free-molecular regime of exchange of molecules between a vapor bubble and a liquid takes place. In contrast, during the decay of a two-component solution besides the free-molecular regime there may be established a diffusive regime when molecules of the mixture components move to or from a bubble in a diffusive way.

We shall consider the gas in the bubble as a perfect one, and as the determining thermodynamic parameters of the bubble we choose its volume, V, and the partial pressures of the components of the mixture, p_1'' and p_2'', in the bubble. Inertia effects in the boiling-up kinetics of the solution will be taken into account by including a dynamic variable into the set of parameters of the bubble, i.e., the rate of change of the bubble volume, $\dot{V} = dV/d\tau$. In dimensionless quantities (x, y, z, v)

$$x = \frac{V - V_*}{V_*}, \quad y = \frac{p_1'' - p_{1*}''}{p_{1*}''}, \quad z = \frac{p_2'' - p_{2*}''}{p_{2*}''}, \quad v = \frac{\tau_r}{V_*}\frac{dV}{d\tau}, \quad (4.25)$$

the potential surface will be written as

$$\Psi(x, y, z, v) = \frac{\Delta\Phi(x, y, z)}{k_B T} + \frac{M_{\text{eff}}}{2k_B T \tau_r^2} v^2. \quad (4.26)$$

Here, the first term on the right-hand side is determined by Eq. (4.4), and τ_r is the characteristic time of establishment of the equilibrium growth rate distribution of bubbles. Expanding Eq. (4.26) into a power series in (x, y, z, v) in the vicinity of the extremum, $\Psi_* = \Psi_*(0, 0, 0, 0)$, including terms of second order, we obtain

$$\Psi(x, y, z, v) = \Psi_* + \frac{1}{2}(\vec{r}\widehat{\Psi}\vec{r}') = G_*\left(1 - \frac{1}{3}x^2 + \frac{1}{b_1}y^2 + \frac{1}{b_2}z^2 + \frac{1}{3\chi}v^2\right), \quad (4.27)$$

where

$$b_i = \frac{2\sigma}{p_{i*}'' R_*}, \quad i = 1, 2, \quad (4.28)$$

holds (for the other notations, see Eq. (3.19)).

The matrix of second-order derivatives, $\widehat{\Psi}$, at the extremum of the potential $\Psi(x, y, z, v)$, which, as a consequence from Eq. (4.27), is the saddle point, will look like

$$\hat{\Psi} = 2G_* \begin{pmatrix} -\frac{1}{3} & 0 & 0 & 0 \\ 0 & \frac{1}{b_1} & 0 & 0 \\ 0 & 0 & \frac{1}{b_2} & 0 \\ 0 & 0 & 0 & \frac{1}{3\chi} \end{pmatrix}. \tag{4.29}$$

The interaction of a bubble with a solution is composed of the hydrodynamic and the absorption–emission interaction. The hydrodynamic interaction is caused by the pressure of the solution, the Laplace force, and the viscous and inertial forces. The absorption–emission interaction includes the free-molecular and the diffusion regime of molecular exchange of a bubble and the phase surrounding it. The hydrodynamic interaction in a solution is described by the equation for a one-component liquid (see Eq. (3.22)), in which ρ_l, η, and σ are mass density, viscosity, and surface tension of the solution, respectively. It can be shown that in this case the relaxation time τ_r, the effective bubble mass, M_{eff}, and the parameter χ in the solution are determined by Eqs. (3.36), (3.64), and (3.65) for a one-component system. In dimensionless quantities, the equation of motion for the bubble volume, Eq. (3.22), is written as follows:

$$\dot{v} = \chi x + 3\frac{\chi}{b_1}y + 3\frac{\chi}{b_2}z - v. \tag{4.30}$$

We now examine the absorption–emission interaction of a bubble with a solution. It changes the number of molecules, (N_1'', N_2''), in the bubble but does not change its volume. In the free-molecular regime, the rates of change of the number of molecules of each of the bubble components are given by

$$\dot{N}_{mi}'' = \frac{\pi \alpha_i v_{ti} R^2}{k_B T}(p_{iR}'' - p_i''). \tag{4.31}$$

The notations used are the same as in Eq. (3.23). The subscript i ($i = 1, 2$) specifies the component of the solution considered.

Besides the free-molecular exchange, the supply of a substance to a growing precritical bubble from the surrounding solution may be realized by diffusion. In the general case, the concentration field around a bubble is determined by the solution of the nonstationary diffusion equation. The characteristic time of change of the bubble radius is proportional to $(R/\dot{R})$. The characteristic time of establishment of a stationary concentration field, $c'(r)$, around a bubble is (R^2/D_g), where D_g is the coefficient of mutual diffusion. In the vicinity of the saddle point of the pass, the rate is $\dot{R}$, and $R\dot{R} \ll D_g$ holds. This inequality makes it possible to consider only the stationary solution of the diffusion equation

$$D_g \frac{1}{r} \frac{\partial^2}{\partial r^2} r c'(r) = 0, \qquad (4.32)$$

with the boundary conditions

$$c'(r)|_{r=R} = c'_R, \qquad c(r)|_{r \to \infty} = c'_s, \qquad (4.33)$$

which has the form

$$c'(r) = c'_s + (c'_R - c'_s) \frac{R}{r}. \qquad (4.34)$$

Using the well-known formula for a diffusion flow of molecules of the ith component in a solution [61] and taking into account the fact that the particle number density of molecules in a solution is ρ', for the rate of change of the number of molecules in a bubble in the diffusion regime we obtain

$$\dot{N}''_{gi} = 4\pi R^2 \rho' D_g \left. \frac{\partial c(r)}{\partial r} \right|_{r=R} = 4\pi R \rho' D_g (c'_R - c'_s). \qquad (4.35)$$

A small change in the concentration around a growing bubble will considerably affect only the values of p''_{iR}. Expanding p''_{iR} into a series in the vicinity of $c' = c'_s$ and restricting ourselves to the first terms of the expansion, we have

$$p''_{iR} = p''_i + \left(\frac{\partial p''_i}{\partial c} \right) (c'_R - c'_s). \qquad (4.36)$$

Substitution of Eq. (4.36) into Eq. (4.35) gives

$$\dot{N}''_{gi} = 4\pi R \rho' D_g H_i (p''_{iR} - p''_i). \qquad (4.37)$$

Here, $H_i = (\partial c / \partial p''_i)$.

The total rate of change in the number of molecules of the components of the solution in a bubble as a result of free-molecular and diffusion exchanges is given by

$$\dot{N}''_i = [(\dot{N}''_{mi})^{-1} + (\dot{N}''_{gi})^{-1}]^{-1}. \qquad (4.38)$$

Linearization of Eqs. (4.31) and (4.37), their substitution into Eq. (4.38) and a subsequent transformation from variables $\dot{N}''_i$ to variables $\dot{p}''_i$ for the emission–absorption interaction of a bubble with its surroundings yields

$$\dot{y} = -\frac{\chi \omega_1}{1 + \gamma_1} y - v, \qquad (4.39)$$

$$\dot{z} = -\frac{\chi \omega_2}{1 + \gamma_2} z - v. \qquad (4.40)$$

Here, the following notations are introduced:

$$\omega_i = \frac{3}{2}\frac{\alpha_i v_{ti} \eta}{\sigma}, \tag{4.41}$$

$$\gamma_i = \frac{\alpha_i v_{ti} R_*}{4 k_B T \rho' D_g H_i}. \tag{4.42}$$

From the equations of motion of the bubbles, Eqs. (4.30), (4.39), and (4.40), and from data on the thermodynamic forces determined by the matrix of the second-order derivatives, $\hat{\Psi}$ (Eq. (4.29)), for the matrix of the generalized diffusion tensor, we obtain

$$\hat{D} = \frac{3}{2}\frac{\chi}{G_*}\begin{pmatrix} 0 & 0 & 0 & -1 \\ 0 & \frac{\omega_1 b_1}{3(1+\gamma_1)} & 0 & 1 \\ 0 & 0 & \frac{\omega_2 b_2}{3(1+\gamma_2)} & 1 \\ 1 & -1 & -1 & 1 \end{pmatrix}. \tag{4.43}$$

Substitution of Eqs. (4.29) and (4.43) into the characteristic equation (2.101) leads to the algebraic equation

$$\tilde{\lambda}^4 + \left(1 + \frac{\omega_1 \chi}{1+\gamma_1} + \frac{\omega_2 \chi}{1+\gamma_2}\right)\tilde{\lambda}^3$$
$$+ \chi\left[\frac{\omega_1 \omega_2 \chi}{(1+\gamma_1)(1+\gamma_2)} + \frac{\omega_1}{1+\gamma_1} + \frac{\omega_2}{1+\gamma_2} + \frac{3}{b_1} + \frac{3}{b_2} - 1\right]\tilde{\lambda}^2$$
$$+ \chi^2\left[\frac{\omega_1 \omega_2}{(1+\gamma_1)(1+\gamma_2)} + \frac{\omega_1}{(1+\gamma_1)}\left(\frac{3}{b_2}-1\right) + \frac{\omega_2}{(1+\gamma_2)}\left(\frac{3}{b_1}-1\right)\right]\tilde{\lambda}$$
$$- \frac{\chi^3 \omega_1 \omega_2}{(1+\gamma_1)(1+\gamma_2)} = 0. \tag{4.44}$$

The positive root $\tilde{\lambda}_0$ of this equation determines the nucleation rate in a supersaturated binary solution with an arbitrary volatility, viscosity, and inertia of the liquid.

As in the case of a one-component system, the nucleation rate in a binary solution is determined by Eq. (2.112). It is usually assumed that a nucleus can form on any molecule of the system. In this case, $\rho = \rho' = \rho'_1 + \rho'_2$ holds, where ρ'_1 and ρ'_2 are the number densities of the solvent and the dissolved substance in the liquid phase, and $z_0 = 1$. Another approach to the determination of ρz_0 is proposed by Deryagin [97]. According to Ref. [97], $z_0 = \rho'/\rho''_*$ holds, where ρ''_* is the density of the mixture in a critical-size bubble. For the factor z_1, which determines the transition from the distribution function of nuclei per volume, V, to the distribution function referred to the number of molecules, N'', in the nuclei, we have $z_1 = dN''/dV|_* = \rho''_*$. Thus, the nucleation

4 Nucleation in Solutions of Liquefied Gases

rate in a superheated binary solution is given by Eq. (2.112), where the kinetic factor is determined by Eq. (3.52) and the positive root of Eq. (4.44).

The final solution of the problem includes four dimensionless parameters

$$b_i = \frac{2\sigma}{R_* p''_{i*}}, \quad \chi = \frac{\rho_l \sigma R_*}{8\eta^2}, \quad \omega_i = \frac{3}{2}\frac{\alpha_i v_{ti}\eta}{\sigma}, \quad \gamma_i = \frac{\alpha_i v_{ti} R_*}{4k_B T D_g \rho' H_i}. \quad (4.45)$$

The parameter b_i characterizes the supersaturation of the solution, χ determines the contribution of inertial forces to the dynamics of nucleus growth, ω_i characterizes the role of the volatility of the components and the viscosity of the medium, and the parameter γ_i determines the role of the free-molecular and the diffusion regime in the absorption–emission interaction of a bubble with the solution.

The diffusion supply of a substance to a bubble may be neglected if the characteristic time of diffusion in the vicinity of the bubble is much shorter than the other time scales connected with the pressure change in the nucleus. In this case, the concentration in the solution is homogeneous up to the bubble surface

$$\gamma_i \ll 1 \quad (4.46)$$

and Eq. (4.44) takes the form

$$\tilde{\lambda}^4 + (1 + \omega_1 \chi + \omega_2 \chi)\tilde{\lambda}^3 + \chi\left(\omega_1 \omega_2 \chi + \omega_1 + \omega_2 + \frac{3-b}{b}\right)\tilde{\lambda}^2$$
$$+ \chi^2\left[\omega_1\omega_2 + \omega_1\left(\frac{3}{b_2} - 1\right) + \omega_2\left(\frac{3}{b_1} - 1\right)\right]\tilde{\lambda} - \chi^3\omega_1\omega_2 = 0, \quad (4.47)$$

where $1/b = 1/b_1 + 1/b_2$. Otherwise, in Eq. (4.46) the diffusion regime is determined and, instead of Eq. (4.44), we have

$$\tilde{\lambda}^4 + \left(1 + \frac{\chi\omega_1}{\gamma_1} + \frac{\chi\omega_2}{\gamma_2}\right)\tilde{\lambda}^3 + \chi\left(\frac{\chi\omega_1\omega_2}{\gamma_1\gamma_2} + \frac{\omega_1}{\gamma_1} + \frac{\omega_2}{\gamma_2} + \frac{3-b}{b}\right)\tilde{\lambda}^2$$
$$+ \chi^2\left[\frac{\omega_1\omega_2}{\gamma_1\gamma_2} + \frac{\omega_1}{\gamma_1}\left(\frac{3}{b_2} - 1\right) + \frac{\omega_2}{\gamma_2}\left(\frac{3}{b_1} - 1\right)\right]\tilde{\lambda} - \chi^3\frac{\omega_1\omega_2}{\gamma_1\gamma_2} = 0. \quad (4.48)$$

If inertia effects of the liquid are neglected (see Eq. (3.41)), then Eq. (4.44) yields

$$\tilde{\lambda}^3 + \chi\left(\frac{\omega_1}{1+\gamma_1} + \frac{\omega_2}{1+\gamma_2} + \frac{3-b}{b}\right)\tilde{\lambda}^2$$
$$+ \chi^2\left[\frac{\omega_1\omega_2}{(1+\gamma_1)(1+\gamma_2)} + \frac{\omega_1}{1+\gamma_1}\left(\frac{3}{b_2} - 1\right) + \frac{\omega_2}{1+\gamma_2}\left(\frac{3}{b_1} - 1\right)\right]\tilde{\lambda} \quad (4.49)$$
$$- \frac{\chi^3\omega_1\omega_2}{(1+\gamma_1)(1+\gamma_2)} = 0.$$

We examine, now, the boiling-up of a nonviscous volatile solution. At a low viscosity, inertia effects of the liquid become essential. Conditions (3.41) and $\omega_1/(1+\gamma_1) \ll 1$, $\omega_2/(1+\gamma_2) \ll 1$ determine the region of superheatings in which the solution may be considered as nonviscous and at the same time as noninertial. Here, at

$$\frac{3-b}{b} \ll 0, \tag{4.50}$$

which corresponds to high stretching pressures, $-p'$, from Eq. (4.49) it follows that

$$\tilde{\lambda}_0 = \chi \tag{4.51}$$

holds. If

$$\frac{3-b}{b} \gg 0, \tag{4.52}$$

the cubic equation, Eq. (4.49), is reduced to a quadratic equation

$$\left(\frac{3-b}{b}\right)\tilde{\lambda}^2 + \chi\left[\frac{\omega_1}{1+\gamma_1}\left(\frac{3}{b_2}-1\right)+\frac{\omega_2}{1+\gamma_2}\left(\frac{3}{b_1}-1\right)\right]\tilde{\lambda} \\ -\frac{\chi^2\omega_1\omega_2}{(1+\gamma_1)(1+\gamma_2)} = 0. \tag{4.53}$$

Substitution of the positive root, $\tilde{\lambda}_0$, of Eq. (4.53) into Eq. (2.112) yields an expression for the nucleation rate in a superheated nonviscous liquid with volatile components at positive and limited negative pressures, p'.

The boiling-up of a gas-filled nonvolatile liquid ($\alpha_1 = 0$, $p''_{1*} = 0$) taking into account inertia effects is described by

$$\tilde{\lambda}^3 + \left(1+\frac{\omega_2\chi}{1+\gamma_2}\right)\tilde{\lambda}^2 + \chi\left(\frac{\omega_2}{1+\gamma_2}+\frac{3}{b_2}-1\right)\tilde{\lambda} - \frac{\chi^2\omega_2}{1+\gamma_2} = 0. \tag{4.54}$$

Neglecting inertia effects and assuming that the volatility of the gas component is caused only by a free-molecular exchange, the cubic equation, Eq. (4.54), is reduced to a quadratic expression

$$\tilde{\lambda}^2 + \chi\left(\omega_2 + \frac{3-b_2}{b_2}\right)\tilde{\lambda} - \chi^2\omega_2 = 0. \tag{4.55}$$

In appropriate notations, Eq. (4.55) coincides with (39) from the paper by Deryagin and Prokhorov [282].

Under the diffusion regime of absorption–emission interaction of a bubble with a solution, Eq. (4.54) takes on the form

$$\tilde{\lambda}^2 + \chi\left(\frac{\omega_2}{\gamma_2}-\frac{3-b_2}{b_2}\right)\tilde{\lambda} - \chi^2\frac{\omega_2}{\gamma_2} = 0. \tag{4.56}$$

4 Nucleation in Solutions of Liquefied Gases

Tab. 4.1 Temperature of attainable superheatings of binary solutions at atmospheric pressure in dependence on the molar fraction of the second component, c', with $J = 10^{11}$ m^{-3} s^{-1}.

c'	T_n (K)	c'	T_n (K)	c'	T_n (K)
		Ethane–propane [127]			
0.04	269.8	0.36	285.1	0.83	312.2
0.14	275.1	0.62	302.1	0.90	317.7
0.21	280.7	0.72	304.6		
		Ethane–n-butane [127]			
0.06	271.4	0.17	282.3	0.74	348.1
0.09	276.7	0.65	333.9	0.88	360.1
		Propane–isobutane [289]			
0.12	329.7	0.32	337.0	0.66	349.2
0.18	331.5	0.45	341.5	0.80	353.2
0.26	334.7	0.56	344.3	0.93	357.1
		Propane–n-butane [127]			
0.12	332.7	0.57	355.7	0.92	372.6
0.23	327.0	0.65	359.4		
0.38	347.7	0.78	369.5		

Substituting the positive root of Eq. (4.56) into Eq. (2.112) gives a result coinciding with the data of the paper by Kuni et al. [283, 284].

4.3
Attainable Superheating of Solutions of Hydrocarbons

Experimental data on the temperatures of attainable superheatings of low-temperature boiling liquid solutions have been obtained by the floating-droplet method [127, 289]. Measurements were made at atmospheric pressure and a nucleation rate $J = 10^{11}$ m^{-3} s^{-1} for binary and ternary solutions with ethane, propane, and butane. The error of determination of the temperature of attainable superheatings is estimated as $\pm(0.5$–$1.0)$ K. The results of measurements are presented in Tables 4.1, 4.2 and in Fig. 4.2.

With respect to their thermodynamic properties, the investigated liquids are close to ideal solutions. For binary solutions, the dependence of the temperature of attainable superheating on the molar concentration is linear within the limit of experimental error (see Fig. 4.2). The values of the temperature of attainable superheatings of the ternary ethane–propane–n-butane system, which are calculated by an additional rule on the basis of the T_n^0-values of pure liquids and their molar fractions in the solution, are close to the experimental

Tab. 4.2 Temperature of attainable superheatings of ethane–propane–n-butane solutions at atmospheric pressure in dependence on the molar fraction of the second (c'_2) and the third (c'_3) component for $J = 10^{11}$ m^{-3} s^{-1}.

c'_2	c'_3	T_n (K)	
		Experiment	Eq. (4.57)
0.155	0.039	278.4–281.2	282.2
0.073	0.121	280.8–283.4	286.4
0.811	0.097	324.0–329.8	322.9
0.771	0.129	326.0–331.0	323.5
0.433	0.332	324.8–344.2	329.7
0.652	0.235	329.4–334.8	331.7
0.299	0.541	337.8–354.4	344.6
0.403	0.521	342.4–353.0	348.4
0.160	0.705	344.6–355.8	354.5

data (see Table 4.2)

$$T_n^{ad} = \sum_{i=1}^{3} c'_i T_{ni}^0. \qquad (4.57)$$

A comparison between theoretical and experimental data in Refs. [127, 289] has not been made because of the absence of data on the thermophysical properties of the investigated solutions which are a prerequisite for the calculations by the homogeneous nucleation theory.

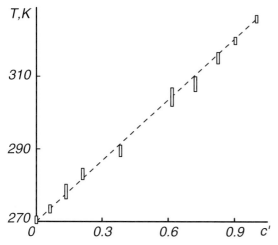

Fig. 4.2 Concentration dependence of the temperature of attainable superheatings of ethane–propane solutions at atmospheric pressure [127].

4.4
Methods of Experimentation on Solutions of Cryogenic Liquids

The equilibrium state of a liquid mixture is characterized by three parameters: temperature, pressure, and composition. A transfer of a mixture into the metastable region may be realized, for example, by changing any one of these parameters retaining the other two constants.

Figure 4.3 shows the (p,c)-diagram of liquid (upper curve)–vapor (lower curve) phase equilibrium for a binary mixture at a constant temperature, $T = T_s$. It is assumed that the components in both phases are mixed in arbitrary proportions. If at constant T_s and concentration in the liquid phase, c'_s, the pressure in the mixture is increased to the value p', the solution will be transferred into a supercompressed state, and the formation of the vapor phase will be impossible. As a result of a rapid pressure decrease, a liquid mixture may be transferred to the point B where it has the same composition c'_s, temperature T_s, but a pressure $p' < p_s$. An equilibrium isotherm, corresponding to a certain temperature T, goes through this point. The liquid state at the point B is superheated as the actual temperature, T_s, is higher than the equilibrium one for this composition, c'_s, and pressure, p'. The state at the point B may also be regarded as supersaturated in the sense that the actual concentration, c'_s, is higher than the equilibrium concentration, c', for the temperature, T_s, and pressure, p' (point C).

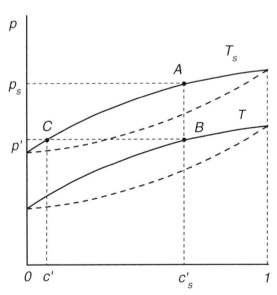

Fig. 4.3 (p,c)-projection of the phase diagram of a binary mixture.

The nucleation rate in superheated solutions of cryogenic liquids was determined by the lifetime measuring method [12, 291–294]. The experimental setup and the experimental techniques are similar to those used in experiments on one-component liquids. The solution under investigation is filled into a glass tube (3) and metallic bellows (7) (Fig. 4.4). In the experiment, a part of the liquid solution with volume $V \simeq 70\,\text{mm}^3$ thermostated ($\pm\,0.002$ K) in a copper block (1) was superheated. The temperature in the block was measured with a platinum resistance thermometer (2). The pressure was created with compressed helium and transferred to the solution through bellows (7). The bellows chamber and the low-temperature valve (6) were thermostated ($\pm\,0.05$ K) in the block (8) at a temperature close to the temperature of the normal boiling of the solution. The measuring device was placed into the vacuum jacket (4) fixed in a Dewar vessel (5). Cooling was realized with liquid nitrogen.

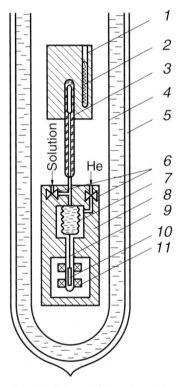

Fig. 4.4 Schema of experimental apparatus.

An experiment begins with the preparation of the gas mixture of the required concentration. The mixture was condensed into the measuring device. Differences in the temperatures of blocks (1) and (8) resulted in the evolution of a concentration gradient throughout the height of the measuring device. To define the solution concentration in a superheated volume before the beginning and after completion of an experiment, the solution saturation pressure was measured. For this purpose, at the temperature T of the experiment (the temperature in the block (1)) the pressure in the solution was decreased below the saturation pressure. The liquid solution in the upper part of the tube (3) then turned into a gaseous state. Subsequently, the pressure was gradually increased, and with the help of a system of detection of changes in the volume of the solution under investigation the pressure was recorded at the moment of the transition of the solution from a gaseous into a liquid state (the saturation pressure, $p_s(T)$). The system of detection of changes in the volume consists of the rod (9) attached to bellows (7) and ending with a core (10) made of a ferromagnetic material. The displacement of the core results in a change of the reactance of the coil (11) connected in the bridge circuit. The error in determining the saturation pressure was ± 0.01 MPa. In calculating the concentration, data on the phase equilibrium were used. The error in determining the concentration is $\pm 0.5\%$.

After preparing a solution of the required concentration c', the initial pressure $p_1 > p_s(T, c')$ was created in the measuring system. The penetration into the metastable region was realized by an abrupt pressure decrease below the saturation pressure. To reduce both the effect of cooling a liquid solution and the hydraulic oscillations in the measuring system, the liquid was brought to a metastable state in two stages. At first the pressure was decreased to the value p_2, which is 0.3–0.5 MPa less than the saturation pressure, $p_s(T, c')$. The probability of boiling-up of a solution at pressure p_2 is sufficiently small. The solution was kept at this pressure, p_2, for approximately 20 s, after which its final release to the value p' was realized. The error of the maintenance of the pressure p' was ± 0.005 MPa. At the moment of the final pressure release, a chronometer was started up. It recorded the time, τ, of the stay of the solution in the given metastable state before its boiling-up (the lifetime of a superheated solution). The moment of boiling-up was recorded by a water hammer in the measuring system. After the boiling-up of a solution, the initial pressure, p_1, was created in the bellows chamber and the measurement process was repeated. Up to $N = 100$ values of τ were measured at every given p' and c', and the mean lifetime, $\bar{\tau} = \sum_i \tau_i / N$, was determined. In processing the results of the experiments, corrections for the liquid cooling in the process of pressure release were made and the delay time, τ_0 ($\tau_0 \simeq 0.5$ s), caused by the inertia effects of the pressure-release system, was subtracted from the measured value of $\bar{\tau}$.

In experiments with solutions as well as with pure liquids [9, 37], in the whole range of (p', T, c')-values investigated, a near identity of the dispersion of the random quantity, τ, and the mean lifetime, $\bar{\tau}$, can be observed. The Poissonian character of the boiling-up of a superheated solution is also reconfirmed by the histograms of experiments (Fig. 4.5), which are in good agreement with the distribution described by Eq. (3.88).

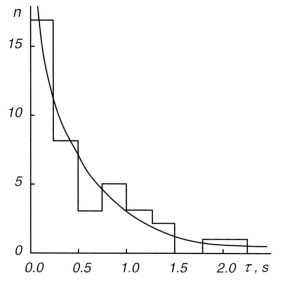

Fig. 4.5 Histogram of exponential data. The parameter values are $c' = 0.428$; $p = 1.6$ MPa; $T = 154.47$ K; $N = 40$; $\bar{\tau} = 0.52$ s. The smooth curve has been calculated from Eq. (3.88).

4.5
Solutions with Complete Solubility of the Components

Solutions of simple classical liquids are convenient objects for the verification of the homogeneous nucleation theory. The spherical symmetry of the force field of the molecules makes it possible to express the free parameters of the functional (4.13) in terms of thermodynamic properties of bulk phases and by doing so to determine the work of nucleus formation without utilizing the macroscopic approximation.

The kinetics of nucleation in superheated solutions Ar–Kr [290,291] and N_2–O_2 [292] has been investigated in experiments measuring the mean lifetime of the respective systems. The dependence of the mean lifetime of liquid Ar–Kr and N_2–O_2 solutions on temperature and concentration has been traced in experiments at two pressures ($p' = 1.0$ and 1.6 MPa for Ar–Kr; $p = 0.5$ and 1.0 MPa for N_2–O_2). The interval of the investigated values of the mean lifetime changes from 0.1 to 1000 s, which corresponds to the variation of the nucleation rate from 10^4 to 10^8 m^{-3} s^{-1}. Figure 4.6 shows in a semilogarithmic scale the temperature dependences of the mean lifetime for several values of the concentration [291]. The vertical size of the dots in the figure corresponds to the statistical error of determination of $\bar{\tau}$. The character of the temperature dependence of $\bar{\tau}$ in solutions is similar to that in pure liquids [12, 37]. Along all lines of constant concentration, after reaching a certain temperature (the boundary of spontaneous boiling) sections of an abrupt decrease of $\bar{\tau}$ can be observed. We associate these sections of curves with homogeneous nucleation. Here, with a temperature increase of 0.1 K the value of $\lg \bar{\tau}$ decreases by 0.6–1.2.

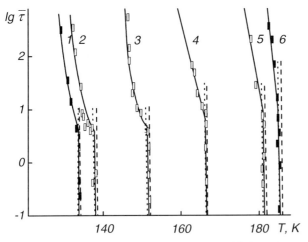

Fig. 4.6 Temperature dependences of the mean lifetime of argon–krypton solutions at $p = 1.0$ MPa for different concentrations. 1: $c' = 0$; 2: 0.109; 3: 0.428; 4: 0.708; 5: 0.938; 6: 1. *Dashed lines* represent calculations using Eqs. (2.112), (3.52), (4.44), (3.8), $z_0 = 1$ and $\sigma = \sigma_\infty$; *dotted lines* employ Eqs. (2.112), (3.52), (4.44), (3.11), and $z_0 = 1$.

The bend of experimental curves at $\bar{\tau} \simeq 8$–10 s is caused by heterogeneous nucleation and the effect of ionizing radiation on nucleation. The source of ionizing radiation in experiments is cosmic rays and the natural radiation background. The increase of the intensity of ionizing radiation results in a shift of the gently sloping sections of curves into a region of smaller values of $\bar{\tau}$ without violating the character of the dependence $\bar{\tau}(T)$. The effect of initiation of nucleation in the series of condensed inert gases correlates with the radiation stability and scintillation properties of the substance, which increases with growing atomic number of the respective element [163]. The strongest resistance to ionizing radiation, which is determined by well-expressed scintillation properties, is found for liquid xenon. As a result, a part of the energy of ionizing radiation is not transformed into thermal energy but fluoresces. The bend of experimental curves for xenon manifests itself at the initiation rate $J_m = (\bar{\tau} V)^{-1} = 2.2 \times 10^4$ m^{-3} s^{-1} [145]. The bend of experimental curves in the case of krypton manifests itself at the initiation rate $J_m \simeq 1.5 \times 10^6$ m^{-3} s^{-1}, and in the case of argon at $J_m \simeq (2$–$3) \times 10^6$ m^{-3} s^{-1}. The increase of the concentration of krypton in argon results in a monotonic decrease of J_m (Fig. 4.7). As distinct from krypton, where at all pressures a linear dependence of $\lg J$ on T can be observed, in argon at low pressures

($p' < 0.3$ MPa, see [37]) a section of approximately constant radiation sensitivity is revealed, which is adjacent to the boundary of liquid spontaneous boiling-up. The presence of such a section may be connected with the fact that the work of formation of a critical nucleus is here much smaller than the average energy of a thermal spike, and the probability of boiling-up on every thermal spike is close to unity [10].

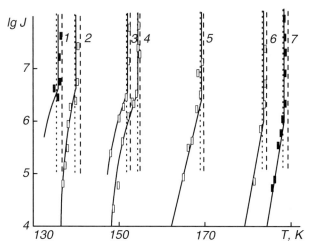

Fig. 4.7 Nucleation rate in superheated argon–krypton solutions at $p = 1.6$ MPa for different concentrations. 1: $c' = 0$; 2: 0.109; 3: 0.382; 4: 0.428; 5: 0.708; 6: 0.938; 7: 1. *Dashed and dotted lines* are determined by the same dependences as explained in the caption to Fig. 4.6.

The details of the process of transformation of the ionizing radiation energy into thermal energy, and subsequently into the work of nucleation, are not finally established to allow for a calculation of the probability of nucleation initiation employing the molecular characteristics of a substance. Besides, at present one cannot be absolutely sure that the action of ionizing particles is the only reason for the observed bends of experimental curves at large waiting times for boiling-up. The similarity of the temperature and pressure dependences of $\bar{\tau}$ obtained at natural conditions and in the presence of a γ-source is only indicative of the close link between these phenomena. As shown in Ref. [264], the value of J_m changes in passing from glass to metallic measuring cells. The character of the dependence $\bar{\tau}(T)$ at $J < J_m$ is affected by easily activated boiling centers [12, 265] and ultrasound fields [258].

Figures 4.8a and b presents the results of experiments on the determination of the mean lifetime and the nucleation rate in the nitrogen–oxygen system [292]. The components of the solution differ with respect to the similarity criterion A (see Table 3.4) by 6%, whereas in the argon–krypton system this difference is only 1%. Lines of constant pressure and concentration values have characteristic bends in the region of large $\bar{\tau}$ and, correspondingly, small values of J. An increase in the concentration of oxygen results in an increase

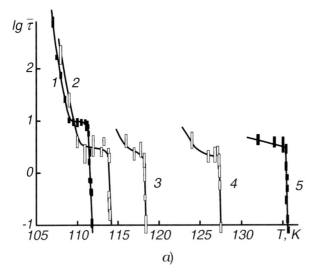

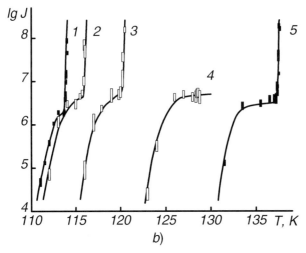

Fig. 4.8 Temperature dependences of the mean lifetime (a, $p = 0.5$ MPa) and the nucleation rate (b, $p = 1.0$ MPa) in the solution nitrogen–oxygen at different concentrations. 1: $c' = 0$, 2: 0.105, 3: 0.300, 4: 0.670, 5: 1.0.

of the temperature of superheating of the solution. However, at small values of c' the opposite effect is also observed. Figure 4.8a shows that, when one dissolves in nitrogen up to 0.105 molar fraction of oxygen ($p = 0.5$ MPa), the temperature of attainable superheating in the region of homogeneous nucleation increases by 2.3 K, whereas in the region of initiated nucleation this value is only 0.5 K. In passing from initiated to homogeneous nucleation, the temperature of superheating for pure nitrogen proves to be higher than that for the solution N_2–O_2 by 1.5 K, i.e., small additions of oxygen into nitrogen can also decrease the superheating. The temperature of attainable superheating, T_n, for the solution N_2–O_2 at a fixed value of the nucleation rate $J = 10^7$ m^{-3} s^{-1} is given in Table 4.3 and Fig. 4.9.

Tab. 4.3 Temperature of attainable superheatings for nitrogen–oxygen solutions of different molar fractions of the second component and two pressures, $J = 10^7$ m^{-3} s^{-1}.

c'	T_s (K)	p_s (MPa)	T_n (K)
		$p = 0.5$ MPa	
0	93.99	1.606	111.58
0.105	94.99	1.695	113.90
0.300	97.01	1.854	118.27
0.670	101.91	2.153	127.36
1.000	108.81	2.278	135.51
		$p = 1.0$ MPa	
0	103.73	1.823	113.86
0.105	104.91	1.889	115.92
0.300	107.25	2.046	120.22
0.670	112.69	2.346	129.21
1.000	119.62	2.494	137.49

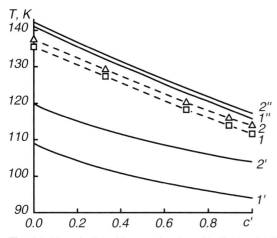

Fig. 4.9 Line of attainable superheatings (1, 2), binodal (1', 2') and diffusion spinodal (1'', 2'') for a nitrogen–oxygen solution at a pressure of $p = 0.5$ MPa (1, 1', 1'') and 1.0 MPa (2, 2', 2'').

Experimental data on the temperature of attainable superheating for the solution N_2–O_2 differ much less from the values calculated by the additivity rule, Eq. (4.57), than for the system Ar–Kr. Thus, at a pressure of $p = 1.0$ MPa the maximum deviation from Eq. (4.57) for the solution N_2–O_2 is $\simeq 0.75$ K ($c' = 0.3$), whereas for the mixture Ar–Kr it reaches 4.3 K ($c' = 0.5$) (see Fig. 4.10).

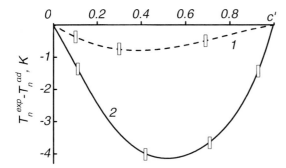

Fig. 4.10 Deviations from the additivity rule, Eq. (4.57), for the temperature of attainable superheatings for nitrogen–oxygen (1) and argon–krypton (2) solutions.

The position of the diffusion spinodal for the nitrogen–oxygen solution is shown in Fig. 4.9. As in the case of the attainable-superheating temperature, the concentration dependence of T_{sp} is close to linear. In the whole investigated range of pressures and concentrations, the ratio of the attainable superheatings of the solution, $\Delta T_n = T_n - T_s$, to $\Delta T_{sp} = T_{sp} - T_s$ is 0.75–0.8.

4.6
Solutions with Partial Solubility of the Components

A supersaturated solution of a gas in a liquid is an example of a two-component metastable system in which nonequilibrium is caused by the excess of the concentration of a dissolved component over its value in the saturated state. The removal of supersaturation here takes place as a result of formation of the gas phase in the form of numerous bubbles ("the effect of champagne"). This phenomenon manifests itself also as Caisson's disease, it is used for degassing liquids, foaming of polymeric materials and in other technological processes.

If at a given temperature and pressure of a supersaturated solution a pure solvent (liquid) and a dissolved substance (gas) are in the stable state, the solvent in the solution is characterized by a low volatility and the dissolved gas by a low solubility. New-phase nuclei in such solutions practically fully consist of molecules of the dissolved substance. The theory of boiling-up of gas-filled nonvolatile liquids was suggested by Deryagin and Prokhorov [282] and Kuni et al. [283–285]. The first experimental investigations of nucleation in nonvolatile gas-filled liquids were made by Hemmingsen [296], Gerth and Hemmingsen [297], Finkelstein and Tamir [298], and Bowers et al. [299]. Supersaturation was created here by an abrupt release of pressure on the liquid (water), which was saturated with a gas at pressures equal to several hundreds of atmospheres. The concentration of a dissolved gas, c', was determined in this case by the saturation pressure, p_s, and the depth of penetration into the metastable region, by the pressure difference $p_n(c', T) = p_s - p_n$, where $p_n(c', T)$ is the pressure of intensive gas emission.

An approach to the investigation of nucleation in gas-supersaturated water solutions, different from that used in Ref. [296], was suggested by Rubin and Noyes [300]. In a solution, which was in a vessel of a fixed volume at controlled values of temperature and pressure, supersaturation was initiated by a chemical reaction. If the rate of gas production during a chemical reaction in a solvent considerably exceeds the rate of its removal through a free interface, considerable supersaturations can be achieved in the volume of a liquid solvent. A decay of the metastable state was accompanied by an abrupt (explosive) gas liberation. The concentration of the dissolved gas in the supersaturated solution was determined by data for the volumes of the solution, gas and liquid phases in a measuring cell and the pressure increase in the system during the initiation of gas liberation by stirring the solution or by sonication. In this case, the supersaturation is characterized by the difference $\Delta c'_n(p, T) = c'_n - c_s$. According to the data of Ref. [300], the concentration, c'_n, of nitrogen dissolved in water before the formation of bubbles was approximately 20–40 times higher than its equilibrium concentration, c_s, at atmospheric pressure.

A comparison of experimental data, obtained by two different methods of creating supersaturation in a gas-filled liquid with homogeneous nucleation theory has shown that the degrees of metastability obtained by experiment are considerably lower than their theoretical values. Thus, in the system oxygen–water the limiting values of concentration, c'_n, calculated from homogeneous nucleation theory at atmospheric pressure and a nucleation rate $J = 10^7$ m^{-3}s^{-1} have proven to be 15 times larger than those achieved by experiment [301]. In this system, at the limiting values of supersaturation registered in experiment, the critical bubble is characterized by a radius, $R_* = 15$ nm, and a pressure, $p''_* = 9.5$ MPa. The Gibbs number, $G_* = W_*/k_B T$, i.e., the relation between the work of formation of a critical nucleus and the energy of thermal motion of molecules, $k_B T$, in this case is $\simeq 1.56 \cdot 10^4$ [302]. Overcoming such a high potential barrier by means of homogeneous nucleation of the gas phase in the characteristic times of experiment is an unlikely event. In the case of spontaneous boiling-up of superheated pure liquids, the value of G_* is $\simeq 72$ [12].

To eliminate contradictions between theory and experiment, a number of authors have suggested models of nucleation in gas-supersaturated solutions different from the classical thermodynamic model [302, 303]. It is postulated, for example, that the formation of gas bubbles in the solution proceeds in two stages. At the first stage, according to Kwak [303], the molecules of a dissolved gas form a cluster, which has no distinct interface and, consequently, no surface energy. In the paper by Bowers et al. [302], the spatial region of increased gas concentration in the solution is called a "blob." The small difference of gas concentrations in a "blob" and the surrounding liquid ensures, according to the opinion of the authors of Ref. [302], a small surface contribution to its excess free energy. The latter proves to be much lower than in a bubble of the same size. Both the "cluster" of Kwak [303] and the "blob" of Bowers et al. [302], after achieving a certain size exceeding the size of a critical nucleus, are transformed into a supercritical gas bubble. The models described in Refs. [302, 303] make it possible, for limiting supersaturations observed by experiment [296, 300], to decrease the height of the nucleation barrier to $\sim 10^2\ k_B T$. Some other approaches to the explanation of discrepancies between the classical homogeneous nucleation theory and experiment are discussed in Refs. [304, 341].

A peculiar situation for a gas-filled liquid is observed at temperatures close to the critical point of the solvent. Here, a two-component system may prove to be "doubly" metastable. At a given temperature, T, and concentration of the dissolved component, c', a solution at a pressure, p', lower than the saturation pressure, p_s, is supersaturated and characterized by the degree of metastability, $\Delta p(c', T) = p_s - p'$, a pure solvent is superheated by the value of $\Delta T(p') = T - T_s(c' = 0)$, where $T_s(c' = 0)$ is the saturation temperature of a pure solvent at a given pressure. In the vicinity of the critical point, the solvent volatility cannot be neglected any more, and viscous and inertia forces and heat exchange at the interface start to play an important role in the bubble-growth dynamics. The theory of boiling-up of such solutions has been treated in detail by Baidakov [286–288].

Solutions of gases in liquids are generally characterized by a higher degree of nonideality than systems with complete solubility of the components. Increasing nonideality is accompanied by an expansion of the region of two-phase equilibrium, an increase in the surface adsorption and deviations from the law of additivity. Studies of the kinetics of boiling-up of gas-saturated liquids are of interest from the viewpoint of both homogeneous and initiated nucleation. Gas saturation may stabilize gaseous inclusions on the vessel walls, in the pores of solid foreign particles, and also initiate their growth at the expense of selective adsorption.

Depending on the chemical nature of substances and thermodynamic state parameters, the water solubility of a gas varies in a sufficiently wide range.

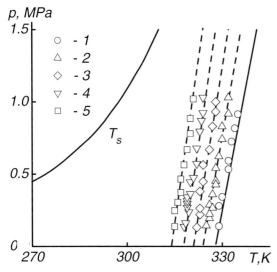

Fig. 4.11 Attainable superheatings of the solution C_3H_8–CO_2 [126]. 1: $c' = 0$, 2: 0.09, 3: 0.15, 4: 0.22, 5: 0.3. *Dashed* and *full lines* show results of calculations by homogeneous nucleation theory ($J = 10^6$ m^{-3}s^{-1}). T_s is the binodal curve.

High solubility is characteristic of carbon dioxide in liquid hydrocarbons of the methane series. The attainable superheatings of solutions C_3H_8–CO_2 and C_4H_{10}–CO_2, and also nitrogen in organic liquids are investigated by Mori et al. [126], Forest and Ward [306] by the method of emerging drops. The error of determination of the temperature of limiting superheatings for a solution is estimated as ± 1.5 K, of the concentration as 1%. The dissolution of carbon dioxide results in a shift of the boundary of spontaneous boiling-up of a liquid toward lower temperatures (Fig. 4.11). The temperature of attainable superheating for liquid isobutane ($p \simeq 0.1$ MPa) decreases by 25 K when one dissolves in it up to 0.33 molar fraction of CO_2. The authors of Ref. [126] compare the data obtained with the results of calculating T_n employing homogeneous nucleation theory of Döring–Volmer (Eqs. (2.112), (3.15), and $z_0 = 1$), in which, according to recommendations of Ward et al. [307], the surface tension of a pure liquid has been replaced by the surface tension of a solution, and the pressure of a vapor–gas mixture in a bubble is determined as the sum of the partial pressures of vapor and gas. At nucleation rates $J = (10^{24}$–$10^{26})$ m^{-3}s^{-1}, gas-filled solutions of high-temperature liquids were investigated by the method of pulse heating of a wire probe [308]. Gas filling of a liquid resulted in a shift of the spontaneous boiling-up boundary toward lower temperatures. The authors of Ref. [309] claim satisfactory agreement between homogeneous nucleation theory and experiment, at least, in the value of the limiting superheatings of the solutions.

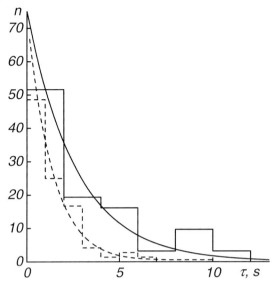

Fig. 4.12 Distribution of the number of boiling-up events, n, observed within the interval $\tau, \tau + \Delta\tau$: $c' = 0.14$ mol%; $p = 1.171$ MPa; $T = 166.66$ K; $N = 31$. *Solid lines*: $\bar{\tau} = 2.66$ s at a "holding" time of 40 min, *dashed lines*: $\bar{\tau} = 1.34$ s at a "holding" time of 3 min.

Gas-saturated cryogenic liquids belong to the class of weak solutions. The solubility of helium in oxygen and nitrogen at pressures lower than 5 MPa does not exceed 2–5 mol% [310–312]. The considerable difference of the parameters of intermolecular interaction of helium atoms and the solvent molecules results in an essential nonideality of the solvent, with helium behaving like a surface-active substance decreasing the excess free energy of the liquid–gas interface [313–317]. We used the method of determination of the mean lifetime to investigate the nucleation kinetics in solutions of O_2–He and N_2–He [293, 294, 318]. Experiments were conducted at two values of pressure and several values of concentration for each of the solutions.

The technique of experiments on the mean lifetime of gas-saturated systems is similar to that used in studying nucleation in solutions with complete solubility of the components. Under fixed external conditions ($T, p, c' = $ const), the appearance of the first viable nucleus in a metastable liquid is an event of the Poisson type. This statement has been confirmed by experiments on one-component liquids [12]. In these experiments, the time of "hold-up" of a liquid under pressure was $\approx$ 3 min. This time interval was sufficient for complete decompression of the vapor phase and the relaxation of thermal inhomogeneities in the measuring system. In experiments on gas-filled liquids, it has been discovered that at times of "hold-up" of about three minutes one can often observe boiling-up with expectation times that are "anomalously"

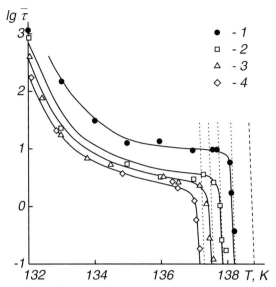

Fig. 4.13 Temperature dependence of the mean lifetime of a metastable mixture of oxygen and helium at a pressure $p = 1.171$ MPa and several concentrations. 1: $c' = 0$ mol%; 2: 0.08; 3: 0.14; 4: 0.20. The *dashed lines* show results of calculations for a one-component system by homogeneous nucleation theory with $\sigma = \sigma_\infty$; *dotted lines* show results of computations by homogeneous nucleation theory with $\sigma = \sigma(R_*)$.

small for the given values of T, p, and c. "Anomalously" small values of τ may appear and disappear in the course of experiments, or be retained during the whole period of measurements. Such a behavior of a gas-filled system is similar to that of a one-component system if the latter contains easily activated nucleation centers [12, 264]. The law of distribution of "anomalously" low expectation times of boiling-up events is also close to the Poisson one (Fig. 4.12, smooth curves). With an increase in the "hold-up" time between measurements up to 30–40 min (i.e., by an order of magnitude) premature liquid boiling-ups did not manifest themselves. In the course of treatment of experimental data, "anomalously" low values of τ were excluded from consideration.

Figures 4.13 and 4.14 show on a semilogarithmic scale the results of determination of the mean lifetime and the nucleation rate in oxygen–helium solutions at two values of pressure. As in the case of one-component liquids, on the curves $\bar\tau(T)$ and $J(T)$ one can distinguish two sections with different characters of the temperature dependences $\bar\tau$ and J. At low superheatings, experimental isobars have characteristic gently sloping sections, which with increasing temperatures give way to sections with an abrupt change of the expectation time of boiling-up of the solution. The first section is connected with initiated and the second with spontaneous nucleation. Dissolution of 0.1 mol% of helium in liquid oxygen decreases the limiting superheatings, $\Delta T = T_n - T_s$, by approximately 10%. In this case, in the region of initiated nucleation the mean lifetime of a solution decreases by three or four times. It is important here that gas saturation manifests itself most noticeably at very low helium concentrations.

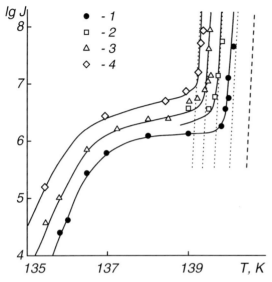

Fig. 4.14 Temperature dependence of the nucleation rate of oxygen–helium mixtures at a pressure $p = 1.667$ MPa and different concentrations. 1: $c' = 0$ mol%; 2: 0.08; 3: 0.14; 4: 0.20. For the way of determination of the *dashed* and the *dotted* curves, see caption to Fig. 4.13.

In sections of an abrupt decrease of τ (increase of J), the value of the derivative of J with respect to temperature $(d \ln J / dT)$ does not depend on concentration within the experimental error. At a pressure of 1.171 MPa, it is about 18, at $p = 1.667$ MPa equal to 21. This value corresponds to an increase of the nucleation rate by approximately 8 or 9 orders of magnitude with an increase in temperature of 1 K. The character of the temperature dependence of $\bar{\tau}$ and J on the lines of constant value of pressure, p, and concentration, c', in the nitrogen–helium system is similar to that in the oxygen–helium mixture (Figs. 4.15 and 4.16).

In the region of homogeneous nucleation, the temperatures of attainable superheatings are, in a first approximation, a linear function of the concentration of the volatile component (Fig. 4.17a). With equal amounts of helium dissolved in nitrogen and oxygen and similar values of the reduced pressure, the value of $T_n^0 - T_n(c')$ in the system N_2–He is smaller than in the system

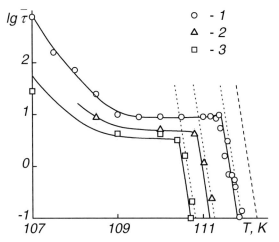

Fig. 4.15 Temperature dependence of the mean lifetime in the system nitrogen–helium at a pressure $p = 0.5$ MPa and different concentrations. 1: $c' = 0$ mol%, 2: 0.051, 3: 0.16, 4: 0.35. For the meaning of the *dashed* and the *dotted* lines, see Fig. 4.13.

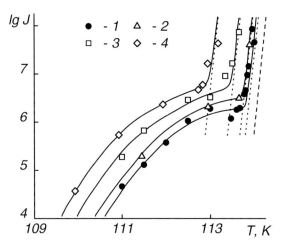

Fig. 4.16 Temperature dependence of the nucleation rate in the system nitrogen–helium at a pressure $p = 1.0$ MPa and different concentrations. 1: $c' = 0$ mol%, 2: 0.051, 3: 0.16, 4: 0.35. For the meaning of the *dashed* and the *dotted* lines, see Fig. 4.13.

O_2–He. This result correlates with the effect of helium on the surface properties of cryogenic liquids [313, 315]. The higher the adsorption of helium at a liquid–gas interface, the smaller the ratio of the critical temperatures of the gas and the liquid dissolved.

In the region of initiated nucleation, the effect of gas saturation on the temperature of attainable superheatings shows up most vividly at small values

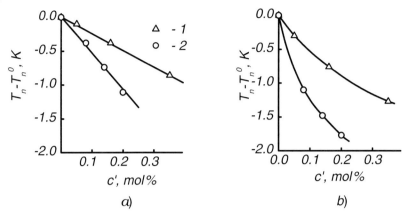

Fig. 4.17 Difference in experimental values of temperatures of attainable superheatings for solutions N_2–He (1, $p = 1.0$ MPa), O_2–He (2, $p = 1.171$ MPa) and a pure solvent at $J = 10^7$ m^{-3}s^{-1} (a) and $J = 10^6$ m^{-3}s^{-1} (b).

of c'. As it follows from Fig. 4.17b, the dependence of T_n on helium concentration in nitrogen and oxygen at $J = 10^6$ m^{-3}s^{-1} is essentially nonlinear. It is rather difficult to explain such a behavior of T_n, when seeing the reason for bends of experimental isobars only in the effect of radiation background.

4.7
Equation of State and Boundaries of Thermodynamic Stability of Solutions

For describing the properties of new-phase nuclei and the determination of the boundary of stability for homogeneous solutions with respect to constantly changing state variables (the spinodal curve), it is necessary to have an equation of state which describes both stable and metastable states. Such an equation of state for Ar–Kr, N_2–He and O_2–He systems has been obtained in the framework of a one-liquid solution model on the basis of experimental data on thermodynamic properties of pure components and phase equilibria in solutions, and for the system N_2–O_2 and three-component solutions in the framework of an extended law of corresponding states [325]. In order to obtain these basic equations, equations of state for argon [319], nitrogen [320], and oxygen [321] were used having a common form for the liquid and the vapor phase. In the region of liquid–gas phase separation, these equations have isotherms of the van der Waals type and satisfy the Maxwell rule. In the general form, the equation of state of the solution is given by

$$\frac{p}{\rho RT} = 1 + \eta \sum_{i,j} b_{ij} \widetilde{\rho}^i \widetilde{T}^{-j}. \tag{4.58}$$

4.7 Equation of State and Boundaries of Thermodynamic Stability of Solutions

Here, $\tilde{\rho} = \rho/\rho_c$, $\tilde{T} = T/T_c$ and (ρ_c, T_c) are the density and the temperature at the critical point, respectively. Both the parameters and the individual parameter η are functions of concentration. The coefficients b_{ij} are determined by experimental data on the thermodynamic parameters of the less volatile solution component (argon, oxygen, nitrogen) and are given in Refs. [319–321].

Three individual parameters of Eq. (4.58), T_c, $\rho_c = 1/v_c$, and η, have been approximated by the expressions

$$T_c(c) = T_{c,1}c^2 + 2\alpha_T T_{c,12} c(1-c) + T_{c,2}(1-c)^2, \tag{4.59}$$

$$v_c(c) = v_{c,1}c^2 + 2\alpha_v v_{c,12} c(1-c) + v_{c,2}(1-c)^2, \tag{4.60}$$

$$\eta(c) = \eta_1 c^2 + 2\alpha_\eta \eta_{12} c(1-c) + (1-c)^2, \tag{4.61}$$

where the cross terms $T_{c,12}$, $v_{c,12}$, and η_{12} satisfy the combination rules

$$T_{c,12} = \delta_T (T_{c,1} T_{c,2})^{1/2}, \tag{4.62}$$

$$v_{c,12} = \delta_v \frac{\left(v_{c,1}^{1/3} + v_{c,2}^{1/3}\right)^3}{8}, \tag{4.63}$$

$$\eta_{12} = \frac{(1+\eta_1)}{2}. \tag{4.64}$$

where $T_{c,i}$ and $\rho_{c,i}$ are the temperature and density, respectively, at the critical point of any of the pure components.

In developing an equation of state for an argon–krypton solution, we started from the fact that for these substances the law of corresponding states is fulfilled with an error close to the experimental one [53]. In reduced coordinates, the (p, ρ, T)-properties of argon and krypton may be described by one equation of state of the form of Eq. (4.58) with coefficients $b_{i,j}$ determined, for example, by the properties of argon, $\eta = 1$. The coefficients $\delta_T = 1.0012$ and $\delta_v = 1.017$ have been found by minimizing the mean-square deviations of experimental data concerning the pressure of saturated vapors [322] and the density of the liquid phase [323] for Ar–Kr solutions calculated from Eqs. (4.58), (4.59), (4.60), (4.62), and (4.63).

Since the region, where oxygen exists in a liquid phase, is located considerably above the helium critical temperature, helium may be regarded here as a gas close to a perfect one. This circumstance makes it possible to use the equation of state for oxygen for a description of helium having introduced into the latter one the correction factor, η. In pure oxygen, $\eta = 1$. At $\eta = 0.431$, the equation of state, Eq. (4.58), with coefficients $b_{i,j}$ of oxygen describes in the best way the (p, ρ, T)-properties of helium in the range of reduced densities $0 < \tilde{\rho} < 1$ and temperatures $0.6 < \tilde{T} < 1$. The free parameters of the equation, α_T, α_v and α_η, may be determined by the pressure of saturated vapor and the composition of the liquid and the vapor phase of the solution. However, the data available so far [324] on the set of parameters (p_s, c', c'') have proven to be insufficient for a reliable representation of the phase diagram in the range of temperatures adjoining the solvent critical point. In the absence of information about the (p, ρ, T, c)-properties of an oxygen–helium system, we used data about the surface tension [314] as an additional information. By including into consideration information about the properties of a liquid–vapor interface it was possible not only to increase the reliability of description of the (p, ρ, T, c)-properties of the solution by the equation of state, Eq. (4.58), but also to determine the influence parameter of pure helium, k_{11}. Calculations were made by employing Eqs. (4.58)–(4.64).

The solvent influence parameter was determined from Eq. (3.119) by data on the surface tension of pure oxygen [156]. The results of such calculations may be presented in the following form:

$$k_{22}\rho_{c,2}^{8/3}/p_{c,2} = 0.5552 + 12.6638\varepsilon - 187.164\varepsilon^2 + 1468.46\varepsilon^3 \\ - 6583.0\varepsilon^4 + 16993\varepsilon^5 - 23465\varepsilon^6 + 13500\varepsilon^7, \quad (4.65)$$

where $\varepsilon = 1 - (T/T_{c,2})$. The temperature dependences of the free parameters in Eqs. (4.59)–(4.61) and also the coefficient k_{11} have been approximated by the following expressions:

$$\alpha_T = 3,$$
$$\alpha_v = 2.88 - 16.8\varepsilon + 54.0\varepsilon^2 - 50.0\varepsilon^3, \quad (4.66)$$
$$\alpha_\eta = -0.238 + 14.95\varepsilon - 61.1\varepsilon^2 + 103.0\varepsilon^3 - 61.0\varepsilon^4,$$

and

$$\frac{k_{11}\rho_{c,2}^{8/3}}{p_{c,2}} = 7.93 - 176.5\varepsilon + 1530\varepsilon^2 - 6248\varepsilon^3 + 12100\varepsilon^4 - 8900\varepsilon^5. \quad (4.67)$$

The coefficient k_{12} was found according to Eq. (4.20).

The use of an extended law of corresponding states for the description of metastable states in solutions [325] also presupposes the availability of unified equations of state for pure fluids describing stable, metastable, and unstable phases. In this case, the equation of state for a solution is written as

$$\frac{F(T, \rho, \vec{c})}{RT} = \tilde{f}(T, \rho, \vec{c}) \\ = \sum_i c_i \left\{ f_i^0(T, \rho) + f_i^r[T_r(\vec{c})/T, \rho/\rho_r(\vec{c})] + \ln c_i \right\} \quad (4.68) \\ + f^E[T_r(\vec{c})/T, \rho/\rho_r(\vec{c}), \vec{c}].$$

Here $\tilde{f}$ is the reduced free energy, $\vec{c}$ is a set of concentrations of the different components, f_i^r is the configurational part of the free energy of the ith component, f_i^0 is the ideal-gas contribution to the free energy, f^E is the correction for nonadditivity, and $T_r(\vec{c})$ and $\rho_r(\vec{c})$ are the reduction parameters. The rules of mixing for $T_r(\vec{c})$ and $\rho_r(\vec{c})$ and the expression for f^E are given in Ref. [325].

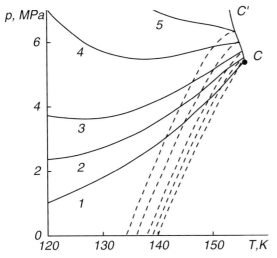

Fig. 4.18 (p, T)-projection of the phase diagram of oxygen–helium mixtures. *Solid lines* show the curves of phase equilibrium, *dashed lines* the diffusion spinodals. 1: $c' = 0$ mol%; 2: 0.5; 3: 1; 4: 2; 5: 3. CC' is the critical line.

The most accurate unified equations of state for pure liquefied gases are presented in Refs. [326–328]. Describing stable one-phase and two-phase states, these equations do not describe metastable and unstable regions in the space of thermodynamic states. In Ref. [329], a procedure of "sewing together" is suggested, which makes it possible to describe metastable and unstable states of pure substances and solutions on the basis of standard equations of state [326–328].

In the region of liquid-gas phase transition, the free energy of a pure fluid is approximated by [329]

$$f = a_0 - a_1 \frac{1}{\tilde{\rho}} + a_2 \frac{\ln \tilde{\rho}}{\tilde{T}} + a_3 \frac{\tilde{\rho}}{\tilde{T}} + \frac{a_4}{2} \frac{\tilde{\rho}^2}{\tilde{T}} + \frac{a_5}{3} \frac{\tilde{\rho}^3}{\tilde{T}} + \frac{a_6}{4} \frac{\tilde{\rho}^4}{\tilde{T}} + \frac{a_7}{5} \frac{\tilde{\rho}^5}{\tilde{T}}. \qquad (4.69)$$

The coefficients $a_1, a_2, \ldots, a_7$ in Eq. (4.69) are functions of temperature. At a given temperature, their values are determined by the following system of equations:

$$p(\rho'_s) = p_s, \qquad p(\rho''_s) = p_s, \qquad (4.70)$$

$$\left(\frac{\partial p}{\partial \rho}\right)_T\bigg|_{\rho'_s} = p'_\rho, \quad \left(\frac{\partial p}{\partial \rho}\right)_T\bigg|_{\rho''_s} = p''_\rho, \tag{4.71}$$

$$\left(\frac{\partial^2 p}{\partial \rho^2}\right)_T\bigg|_{\rho'_s} = p'_{\rho\rho}, \quad \left(\frac{\partial^2 p}{\partial \rho^2}\right)_T\bigg|_{\rho''_s} = p''_{\rho\rho}, \tag{4.72}$$

$$\int_{\rho''_s}^{\rho'_s} \frac{p}{\rho^2} d\rho = p_s \left(\frac{1}{\rho''_s} - \frac{1}{\rho'_s}\right). \tag{4.73}$$

The orthobaric densities ρ'_s and ρ''_s, the saturation pressure p_s, the first-order, p_ρ, and the second-order, $p_{\rho\rho}$, derivatives of pressure along the saturation line are determined by the standard equations of state, and Eq. (4.73) expresses the Maxwell rule of equal areas. The coefficient a_0 is determined by the free energy on the saturation line. Thus, the polynomial, Eq. (4.69), proves to be rigidly related to the binodal of the standard equation of state both in the value of the free energy and in its derivatives with respect to the density.

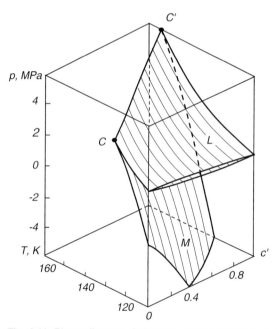

Fig. 4.19 Phase diagram of nitrogen–oxygen solution. L: phase-equilibrium surface, M: diffusion-spinodal surface, CC': critical line.

The phase diagram of an oxygen–helium system, calculated by Eq. (4.58), is given in Fig. 4.18. The lines of phase equilibrium separate the regions of stable and metastable phases on the thermodynamic surface of the homogeneous states. The metastable phase of a two-component system retains the restoring reaction to infinitesimal changes of state variables up to the diffusion spinodal determined by the condition, Eq. (2.22). In a one-component system, the stability of the metastable phase is disturbed, if Eq. (2.26) holds. In pure oxygen at a temperature $T = 140$ K a stretch of the liquid phase by $\Delta p_{sp}(T) = p_s - p_{sp} = 2.65$ MPa corresponds to the spinodal. Dissolution in oxygen of 1 mole% helium results in a rising limiting value of the tensile stress up to $\Delta p_{sp}(T,c) = p_s - p_{sp} \simeq 3.17$ MPa. Thus, with increasing concentration of a dissolved gas, the width of the metastable region increases. It should be mentioned that at high gas concentrations ($c' > 3$ mole%) penetration into the metastable region at a fixed external pressure is already connected with decreasing rather than increasing temperatures (see Fig. 4.18).

The phase diagram of the nitrogen–oxygen solution calculated by the equation of state, Eq. (4.68), with the use of experimental data on thermodynamic parameters and phase equilibrium from Refs. [330, 331], is presented in Fig. 4.19. The surfaces of phase equilibrium and the diffusion spinodal intersect on the line of critical points, $C_N C_O$. The width of the metastable region increases in pressure and temperature with increasing concentration of oxygen in the mixture.

4.8
Properties of Critical Bubbles in Binary Solutions

A binary two-phase system containing a curved surface of discontinuity possesses three degrees of freedom. Therefore, at a given temperature the surface tension of a nucleus is not only a function of the radius of curvature of the dividing surface, but also a function of the composition [332]. The properties of bubbles in solutions of simple liquids may be described in the framework of the van der Waals square gradient representation for the functional of the Helmholtz free energy, Eq. (4.13).

According to the Gibbs method of dividing surfaces [1], during the formation of a nucleus of radius R_* in a metastable system, the change in the thermodynamic potential is given by Eq. (4.1). For bubbles which are in equilibrium with the surrounding liquid phase from Eqs. (4.1) and (3.8) we have

$$W_* = 4\pi\sigma R^2 + \frac{4\pi}{3}(p' - p'')R_*^3. \tag{4.74}$$

4 Nucleation in Solutions of Liquefied Gases

By solving the system of equations, Eqs. (4.21), (4.24), and (4.74), with respect to the surface tension, we obtain

$$\sigma = \int_0^R (p'' - p) \frac{r^2}{R^2} dr + \int_R^\infty (p' - p) \frac{r^2}{R^2} dr \qquad (4.75)$$
$$+ \int_0^\infty \left[\rho_\beta(\mu_2 - \mu_2') + k_{22} \left(\frac{d\rho_\beta}{dr} \right)^2 \right] \frac{r^2}{R^2} dr.$$

Equations (4.75), (4.22) and (4.23) determine, employing the approximation of Eq. (4.20), the value of the surface tension for a binary solution at a curved interface.

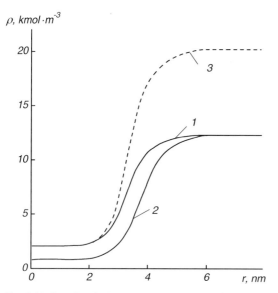

Fig. 4.20 Density distributions of the first (1) and the second (2) component of argon–krypton solutions, and also of the generalized density ρ_β (3) in a critical bubble at $T = 155$ K and $c' = 0.5$.

At small curvatures of the separating surface, the surface tension, σ, may be presented as a series in terms of $1/R_*$, in which the first expansion term is the surface tension at a planar interface. Restricting ourselves to the terms that are squared in curvature, we have

$$\sigma = \sigma_\infty + \frac{\sigma_1}{R_*} + \frac{\sigma_2}{R_*^2}. \qquad (4.76)$$

Now moving over to a new variable, $z = r - R$, and expanding the quantities in Eqs. (4.22), (4.23), and (4.75) into a Taylor series, we obtain [333]

$$\sigma_0 = 2k_{22} \int_{-\infty}^{+\infty} \left(\frac{d\rho_{0,0}}{dz} \right)^2 dz, \qquad (4.77)$$

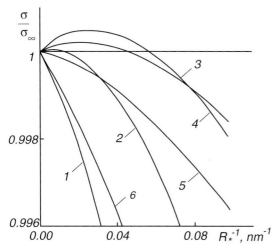

Fig. 4.21 Dependence of the reduced surface tension of critical bubbles of the solution argon–krypton on the curvature of the dividing surface at $T = 145$ K for different concentrations. 1: $c' = 0$; 2: 0.1; 3: 0.3; 4: 0.5; 5: 0.7; 6: 1.

$$\sigma_1 = \sum_{i=1}^{2} \mu'_{i,1} \Delta \rho_{i,s} \delta_{i,\infty}, \qquad (4.78)$$

$$\sigma_2 = I_1 + 2I_2 + I_3 + I_4 + \sigma_\infty z_*^2 + \frac{1}{2} \sum_{i=1}^{2} \mu'_{i,2} \Delta \rho_{i,s} \delta_{i,\infty}, \qquad (4.79)$$

where $\Delta \rho_{i,s} = \rho'_{i,s} - \rho''_{i,s}$ is the difference of the densities of the components at a planar interface, and $\delta_{i,\infty} = z_{i,e} - z_*$ is the distance between the equimolecular dividing surfaces and the surface of tension at a planar interfacial layer,

$$z_* = \frac{2k_{22}}{\sigma} \int_{-\infty}^{+\infty} \left(\frac{d\rho_{0,0}}{dz} \right)^2 z \, dz, \qquad z_{i,e} = \frac{1}{\Delta \rho_{i,s}} \int_{-\infty}^{+\infty} \frac{d\rho_{i,0}}{dz} z \, dz, \qquad (4.80)$$

$$I_1 = 2k_{22} \int_{-\infty}^{+\infty} \left(\frac{d\rho_{0,0}}{dz} \right)^2 (z - z_*)^2 dz, \qquad (4.81)$$

$$I_2 = \sum_{i=1}^{2} \frac{\mu'_{i,1}}{2} \int_{-\infty}^{+\infty} \frac{d\rho_{i,0}}{dz} (z - z_*)^2 dz, \qquad (4.82)$$

$$I_3 = \sum_{i=1}^{2} \frac{\mu'_{i,1}}{2} \int_{-\infty}^{+\infty} \frac{d\rho_{i,1}}{dz} (z - z_*)^2 dz, \qquad (4.83)$$

$$I_4 = 2k_{22} \int_{-\infty}^{+\infty} \frac{d\rho_{0,0}}{dz} \rho_{0,1} dz. \qquad (4.84)$$

Equations (4.77)–(4.84) express the parameters of Eq. (4.76) solely in terms of characteristics of a planar interface. This method of calculating the functions $\rho_{i,1}(z)$ is described in Ref. [333].

Tab. 4.4 Characteristic parameters of critical bubbles in an argon–krypton solution at a pressure 1.0 MPa.

Parameter	c'		
	0	0.5	1.0
T (K)	134	155	186
p_s (MPa)	2.440	2.906	2.738
$W_*/k_B T$ (Eq. (3.8))	78.0	82.9	71.1
$W_*/k_B T$ (Eq. (4.24))	64.1	76.7	57.9
lg J (Eqs. (2.86), (3.8))	5.5	4.1	8.3
lg J (Eqs. (2.86), (4.24))	11.5	6.8	14.0
σ_∞ (mN/m)	2.402	3.065	2.952
δ_∞ (nm)	0.035	−0.023	0.036
$\delta_{\infty,1}$ (nm)	0.035	−0.210	–
$\delta_{\infty,2}$ (nm)	–	0.182	0.036
$\sigma(R_*)$ (mN/m)	2.249	2.987	2.758
R_* (nm)	3.55	3.62	3.59
N_*''	319	330	265
N_{2*}''	0	75	265
δ (nm)	0.207	0.095	0.219
δ_1 (nm)	0.207	−0.067	–
δ_2 (nm)	–	0.329	0.219

The values of the coefficients of decomposition for the chemical potential $\mu'_{i,1}$ and $\mu'_{i,2}$ depend on how metastability is generated. At a constant temperature, the penetration of a binary solution into a metastable region may be realized in different ways. The particular way of penetration is determined by a trajectory in the space of two independent variables whose role can be played by densities, ρ_1 and ρ_2, chemical potentials, μ_1 and μ_2, etc. We shall examine two ways of penetration: along the line determined by the condition

$$c' = \frac{\rho'_2}{(\rho'_1 + \rho'_2)} = \text{const} \qquad (4.85)$$

(c-penetration) and along the line

$$(\mu'_1 - \mu'_{1,0}) = \left(\frac{k_{11}}{k_{22}}\right)^{1/2} (\mu'_2 - \mu'_{2,0}) \qquad (4.86)$$

(β-penetration). The first way mentioned is traditionally employed for experimental investigations of nucleation in creating phase metastability in a binary system [12, 291]. The second possibility is interesting as it allows one establishing the isomorphism in the behavior of the surface tension of nuclei in a binary and a one-component system.

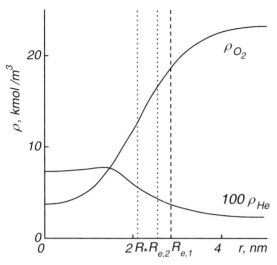

Fig. 4.22 Density distributions of the components of an oxygen–helium solution in a critical bubble at the following values of the parameters: $T = 140$ K, $c' = 0.1$ mol%, $p = 1.0$ MPa.

In fulfilling the condition expressed by Eq. (4.85), we obtain

$$\rho'_{2,1} = \rho'_{1,1}\frac{c'}{(1-c')}, \qquad \rho'_{2,2} = \rho'_{1,2}\frac{c'}{(1-c')}. \tag{4.87}$$

As a result, for the decomposition coefficients of the chemical potential we can write [333]

$$\mu'_{1,1} = \mu'_{2,1}a, \quad \mu'_{1,2} = \mu'_{2,2}a + \rho_1^{2'}a, \tag{4.88}$$

$$\mu'_{2,1} = \frac{-2\sigma_0}{\Delta\rho_{1,0}a + \Delta\rho_{2,0}}, \tag{4.89}$$

$$\mu'_{2,2} = \frac{\rho_1^{2'}\Delta\rho_{1,0}a + \sum_{i=1}^{2}\Delta\rho_{i,1}\mu'_{i,1} + 4\sum_{i=1}^{2}\Delta\rho_{i,0}\mu_{i,1}\delta_{i,0}}{\Delta\rho_{1,0}a + \Delta\rho_{2,0}}, \tag{4.90}$$

$$a = \left(\frac{\partial\mu_{1,0}}{\partial\rho}\right)_{c=c'} \bigg/ \left(\frac{\partial\mu_{2,0}}{\partial\rho}\right)_{c=c'}, \tag{4.91}$$

$$b = \left(\frac{\partial^2\mu_{1,0}}{\partial\rho^2}\right)_{c=c'} - \left(\frac{\partial^2\mu_{2,0}}{\partial\rho^2}\right)_{c=c'}a, \tag{4.92}$$

$$\rho = \rho_1 + \rho_2, \quad \rho_1 = \rho_{1,1} + \rho_{2,1}. \tag{4.93}$$

At β-penetration, we have

$$\mu'_{1,1} = \left(\frac{k_{11}}{k_{22}}\right)^{1/2} \mu'_{2,1}, \quad \mu'_{1,2} = \left(\frac{k_{11}}{k_{12}}\right)^{1/2} \mu'_{2,1}. \tag{4.94}$$

As a result of isomorphism, the coefficients in Eq. (4.76), and also the quantities determining them, prove to be similar to a one-component system [333]

$$\sigma_1 = -2\sigma_\infty \delta_{0,0}, \tag{4.95}$$

$$\sigma_2 = I_1 + 2I_2 + I_3 + I_4 + \sigma_\infty \frac{\Delta\rho_{0,1}}{\Delta\rho_{0,0}} \delta_{0,0} + 3\sigma_0 \delta_{0,0}^2 + \sigma_0 z_{0,e}^2, \tag{4.96}$$

where

$$\Delta\rho_{0,0} \mu'_{1,1} = -2\sigma_\infty, \tag{4.97}$$

$$\Delta\rho_{0,0} \mu'_{1,2} = -\Delta\rho_{0,1} \mu'_{1,1} - 4\mu'_{1,1}\Delta\rho_{0,0} z_{0,e} - 4\sigma_\infty z_*, \tag{4.98}$$

$$I_2 = \frac{\mu'_{1,1}}{2} \int_{-\infty}^{+\infty} \left(\frac{d\rho_{0,0}}{dz}\right) z^2 dz, \quad I_3 = \frac{\mu'_{1,1}}{2} \int_{-\infty}^{+\infty} \left(\frac{d\rho_{0,1}}{dz}\right) z dz. \tag{4.99}$$

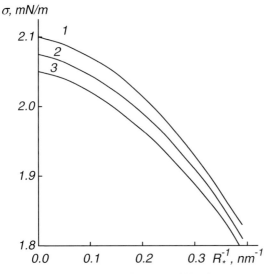

Fig. 4.23 Surface tension of critical bubbles in an oxygen–helium solution along the line $c' = $ const at $T = 140$ K. 1: $c' = 0$ mol%, 2: 0.1; 3: 0.2.

In the Gibbs method of dividing surfaces, the differential equation determining the size dependence of the surface tension takes the form [334]

$$\frac{1}{\sigma}\frac{d\sigma}{d(1/R_*)} = \frac{2\psi}{1 + 2\psi/R_*}, \tag{4.100}$$

where

$$\psi = \left(\Gamma_1 \frac{d\mu'_1}{d(1/R_*)} + \Gamma_2 \frac{d\mu'_2}{d(1/R_*)}\right) \bigg/ \left(\Delta\rho_1 \frac{d\mu'_1}{d(1/R_*)} + \Delta\rho_2 \frac{d\mu'_2}{d(1/R_*)}\right). \quad (4.101)$$

Here, Γ_i is the absolute adsorption of the ith component. At a small curvature of the dividing surface, the function $\psi(R_*)$ may be presented as

$$\psi = \psi_\infty + \frac{\psi_1}{R_*}, \quad (4.102)$$

where ψ_1 is a parameter depending on temperature and concentration. Performing a substitution of Eq. (4.102) into Eq. (4.100), we arrive at Eq. (4.76), where

$$\sigma_1 = -2\psi_\infty, \qquad \sigma_2 = 4\psi_\infty^2 - \psi_1. \quad (4.103)$$

If $\psi_\infty < 0$ and $\psi_1 \gg 4\psi_\infty^2$, then Eq. (4.86) gives a nonmonotonic dependence of $\sigma(R)$ on R with a maximum at

$$\frac{\sigma_{max}}{\sigma_\infty} = 1 + \frac{\psi_\infty^2}{\psi_1}, \qquad R_{*\,max} = \frac{\psi_1}{(-\psi_\infty)}. \quad (4.104)$$

In pure substances, the maximum manifests itself only in liquid droplets [186]. In a binary system, the presence or the absence of a maximum on the dependence $\sigma(R_*)$ at constant temperature depends on how metastability is created, i.e., on the trajectory of motion in the space of independent variables. Thus, if during the process of penetration of a liquid into the region of metastable states a constant concentration is maintained, then a maximum at the dependence $\sigma = \sigma(R_*)$ of vapor bubbles will be observed [334] for the solution of equimolecular composition. On the line $c' = $ const, vapor bubbles and liquid droplets are characterized by two different parameters, ψ. For the β-line, this is the parameter

$$\psi = \psi_\beta = \frac{\Gamma_\rho}{\rho''_\rho - \rho'_\rho}, \quad (4.105)$$

where $\Gamma_\rho = \Gamma_1 + (k_{11}/k_{22})^{1/2}\Gamma_2$ is the absolute adsorption determined by the density, ρ_β. Distinct from the parameter ψ_∞ on the line $c' =$const, the parameter ψ_∞ on the β-line does not change its sign in passing from pure substances to mixtures, and the character of the dependence, $\sigma(R)$, in a binary solution on the β-line is similar to the dependence, $\sigma(R)$, in a one-component system in the whole range of state variables from the binodal to the spinodal.

The functional, Eq. (4.13), is given if the equation of state of a homogeneous solution and the matrix of the influence coefficients, κ_{ij}, are known. Equations of state for binary solutions are presented in Section 4.7. The influence coefficients, κ_{11} and κ_{22}, may be obtained from Eq. (3.119) by experimental data on the surface tension of pure substances at a planar interface [57]. The coefficient κ_{12} is related to the influence coefficient for pure components by the combination rule

$$\kappa_{12} = \delta_\kappa (\kappa_{11}\kappa_{22})^{1/2}. \tag{4.106}$$

From the conditions of stability, Eq. (4.15), we have $\delta_\kappa^2 \leq 1$. The value of δ_κ may be determined in the framework of the van der Waals capillarity theory by experimental data on the macroscopic value of the surface tension, σ_∞, of the solution [276]

$$\sigma_\infty = 2 \int_{-\infty}^{+\infty} \sum_{i,j=1}^{2} \kappa_{ij} \left(\frac{d\rho_i}{dz}\right)\left(\frac{d\rho_j}{dz}\right) dz, \tag{4.107}$$

where the density distributions of the components of the mixture in a planar interfacial layer, $\rho_i(z)$, are found by solving the system of differential equations to which Eqs. (4.17) are reduced when the radius of curvature of the dividing surface is increased to infinity.

The coefficients κ_{ij} (Eq. (4.107)) are functions of temperature and exhibit a weak dependence on density [57, 334]. If we neglect this weak dependence of κ_{ij} on ρ_i, the value of κ_{ii} for a pure substance in solutions with complete solubility of the components may be determined by data on the surface tension at a planar interface via Eq. (3.119). Numerical calculations of the values of $\kappa_{11}(T)$ and $\kappa_{22}(T)$ by data on the surface tension of argon and krypton from Refs. [153, 154], and also of the coefficient δ_κ by the results of measuring σ_∞ of an argon–krypton solution [335–338] have shown that in this solution $\delta_\kappa \simeq 1$. Thus, the properties of the nuclei in an argon–krypton system can be described with good accuracy by Eqs. (4.21)–(4.23).

The results of numerical solution of the system of equations (4.21)–(4.23) for states close to the boundary of spontaneous boiling-up of argon–krypton solutions are presented in Fig. 4.20. At small supersaturations, a homogeneous core and a transition layer can be distinguished in the nucleus. Hereby, the thickness of the transition layer is small as compared with the bubble size. With further penetration into a metastable region, the homogeneity core becomes smaller, and the functions $\rho_1(r)$ and $\rho_2(r)$ take a bell-like shape. A shift in the distribution of $\rho_1(r)$ with respect to $\rho_2(r)$ means excess adsorption of the first component in the surface layer of a nucleus.

By solving the system of Eqs. (3.4), (3.8), (4.2), and (4.24) with respect to σ and R_*, the dependence of the surface tension on the radius of curvature of the surface of tension can be determined. The results of such a calculation at a fixed temperature and several values of solution concentration are given in Fig. 4.21. In contrast to a one-component system, where σ is a monotonically increasing function of R_* [184, 186, 187], for a solution with $c' \simeq 0.5$ the dependence $\sigma(R_*)$ has a characteristic maximum [291]. The value of σ_{max} exceeds that of σ_∞ only by 0.5–0.7%, which is close to the error of experimental determinations of σ_∞. The results of numerical calculations of the dependence $\sigma(R_*)$ in a wide range of nucleus radii may be described by Eq. (4.76), where the parameters σ_1 and σ_2 are determined by Eq. (4.103). Bubbles, responsible for the spontaneous boiling-up of a liquid up to nucleation rates $J = 10^{15}$ m^{-3}s^{-1}, are correctly described by Eq. (4.76). The results of calculating the properties of such bubbles in an argon–krypton system are given in Table 4.4.

Distributions of the densities of components in a critical bubble of an oxygen–helium solution are presented in Fig. 4.22. As distinct from solutions with complete solubility of the components, the density of the first gas component in a bubble is higher than in the solvent. The presence of a weak maximum on the density profile on the side of the gas phase points to the excess adsorption of the first component in the surface layer of a nucleus.

The Tolman parameters of the components of a binary solution ($\delta_i = R_{e,i} - R_*, i = 1, 2$) behave similarly to the Tolman parameter, δ, of pure substances; they increase monotonically with increasing curvature (in the region of weak metastability where the dependence $\delta_i(1/R_*)$ is linear). At small radii of the bubble, the Tolman parameter diverges proportionally to $R_*^{-1/3}$. As distinct from σ_1 and σ_2, the value of $\delta_{1(2)} = \delta_1 - \delta_2 = R_{e,1} - R_{e,2}$ shows a weak dependence on both the curvature of the separating surface and the concentration.

Figure 4.23 shows the surface tension of critical bubbles of O_2–He solutions as a function of the curvature of the surface of tension at a constant temperature for several compositions of the ambient phase. Small additions of helium in liquid oxygen do not change the character of the size dependence of the surface tension, evenly shifting the whole dependence toward smaller values of δ. Taking into account the dependence $\delta(R_*)$ in the work of formation of a critical nucleus leads to a decrease in the height of the activation barrier from $G_* = W_*/k_B T = 85$ to 70 and facilitates the boiling-up of a gas-filled solution.

The characteristic parameters of critical bubbles corresponding to the Gibbs number $G_* = 70$ are given in Table 4.5. It follows from the results presented in the table that at $T = 138$ K and $p = 1.171$ MPa the radius of a critical bubble, R_*, in superheated pure oxygen equals 3.732 nm, and the number of molecules in the bubble equals $N_*'' \cong 457$. Dissolution in oxygen of 0.2 mol% helium decreases the size of a critical bubble to $R_* = 3.615$ nm. This decrease is balanced by increasing $\sigma(R_*)$-values from 2.288 to 2.422 mN/m. In this case, the helium concentration in a bubble is equal to $c_*'' \cong 6$ mol%, the number of molecules $N_*'' \cong 428$ and the bubble is practically free of helium atoms, $N_{*1}'' \cong 31$. Before the formation of a bubble, its volume was occupied

4 Nucleation in Solutions of Liquefied Gases

Tab. 4.5 Characteristic parameters of critical bubbles of pure oxygen and oxygen–helium mixtures at the pressure $p = 1.171$ MPa and the Gibbs number $G_* = 70$.

Parameter		c' (mol%)	
	0	0.1	0.2
T (K)	138.149	137.650	137.150
p_s (MPa)	2.567	2.665	2.754
B (10^8 s^{-1})	1.969	0.855	0.556
$\log J$ (Eqs. (2.86), (3.8))	6.09	5.72	5.60
σ_∞ (mN/m)	2.44	2.51	2.58
$\delta_{\infty,1}$ (nm)	–	0.301	0.297
$\delta_{\infty,2}$ (nm)	0.0363	0.0367	0.0371
R_* (nm)	3.732	3.673	3.615
$\sigma(R_*)$ (mN/m)	2.288	2.354	2.422
N''_*	457	442	428
N''_{1*}	–	16	31
δ_1 (nm)	–	0.463	0.458
δ_2 (nm)	0.215	0.213	0.210
c''_* (mol%)	0	3.03	6.02
N'_1	0	5	9
N'_2	3894	3731	3575

by $N'_2 \cong 3575$ oxygen molecules and $N'_1 \cong 9$ helium atoms. It is evident that the supply of $\cong 22$ atoms of a dissolved gas to the growing bubble does not require a diffusion process and may be caused by molecular exchange. Thus, the process of nucleation in a gas-filled solution at temperatures close to the temperature of the attainable superheating of the solvent differs only slightly from the process of nucleation in a superheated pure liquid. The decrease of the temperature of attainable superheating of the liquid, observed as a gas component is dissolved in it, is first connected with a shift of the line of equilibrium coexistence of phases and, to a lesser degree, with the change of the surface tension. All these features are responsible for the differences of the behavior of a gas-filled solution in the vicinity of the critical point of the solvent and a gas-filled liquid at low temperatures.

4.9
Comparison of Theory and Experiment for Binary Solutions

In the present section, experimental data on the boiling-up kinetics of binary solutions of cryogenic liquids are analyzed in the framework of the homogeneous nucleation theory employing the expression for the steady-state nucleation rate. An evaluation of the time, τ_l, required for establishing a stationary flux of critical nuclei in solutions is given in Ref. [286]. In the region of state variables, where the experiments have been carried out, the time τ_l does not exceed 10^{-8} s, which justifies the neglect of nonsteady state effects.

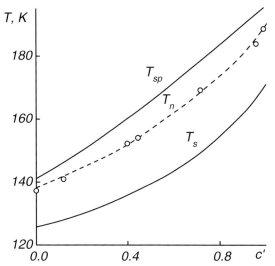

Fig. 4.24 Temperature of attainable superheatings of an argon–krypton solution at $p = 1.6$ MPa. Circles are experimental data. T_n has been calculated using Eqs. (2.86), (4.49), (3.8), and $\sigma = \sigma_\infty$, $J = 10^7$ m^{-3}s^{-1}. T_{sp} specifies the location of the spinodal curve and T_s the binodal curve of the liquid.

In the expression for the steady-state nucleation rate, the exponential term is dominant. The kinetic factor, B, depends weakly only on temperature, pressure, and concentration. A very strong dependence of the nucleation rate on the exponential term makes differences in evaluations of the factor B insignificant. Thus, an uncertainty in the value of B of 1–2 orders of magnitude results in an uncertainty in the temperature of the attainable superheatings of the liquid equal to 0.1–0.2 K. This uncertainty does not exceed the error of determining this value by experiment. A change from the normalization constant $C_0 = \rho'$ ($z_0 = 1$) to $C_0 = \rho'^2/\rho''_*$ ($z_0 = \rho'/\rho''_*$) decreases the temperature of attainable superheatings for oxygen–helium solutions by approximately 0.2–0.3 K. Comparing theory and experiment, we use $C_0 = \rho'$ as the normalization constant for the distribution function.

In Table 4.6 and Figs. 4.6, 4.7, and 4.24 experimental data on the temperature of attainable superheatings for Ar–Kr solutions are compared first with the results of their calculation by homogeneous nucleation theory in a macroscopic approximation and then incorporating the dependence $\sigma = \sigma(R_*)$ into the computations. In a macroscopic approximation, the discrepancies of the-

Tab. 4.6 Temperature of attainable superheating in argon–krypton solutions for $J = 10^7$ m^{-3}s^{-1}.

c'	T_s (K)	p_s (MPa)	T_n (K) Experiment	T_n (K) Theory[a]	T_n (K) Theory[b]
		$p = 1.0$ MPa			
0.0	116.55	2.431	133.9	134.12	133.65
0.109	118.60	2.559	138.1	138.29	137.88
0.428	125.86	2.887	151.8	151.81	151.50
0.708	136.38	2.973	166.5	166.73	166.30
0.938	152.72	2.784	181.1	181.42	180.81
1.0	159.13	2.673	185.3	185.84	185.17
		$p = 1.6$ MPa			
0.0	125.18	2.698	136.2	136.41	136.05
0.109	127.63	2.823	140.5	140.63	140.33
0.382	134.76	3.098	152.1	152.19	151.97
0.428	136.21	3.151	154.4	154.37	154.14
0.708	148.09	3.224	169.2	169.42	169.10
0.938	164.71	3.034	183.8	184.21	183.72
1.0	170.70	2.934	188.1	188.63	188.10

[a] Calculations performed with $\sigma = \sigma_\infty$,
[b] Calculations with $\sigma = \sigma(R_*)$.

ory and experiment in T_n do not exceed 0.6 K. They have the largest values for pure substances and the lowest ones for solutions of equimolar composition. Taking into account the dependence $\sigma(R_*)$ in the work of formation of a critical bubble improves the agreement of theory and experiment for pure substances and weak solutions. The nonmonotonic character of the dependence $\sigma(R_*)$ in a solution in the vicinity of equimolar composition leads to the convergence of the values of σ for critical bubbles and σ_∞. This property manifests itself in a decreasing disagreement between experiment and the theoretical values of the temperature of attainable superheatings calculated in a macroscopic approximation.

Temperature dependences of the nucleation rate are also close to the theoretical ones (Fig. 4.7). A shift in the temperature of 1 K results in the change of J by 6–12 orders of magnitude. The derivative $d \ln J / dT$ becomes larger with increasing pressure. The theory also provides for a qualitatively correct representation of the concentration dependence of the temperature of attainable superheatings observed experimentally (Fig. 4.24). At a pressure $p = 1.0$ MPa and concentration $c' = 0.5$, the attainable superheating, achieved experimentally, was given by $\Delta T = T_n - T_s = 27.2$ K. Under the same conditions, the superheating $\Delta T_{sp} = 36.2$ K corresponds to the diffusion spinodal. The spin-

Tab. 4.7 Temperature of attainable superheatings of helium–oxygen mixtures for nucleation rates, $J = 10^7$ m^{-3}s^{-1}.

c', mol %	T_s (K)	p_s (MPa)	T_n (K) Experiment	T_n (K) Theory[a]	T_n (K) Theory[b]
		$p = 1.171$ MPa			
0	122.40	2.570	138.17	138.73	138.22
0.08	118.96	2.640	137.79	138.38	137.85
0.14	114.55	2.686	137.43	138.11	137.56
0.20	–	2.732	137.06	137.85	137.27
		$p = 1.667$ MPa			
0	129.05	2.786	140.00	140.53	140.13
0.08	127.12	2.860	139.72	140.18	139.76
0.14	125.34	2.910	139.46	139.91	139.47
0.20	123.04	2.963	139.23	139.64	139.19

[a] Calculations performed with $\sigma = \sigma_\infty$,
[b] Calculations with $\sigma = \sigma(R_*)$.

odal has been calculated from Eq. (4.58) according to its definition (Eq. (2.22)).

Table 4.7 shows data on the temperature of attainable superheatings for an oxygen–helium solution corresponding to a nucleation rate, $J = 10^7$ m^{-3}s^{-1}. The experimental results are compared with calculations of T_n by homogeneous nucleation theory in a macroscopic approximation ($\sigma = \sigma_\infty$) and accounting for the dependence of the bubble surface tension on the curvature of its separating surface, $\sigma = \sigma(R_*)$. The kinetic factor, B, was determined from Eqs. (4.33) and (4.44). In addition, Table 4.7 shows the values of the saturation temperature, T_s, at a given pressure, p, and the values of the pressure, p_s, corresponding to the experimental temperature of attainable superheatings, T_n. In Figs. 4.13 and 4.14, the temperature dependences of $\bar{\tau}$ and J are shown, calculated by Eq. (3.10) and taking into account the dependence $\sigma(R_*)$. The results of calculation of the functions $\bar{\tau}(T)$ and $J(T)$ for pure oxygen in a macroscopic approximation are also presented. The data from Table 4.7, Figs. 4.13 and 4.14 show a good agreement between experiment and classical homogeneous nucleation theory if the latter takes into account the size dependence of the properties of critical nuclei. The disagreement between theoretical and experimental values of the temperature of attainable superheatings is within the limits of experimental error and accuracy of determining the thermophysical parameters in Eqs. (3.10) and (3.11). The experimental data do not reveal any peculiarities with respect to the nucleation kinetics in the vicinity of $c' = 0$ and $c' = 1$, where the resultant flux of nuclei can develop by-passing the saddle point. This result confirms the assumption of the equilibrium character of the composition in a critical bubble used in the theory of binary nucleation.

Figure 4.25 shows the line of attainable superheatings and the spinodal in the oxygen–helium system at two investigated pressure values. The dissolution of helium in oxygen decreases both the temperature of superheating and the temperature of stability loss of the system with respect to infinitely small perturbations of concentration. At a helium concentration in oxygen of 0.3 mol%, the temperature of spontaneous boiling-up of the solution is approximately 1.2–1.5 K lower than that of oxygen. At a concentration of 0.1 mol% and a pressure of 1.171 MPa, the superheating is $\Delta T \simeq 20$ K, and the loss of stability of the homogeneous solution takes place at $\Delta T_{sp} \simeq 25$ K. Close to

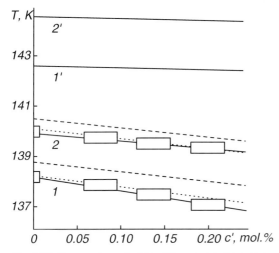

Fig. 4.25 Temperature of attainable superheatings of oxygen–helium solutions (1, 2; $J = 10^7$ m^{-3}s^{-1}) and spinodal (1′, 2′) at pressures $p =$ 1.171 (1, 1′) and 1.667 MPa (2, 2′). The *dashed line* shows the results of the calculation by homogeneous nucleation theory in a macroscopic approximation, $\sigma = \sigma_\infty$, the *dotted line* accounts for the dependence, $\sigma(R_*)$. The width and length of the rectangles determine the error in the determination of temperature and concentration.

the temperature of spontaneous boiling-up of a pure solvent ($T > 0.9 T_c$), a critical bubble of a gas-filled solution contains about an order of magnitude fewer molecules of a dissolved substance than of a solvent. If we assume that an easily volatile component, which is to be dissolved, is distributed uniformly in a solvent, in the process of nucleation the solvent molecules will then predominantly leave the volume where a critical bubble forms. To obtain an equilibrium composition of a dissolved component in a bubble, diffusion may not be required as the inflow of the lacking number of gas molecules may take place at the expense of the thermal motion of the molecules.

The situation in a gas-filled solution changes cardinally if nucleation proceeds at temperatures much lower than the critical temperature of the solvent. Thus, in an oxygen–helium solution at a temperature of 100 K, a pressure of

1.171 MPa, a helium concentration $c' \cong 3$ mol%, the solution is supersaturated (the Gibbs number equals $G_* \simeq 80$) and the radius of a critical nucleus will be $\cong 1.7$ nm, the number of helium molecules in a bubble $N''_{1*} \cong 176$, and the number of oxygen molecules $N''_{2*} \cong 3$. Before the formation of a nucleus took place, in the volume of critical size about 12 helium atoms were contained in it. To provide an equilibrium concentration in a bubble, helium atoms should diffuse via distances approximately 2.5 times exceeding the bubble radius. The characteristic times of diffusion in this case are $\tau_D \cong 2 \cdot 10^{-8}$ s, which is comparable with the time of establishment of a stationary nuclei size distribution.

At low temperatures, a stronger size dependence of the surface tension of critical bubbles may be observed [339–341]. The character of the dependence $\sigma(R_*)$ differs not only quantitatively, but also qualitatively from that revealed at elevated temperatures. At temperatures $T = 0.6T_c$, the radius of a critical bubble is 2–2.5 times smaller than at $T = 0.9T_c$. By this reason, at low temperatures a stronger sensitivity of supersaturated solutions is found with respect to inhomogeneities and accidental inclusions in a system and on the walls of the surrounding vessel. An increase in the number of regions where a viable bubble is easily activated hinders the experimental realization of the homogeneous nucleation mechanism. However, even with existing boiling centers and free liquid surfaces, the temperature of intensive fluctuation nucleation in a supersaturated solution can be achieved in nonstatic processes with a rapid introduction of heat into the system or an abrupt removal of the external pressure [9, 10]. For this purpose, it is necessary that the speed of penetration into a metastable region satisfies a certain criterion which would be ensured by liquid superheatings under an intensive heat flow into existing evaporation centers. At low temperatures, the conditions of realization of such a shock liquid boiling-up regime prove to be more rigid than at temperatures close to the critical point. Practically all experiments on the investigation of nucleation in gas-filled solutions at low temperatures were carried out using quasistatic methods [296–301] with a sufficiently slow transfer of the system into a metastable state. As done in the case of cavitation [12], the achievement of conditions of homogeneous nucleation here evidently requires the use of nonstatic (pulse) methods.

4.10
Kinetics and Thermodynamics of Nucleation in Three-Component Solutions

Let us now examine some aspects of nucleation in ternary solutions. If nucleation proceeds under constant external pressure and temperature, the Gibbs excess thermodynamic potential is written as Eq. (4.1), where $\nu = 3$. Substitution of the conditions of mechanical, Eq. (3.4), and diffusion, Eq. (4.2), equilibria into Eq. (4.1) gives the work of formation of a critical nucleus, which has the form of Eq. (3.8).

In a region of weak metastability ($G_* \gg 1$), approximate relations can be obtained relating the parameters of a critical bubble in a ternary system with those of the line of phase equilibrium at a planar interface. Assuming that the liquid solution is incompressible and the vapor in a bubble is ideal, we have

$$c''_{2*} = c''_{2s} \left\{ c''_{1s} + c''_{2s} \exp\left[\frac{(p_s - p')(v'_2 - v'_3)}{p_s} \frac{1}{v''}\right] \right. \\ \left. + (1 - c''_{2s} - c''_{3s}) \exp\left[\frac{(p_s - p')(v'_2 - v'_3)}{p_s} \frac{1}{v''}\right] \right\}^{-1}. \tag{4.108}$$

$$c''_{3*} = c''_{3s} \left\{ c''_{3s} + c''_{2s} \exp\left[\frac{(p_s - p')(v'_3 - v'_2)}{p_s} \frac{1}{v''}\right] \right. \\ \left. + (1 - c''_{1s} - c''_{2s}) \exp\left[\frac{(p_s - p')(v'_2 - v'_3)}{p_s} \frac{1}{v''}\right] \right\}^{-1}. \tag{4.109}$$

$$p''_* = p_s \left\{ c''_{2s} \exp\left[-\frac{(p_s - p') v'_2}{p_s} \frac{1}{v''}\right] + c''_{3s} \exp\left[-\frac{(p_s - p') v'_3}{p_s} \frac{1}{v''}\right] \right. \\ \left. + (1 - c''_{2s} - c''_{3s}) \exp\left[-\frac{(p_s - p') v'_1}{p_s} \frac{1}{v''}\right] \right\}. \tag{4.110}$$

Here, c''_{2s} and c''_{3s} are the concentrations of the second and third components of the solution at a planar interface at temperature, T, and pressure, p_s, v'_i is the partial molar volume in the liquid phase, and v'' is the molar volume of the gas phase. If $v'_i \ll v''$ ($i = 1, 2, 3$), then expanding the exponential terms in Eqs. (4.108)–(4.110) into a series and restricting ourselves to the first terms of the expansion, after substituting the results into Eq. (3.8) for the work of formation of a critical nucleus in a ternary mixture, we obtain

$$W_* = \frac{16\pi}{3} \frac{\sigma^3}{(p_s - p')^2 \left[1 - c''_{2s}\frac{v'_2}{v''} - c''_{3s}\frac{v'_3}{v''} - (1 - c''_{2s} - c''_{3s})\frac{v'_1}{v''}\right]^2}. \tag{4.111}$$

According to the van der Waals theory of capillarity, the change of the Helmholtz free energy at the formation of a spherical inhomogeneity with the properties of a competitive phase in a ternary mixture is given by Eq. (4.13), where $\nu = 3$. It follows from the conditions of stability of a two-phase three-component system that the inequalities

$$\begin{pmatrix} \kappa_{11} & \kappa_{12} & \kappa_{13} \\ \kappa_{21} & \kappa_{22} & \kappa_{23} \\ \kappa_{31} & \kappa_{32} & \kappa_{33} \end{pmatrix} > 0, \quad \begin{pmatrix} \kappa_{11} & \kappa_{12} \\ \kappa_{21} & \kappa_{22} \end{pmatrix} > 0, \quad \kappa_{11} > 0, \tag{4.112}$$

4.10 Kinetics and Thermodynamics of Nucleation in Three-Component Solutions

have to be fulfilled. If, for the cross coefficients κ_{12}, κ_{23}, and κ_{13}, we take

$$\kappa_{12} = (\kappa_{11}\kappa_{22})^{1/2}, \qquad \kappa_{23} = (\kappa_{22}\kappa_{33})^{1/2}, \qquad \kappa_{13} = (\kappa_{11}\kappa_{33})^{1/2}, \qquad (4.113)$$

the first and the second of inequalities (4.112) transform to an equality, which corresponds to an indifferent equilibrium of the system with respect to density gradients opposite in sign and determines the limiting value of the coefficients κ_{12}, κ_{23}, and κ_{13}. With Eq. (4.113), Eq. (4.13) can be written in the more simple form

$$\Delta F[\vec{\rho}] = 4\pi \int_0^\infty \left[\Delta f + \kappa_{33} \left(\frac{d\rho_\beta}{dr} \right)^2 \right] r^2 dr. \qquad (4.114)$$

Here, $\rho_\beta = \rho_3 + (\kappa_{11}/\kappa_{33})^{1/2}\rho_1 + (\kappa_{22}/\kappa_{33})^{1/2}\rho_2$. Density distributions, ρ_i ($i = \beta, 1, 2, 3$), in a critical bubble can be found from the solution of the Euler system of equations for the functional equation. (4.114), which includes one differential equation

$$\frac{d^2\rho_\beta}{dr^2} + \frac{2}{r}\frac{d\rho_\beta}{dr} = \frac{\mu_3 - \mu_3'}{2\kappa_{33}} \qquad (4.115)$$

with the boundary conditions $\rho_\beta \to \rho_\beta'$ at $r \to \infty$, $d\rho_\beta/dr \to 0$ at $r \to 0$ and $r \to \infty$ and two algebraic equations

$$\mu_1 - \mu_1' = \left(\frac{\kappa_{11}}{\kappa_{33}} \right)^{1/2} (\mu_3 - \mu_3'), \qquad (4.116)$$

$$\mu_2 - \mu_2' = \left(\frac{\kappa_{22}}{\kappa_{33}} \right)^{1/2} (\mu_3 - \mu_3'). \qquad (4.117)$$

In this case, the work of formation of a critical bubble is given by Eq. (4.24).

Let us examine the kinetics of nucleation in a three-component system assuming that the nucleation process is isothermal, the initial supersaturation is small ($G_* \gg 1$), and the gas in a bubble is ideal. The inertial properties of the liquid in the process of bubble growth are neglected. In the approximations mentioned above after introducing dimensionless quantities (see Section 4.2) for the matrix of second-order derivatives of the thermodynamic potential, $\Psi(x, y, z, \omega)$, at the saddle point of the potential barrier we have

$$\widehat{\Psi} = 2G_* \begin{pmatrix} -\frac{1}{3} & 0 & 0 & 0 \\ 0 & \frac{1}{b_1} & 0 & 0 \\ 0 & 0 & \frac{1}{b_2} & 0 \\ 0 & 0 & 0 & \frac{1}{b_3} \end{pmatrix}. \qquad (4.118)$$

The exchange of molecules of the mixture components between a growing bubble and a liquid solution may take place in the free-molecular (Eq. (4.31)) and the diffusion (Eq. (4.37)) regimes. The total rate of change of the number of molecules of each of the components in a bubble is then given by Eq. (4.38).

The rate of change of the bubble volume is determined by Eq. (3.22). Neglecting the liquid inertia in the process of the nucleus growth implies, when one restricts the attention to processes in the vicinity of the saddle point of the pass, the elimination from Eq. (3.22) besides the term containing $\ddot{R}^2$ and the term containing $\ddot{R}$. The condition of such neglect is expressed by inequality (4.43). Now moving to the equation of motion for the bubble volume and the partial pressures of the solution components in the bubble to dimensionless variables defined by Eq. (4.47), we obtain

$$\dot{x} \equiv \frac{dx}{dt} = \frac{1}{3}x + \frac{1}{b_1}y + \frac{1}{b_2}z + \frac{1}{b_3}\omega, \qquad (4.119)$$

$$\dot{y} \equiv \frac{dy}{dt} = -\frac{1}{3}x - \left[\frac{1}{b_1} + \frac{\omega_1}{3(1+\gamma_1)}\right]y - \frac{1}{b_2}z - \frac{1}{b_3}\omega, \qquad (4.120)$$

$$\dot{z} \equiv \frac{dz}{dt} = -\frac{1}{3}x - \frac{1}{b_1}y - \left[\frac{1}{b_2} + \frac{\omega_2}{3(1+\gamma_2)}\right]z - \frac{1}{b_3}\omega, \qquad (4.121)$$

$$\dot{\omega} \equiv \frac{d\omega}{dt} = -\frac{1}{3}x - \frac{1}{b_1}y - \frac{1}{b_2}z - \left[\frac{1}{b_3} + \frac{\omega_3}{3(1+\gamma_3)}\right]\omega. \qquad (4.122)$$

Here $t = \tau/\tau_m$ is a dimensionless time and $\tau_m = 4\eta/3bp''_*$. By using the relation between the rate of change of the characteristic variables of the nucleus and the thermodynamic forces, Eq. (3.79), for the tensor of generalized diffusion we have

$$\hat{D} = \frac{1}{2G_*}\begin{pmatrix} 1 & -1 & -1 & -1 \\ -1 & 1+\dfrac{\omega_1 b_1}{3(1+\gamma_1)} & 1 & 1 \\ -1 & 1 & 1+\dfrac{\omega_2 b_2}{3(1+\gamma_2)} & 1 \\ -1 & 1 & 1 & 1+\dfrac{\omega_3 b_3}{3(1+\gamma_3)} \end{pmatrix}. \qquad (4.123)$$

Substitution of Eqs. (4.118) and (4.123) into the characteristic equation, Eq. (2.101), yields the algebraic equation

$$\tilde{\lambda}^4 + \frac{1}{3}\left(\frac{\omega_1}{1+\gamma_1} + \frac{\omega_2}{1+\gamma_2} + \frac{\omega_3}{1+\gamma_3} + \frac{3-b}{b}\right)\tilde{\lambda}^3 + \frac{1}{9}\left[\frac{\omega_1}{1+\gamma_1}\left(\frac{3}{b_2}+\right.\right.$$

$$+ \frac{3}{b_3} - 1\right) + \frac{\omega_2}{1+\gamma_2}\left(\frac{3}{b_1}+\frac{3}{b_3}-1\right) + \frac{\omega_3}{1+\gamma_3}\left(\frac{3}{b_1}+\frac{3}{b_2}-1\right) +$$

$$+ \frac{\omega_1\omega_2}{(1+\gamma_1)(1+\gamma_2)} + \frac{\omega_2\omega_3}{(1+\gamma_2)(1+\gamma_3)} + \frac{\omega_1\omega_3}{(1+\gamma_1)(1+\gamma_3)}\Bigg]\tilde{\lambda}^2 +$$

$$+ \frac{1}{27}\left[\frac{\omega_1\omega_2}{(1+\gamma_1)(1+\gamma_2)}\left(\frac{3}{b_3}-1\right) + \frac{\omega_2\omega_3}{(1+\gamma_2)(1+\gamma_3)}\left(\frac{3}{b_1}-1\right) +\right.$$

$$+ \frac{\omega_1\omega_3}{(1+\gamma_1)(1+\gamma_3)}\left(\frac{3}{b_2}-1\right) + \frac{\omega_1\omega_2\omega_3}{(1+\gamma_1)(1+\gamma_2)(1+\gamma_3)}\Bigg]\tilde{\lambda} -$$

$$- \frac{1}{81}\frac{\omega_1\omega_2\omega_3}{(1+\gamma_1)(1+\gamma_2)(1+\gamma_3)} = 0. \quad (4.124)$$

The positive root, $\tilde{\lambda}_0$, of this equation determines the nucleation rate in a supersaturated three-component viscous solution with allowance for the diffusional and the molecular supply of a substance to a growing bubble.

The nucleation rate is determined by Eq. (2.112), where ρ' is the particle number density of a metastable solution, and the kinetic factor is given by

$$B = \lambda_0 \rho''_* R_*^2 \left(\frac{k_B T}{\sigma}\right)^{1/2}. \quad (4.125)$$

The diffusional mechanism will be decisive if $\gamma_i \gg 1$. Otherwise the concentration of the solution will be homogeneous up to the bubble surface, and the molecular exchange will be determined only by the volatility of the components of the mixture. In the limiting case $\omega_3 \to 0$, $b_3 \to \infty$, Eq. (4.124) is transformed into Eq. (4.51), which describes the boiling-up of an inertialess binary solution.

4.11
Attainable Superheatings of Ternary Solutions of Cryogenic Liquids

In the present section, results of experimental investigations of nucleation in ternary systems are presented. On the one hand, the increase of the number of components in a mixture considerably complicates the description of processes proceeding in it, but, on the other hand, makes it possible to hope for revealing new phenomena connected with nucleation.

The first experimental data on the temperatures of attainable superheatings for ternary solutions of cryogenic liquids were obtained for nitrogen–

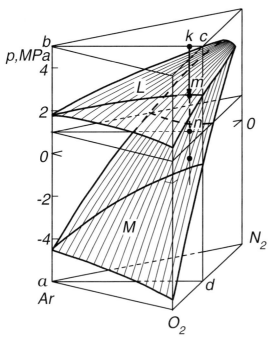

Fig. 4.26 Phase diagram of the system nitrogen–oxygen–argon. L: phase equilibrium, M: diffusion-spinodal surface.

oxygen–argon and nitrogen–oxygen–helium systems [342, 343]. The quasistatic method of measuring the lifetime, τ, of superheated solutions in glass capillaries was used. The procedure of determining $\bar{\tau}$ in ternary systems is similar to the one used previously for binary solutions (see Section 4.4). All measurements were conducted at the same pressure, $p' = 1.0$ MPa and concentrations, (c'_2, c'_3). For determining the composition of the solutions under investigation, manometric, piezometric, and weight methods were employed.

The piezometric method, as distinct from the manometric and the weight methods, allows one to determine the composition of the solution directly in the measuring cell at the test temperature. For this purpose, the pressure dependence of the specific volume of the solution is measured in the course of the experiment, after completion of the cycle of measurements of the lifetime, during the process of condensation of the mixture ($T = $ const). The data obtained on the phase-equilibrium pressure and data from the literature on the phase diagram of a solution are used to determine the concentration of the components in a mixture. For the system N_2–O_2–Ar, data on phase equilibrium from Ref. [325] were used, the phase diagram of the solution N_2–O_2–He was not investigated by experiment. Figures 4.26 and 4.27 present the phase diagrams of solutions in the region of the state variables at which the investi-

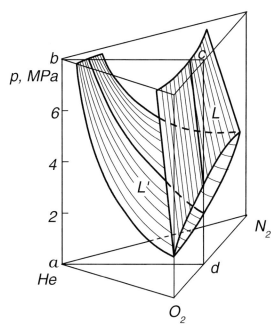

Fig. 4.27 Phase diagram of the system nitrogen–oxygen–helium.
L, L': phase-equilibrium surfaces.

gations of nucleation kinetics were carried out. The results of determination of the temperature dependence of the nucleation rate, J, in solutions N_2–O_2–Ar and N_2–O_2–He at fixed values of pressure and composition are presented in Figs. 4.28 and 4.29. The character of the dependence $J(T)$ is similar to that of pure liquids and binary solutions. Along the experimental curves, there are sections of an abrupt increase in the nucleation rate, which correspond to spontaneous boiling-up of the solution. At given state variables (temperature, pressure, composition), after the cut-off of premature acts of liquid boiling-up, the distribution of expectation times of boiling-up for a metastable solution is close to the Poissonian one (Eq. (3.88)).

For the solution N_2–O_2–Ar containing 0.442 molar fraction of oxygen (c'_2) and 0.166 molar fraction of argon (c'_3) at a pressure $p = 1.0$ MPa and $J = 10^7$ m^{-3}s^{-1}, the temperature of attainable superheating was equal to $T_n = 127.72$ K. In this case, the value of the derivative $(d \lg J/dT)$ is equal to 15.2. The value of the temperature of attainable superheatings, calculated by the additivity rule Eq. (4.54) from the data on T_n for pure liquids, is 127.62 K, which coincides with the experimental value within the limits of error of measurements. The temperatures of attainable superheatings for a binary solution containing 0.44 molar fraction of nitrogen (c'_1) and 0.56 molar fraction of oxygen (c'_2) at a pressure $p = 1.0$ MPa and $J = 10^7$ m^{-3}s^{-1} is 126.46 K. The disso-

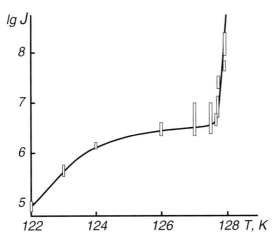

Fig. 4.28 Nucleation rate in the system N_2–O_2–Ar at $p' = 1.0$ MPa, $c_2' = 0.442$, $c_3' = 0.166$.

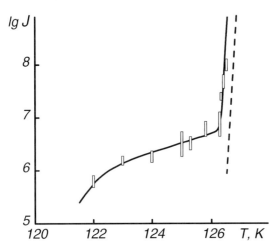

Fig. 4.29 Nucleation rate in the system N_2–O_2–He at $p' = 1.0$ MPa, $c_2' = 0.56$, $c_3' = 0.05$ mol%. The *dashed line* shows results of calculations by the homogeneous nucleation theory in a macroscopic approximation.

lution in a mixture N_2–O_2 of such a composition of 0.05 mol% helium leads to a decrease in the temperature of attainable superheatings to $T_n - 126.30$ K.

The comparison of homogeneous nucleation theory and experiment for ternary systems is hampered by the lack of data on a number of thermophysical parameters. For the system N_2–O_2–Ar at temperatures above 115 K, data on the surface tension and the interdiffusion coefficients are not available. The surface tension of the solution N_2–O_2–He at temperatures close to the temperature of its attainable superheatings is measured in Ref. [344], but the absence of a single equation of state for such a system did not allow us to perform calculations of the work of formation of a critical bubble in the framework of the van der Waals approach (see Section 4.8). The temperatures of attainable superheatings for the system N_2–O_2–He were calculated in the framework of the model described in Section 4.10.

The work of formation of a critical nucleus was determined by Eq. (4.11) using the macroscopic approximation, $\sigma(R_*) = \sigma_\infty$. Due to the lack of data on the interdiffusion coefficients of the components it was assumed that the molecular exchange between a nucleus and the solution was determined only by the volatility of the mixture components. Then, Eq. (4.124) takes the form

$$\tilde{\lambda}^4 + \frac{1}{3}\left(w_1 + w_2 + w_3 + \frac{3-b}{b}\right)\tilde{\lambda}^3 + \frac{1}{9}\left[w_1\left(\frac{3}{b_2} + \frac{3}{b_3} - 1\right) + \right.$$
$$+ w_2\left(\frac{3}{b_1} + \frac{3}{b_3} - 1\right) + w_3\left(\frac{3}{b_1} + \frac{3}{b_2} - 1\right) + w_1 w_2 + w_2 w_3 + w_1 w_3 \left.\right]\tilde{\lambda}^2 +$$
$$+ \frac{1}{27}\left[w_1 w_2 \left(\frac{3}{b_3} - 1\right) + w_2 w_3\left(\frac{3}{b_1} - 1\right) + w_1 w_3\left(\frac{3}{b_2} - 1\right) + w_1 w_2 w_3\right]\tilde{\lambda} -$$
$$- \frac{1}{81}w_1 w_2 w_3 = 0. \qquad (4.126)$$

The results of calculations of the nucleation rate in the system N_2–O_2–He are compared with the experimental data in Fig. 4.29. Within the limits of experimental error between theoretical and experimental values of the derivative $(\partial \ln J/\partial T)_{p,c_2',c_3'}$, theoretical values of the temperature of attainable superheatings are 0.2–0.3 K higher than the experimental data.

5
Nucleation in Highly Correlated Systems

5.1
Introduction

The critical point of a thermodynamic system is a singular point where the curve enclosing the region of essential instability (the spinodal) divides unstable and stable states. In its vicinity both the correlation length, ξ, and the relaxation time, τ, of fluctuations are large. For describing the kinetics of first-order phase transitions in systems with a large correlation length, the methods of the theory of second-order phase transitions are employed here. Such a description is based on the concept of an order parameter, φ, which has different values in both co-existing phases and, therefore, undergoes a finite jump in the process of phase transition [20]. We are interested in the formation and growth of nuclei of a new phase, and also in the processes of decay of unstable states in the vicinity of singular points of a thermodynamic system. Of all the degrees of freedom of the system, we explicitly take into account only slowly changing variables $a_i(\vec{r}, \tau)$, $(i = 1, 2, \ldots, M)$, assuming that the remaining degrees of freedom adjust themselves adiabatically to such large-scale (hydrodynamic) modes. The variables a_i either include the order parameter as an independent mode or the order parameter is their certain linear combination.

The dynamics of the fields, a_i, is described by a system of nonlinear equations, Eq. (2.64). Later on, unless not specified otherwise, it is assumed that the system has only one large-scale mode described by the order parameter, $\varphi(\vec{r}, \tau)$. For a one-component system it is the density field, $\varphi(\vec{r}) = \rho(\vec{r}) - \rho_c$, for a binary solution, the concentration field, $\varphi(\vec{r}) = c(\vec{r}) - c_c$, where ρ_c and c_c are the values of density and concentration at the corresponding critical points. In the vicinity of a critical point, the Helmholtz free energy, F, may be expanded into a series in powers of the order parameter. In the space of dimensionality, d, assuming the invariance of the thermodynamic potential, F, with respect to the inversion of the sign of φ, we have [20, 62]

$$F[\varphi] = \int \left[\frac{1}{2} a \varphi^2 + \frac{1}{4} b \varphi^4 - h\varphi + \kappa (\nabla \varphi)^2 \right] d^d r. \tag{5.1}$$

Explosive Boiling of Superheated Cryogenic Liquids. Vladimir G. Baidakov
Copyright © 2007 WILEY-VCH Verlag GmbH & Co. KGaA, Weinheim
ISBN: 978-3-527-40575-6

The coefficients a, b, and κ are functions of temperature and pressure; the quantity h plays the role of an external field. The subsequent terms of the expansion, Eq. (5.1), and the asymmetry may be taken into account in the framework of the well-known models of the theory of critical phenomena [63, 345].

The properties of the system described by the functional equation (5.1) are well-known [20]. At $h = 0$ and $a = 0$, a critical point is found in the system; the line $h = 0$, $a < 0$ is the line of a first-order phase transition (binodal)

$$\varphi_s = \pm \left(-\frac{a}{b}\right)^{1/2}, \qquad b > 0, \tag{5.2}$$

where

$$a = a_o t, \qquad t = \frac{(T_c - T)}{T_c}, \qquad a_o < 0. \tag{5.3}$$

The condition $(\partial h/\partial \varphi)_T = 0$ determines the boundary of essential instability of the metastable phase (spinodal)

$$\varphi_{sp} = \pm \left(-\frac{a}{3b}\right)^{1/2}, \qquad h_{sp} = \left(\frac{2}{3}\right)\left(-\frac{a^3}{3b}\right)^{1/2}. \tag{5.4}$$

In the absence of random forces, $\zeta_i(\vec{r}, \tau)$, the system of equations, Eq. (2.68), has stationary homogeneous solutions which are stable with respect to infinitely small perturbations, and also a class of quasistationary solutions describing critical heterophase fluctuations in the initial metastable phase. At small values of the random forces, the solutions of Eqs. (2.68) are always perturbations of quasistationary configurations of the fields, $a_i(\vec{r}, \tau)$ [65]. The quantity ζ may be regarded as a perturbation if for regions with a linear dimension of the order of the correlation radius, ξ, the amplitude of the response of the field, $\varphi(\vec{r}, \tau)$, is small compared to φ_s, i.e., [20]

$$\langle [\delta \varphi(\vec{r})]^2 \rangle \ll \varphi_s^2. \tag{5.5}$$

The correlation radius of fluctuations of the order parameter, ξ, is determined by the expression

$$\xi = \left(\frac{\kappa}{|a|}\right)^{1/2}. \tag{5.6}$$

The region of values for the parameters of the potential, Eq. (5.1), at which condition (5.5) is fulfilled, is given by the inequality ($d = 3$) [62, 63]

$$1 \gg |t| \gg \text{Gi} = \frac{b^2 (k_B T_c)^2}{|a_o| \kappa^3}, \tag{5.7}$$

where the quantity Gi is known as the Ginzburg number. In a space of arbitrary dimensionality, d, inequality (5.7) is written as

$$\widetilde{Gi}^{-1} = \frac{\varphi_s^2}{\langle[\delta\varphi(\vec{r})]^2\rangle} \simeq \frac{\varphi_s^2}{\xi^{2-d}r_0^{-2}v_m^{d-4}} = \frac{\kappa^{d/2}\varphi_s^{4-d}}{k_B T_c b^{d/2-1}} \gg 1. \quad (5.8)$$

Here, v_m is the molecular volume and r_0 is the radius of molecular interaction. The criteria (5.7) and (5.8) determine the range of applicability of the approach discussed above, i.e., of the Landau mean-field theory [20].

In the case of strong fluctuations, the field $\varphi(\vec{r},\tau)$ should be averaged over perturbations with scales smaller than the characteristic linear dimensions of space inhomogeneities. The relaxation equation for a smoothened field is found by renormalization group methods [62, 63]. As long as the characteristic size of a nucleus is larger than the correlation length $R_* \gg \xi$, the relaxation equation and the thermodynamic potential have the same form like in the case of weak fluctuations (Eqs. (2.68), (5.1)), but with renormalized coefficients, $\hat{\Gamma}_*, \kappa_*, a_*, b_*$ [65]. The coefficients κ_*, a_*, b_* are expressed in terms of the measured characteristics of the system, such as the correlation radius ξ, the susceptibility (isothermal compressibility) $(\partial\varphi/\partial h)_T$, or the equilibrium value of the order parameter, φ_s. In the asymptotic vicinity of the critical point, these quantities are described by simple power laws

$$\varphi_s = \pm B_0 t^\beta, \quad (5.9)$$

$$\left(\frac{\partial\varphi}{\partial h}\right)_T = \chi_0|t|^{-\gamma}, \quad (5.10)$$

$$\xi = \xi_0|t|^{-\nu}. \quad (5.11)$$

Between the critical exponents of thermodynamic quantities there exists the relation

$$\alpha + 2\beta + \gamma = 2, \quad \beta(\delta - 1) = \gamma, \quad (5.12)$$

where α and δ are the critical exponents of the heat capacity and the critical isotherm, respectively. In the mean-field theory, one obtains $\beta = \nu = 1/2$, $\gamma = 1, \alpha = 0, \delta = 3$, whereas in the fluctuation theory of critical phenomena, the relations $\beta \simeq 0.33, \gamma \simeq 1.23, \nu \simeq 0.63, \alpha \simeq 0.11, \delta \simeq 4.6$ hold [345]. Besides, in a space of arbitrary dimensionality, d, the following relationships are true:

$$\nu d = 2 - \alpha, \quad (d-1)\nu = \mu, \quad (5.13)$$

where μ is the critical exponent of surface tension. Critical exponents depend only on the dimensionality of space, d, and the number of components of the order parameter. The coefficients of proportionality in the power laws,

Eqs. (5.9)–(5.11), are individual parameters of a substance and depend on the direction of approach to the critical point.

Equations (5.12) make it possible to postulate the form of the singular part of the density of the Helmholtz free energy in the vicinity of the critical point as a homogeneous function of t and φ. The homogeneity hypothesis does not determine the explicit form of $f(t, \varphi)$ pointing only to the asymptotic properties of this function. This problem is solved in the method of renormalization group by a successive decrease of the number of degrees of freedom by changing the scale.

5.2
Critical Configuration and its Stability

Let us determine the work of formation of an equilibrium nucleus in the vicinity of the critical point of an isotropic system in a space of arbitrary dimensionality, d. According to Eq. (5.1), at a given configuration of the field of the order parameter, $\varphi(\vec{r})$, the Helmholtz excess free energy may be written as

$$\Delta F[\varphi] = \int [\Delta f + \kappa (\Delta \varphi)^2] d^d r, \tag{5.14}$$

where

$$\Delta f = \tilde{f} - \tilde{f}_o - (\varphi - \varphi_o) h_o, \tag{5.15}$$

$$\tilde{f} = \frac{1}{2} a \varphi^2 + \frac{1}{4} b \varphi^4, \tag{5.16}$$

$$h_o = (\mu_o - \mu_s)/\rho_c = a \varphi_o + b \varphi_o^3. \tag{5.17}$$

Here, μ_s is the chemical potential on the saturation line. The subscript o refers to the initial state of a homogeneous metastable phase.

We introduce dimensionless variables

$$\Psi = \frac{\varphi}{\varphi_s}, \quad x = \left(\frac{b \varphi_s^2}{4\kappa}\right)^{1/2} r, \quad r = \frac{1}{2\xi} r, \quad H_o = \frac{4 h_o}{|a| \varphi_s}. \tag{5.18}$$

Now moving to dimensionless variables, Eq. (5.14) yields

$$\Delta F[\Psi] = \Delta F_o \int [(1 - \Psi^2)^2 - (1 - \Psi_o^2)^2 + (\Psi - \Psi_o) H_o + (\nabla \Psi)^2] d^d x, \tag{5.19}$$

where

$$\Delta F_o = \frac{\kappa^{d/2} \varphi_s^{4-d}}{b^{d/2 - 1}}, \quad H_o = 4(\Psi_o^3 - \Psi_o). \tag{5.20}$$

The equilibrium configuration of the field of the order parameter, $\Psi_*(\vec{x})$, is determined by the Euler equation

$$\frac{d^2\Psi}{dx^2} + \frac{d-1}{x}\frac{d\Psi}{dx} + 2(\Psi - \Psi^3) - \frac{1}{2}H_o = 0 \tag{5.21}$$

with the boundary conditions $\Psi \to \Psi_o$ at $x \to \infty$, $d\Psi/dx = 0$ at $x = 0$ and $x \to \infty$.

The form of the solutions of Eq. (5.21) depends on supersaturation and dimensionality of space, d. We shall restrict our consideration to small supersaturations (the region of weak metastability), when

$$h_o \ll h_{sp} \quad \text{or} \quad G_* = \frac{\Delta F[\Psi_*]}{k_B T} \gg 1. \tag{5.22}$$

In the one-dimensional case at $H_o = 0$, from Eq. (5.21), we have

$$\Psi(x) = \pm\tanh(x - x_o), \tag{5.23}$$

where x_o is the value of the coordinate x determining the position of the interface. If $H_o \neq 0$, and $d = 1$, Eq. (5.21) is reduced to

$$\left(\frac{d\Psi}{dx}\right)^2 = 2[(1-\Psi^2)^2 - (1-\Psi_o^2)^2 + (\Psi - \Psi_o)H_o] = 0. \tag{5.24}$$

The characteristic feature of this equation is that the value of the order parameter at the center of the equilibrium configuration, $\Psi_\diamond(0)$, coincides with the physically undistinguished value of Ψ at which Δf becomes equal to zero (Fig. 5.1). Even in the vicinity of the binodal, the value of $\Psi_\diamond(0)$ is much smaller than the equilibrium value of Ψ_+ determined from the condition of equality of the chemical potentials of the initial and the newly evolving phase. With increasing dimensionality of space, $\Psi_\diamond(0)$ tends to Ψ_+ and at $d \geq 3$ practically coincides with Ψ_+ (note that an exact equality $\Psi_\diamond(0) = \Psi_+$ is unattainable as Ψ_+ is the fixed point of Eq. (5.21)).

In a three-dimensional space the solution of Eq. (5.21), describing a planar interface at ($H_o = 0$), will be similar to the solution of a one-dimensional problem. At small supersaturations, the nucleus is sufficiently large, and its boundary may be considered as being quasiplanar. Substituting into Eq. (5.21) the factor $(d-1)/x$ for $(d-1)/x_*$, where x_* is the effective nucleus radius, we get [65]

$$\Psi_*(x) \simeq \pm\tanh(x - x_*) - \frac{1}{8}H_o, \tag{5.25}$$

$$x_* = \frac{8}{3|H_o|}. \tag{5.26}$$

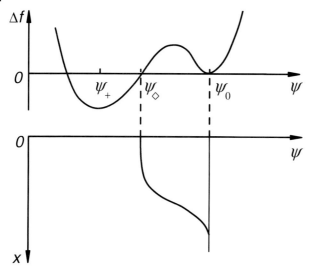

Fig. 5.1 Excess density of the Helmholtz free energy as a function of the reduced order parameter and the dependence of the order parameter on the space coordinate in a one-dimensional system.

In the case of arbitrary values of H_o, Eq. (5.21) can only be solved numerically. The algorithm of finding the distribution of the order parameter in the critical configuration based on the Runge–Kutta method is described in Ref. [103].

The stability of the equilibrium distribution of the function $\Psi_*(\vec{x})$ with respect to perturbations, $\delta\Psi(\vec{x}) = \Psi(\vec{x}) - \Psi_*(\vec{x})$, is determined by the second variation of the functional $\Delta F[\Psi]$ (Eq. (2.41), see Section 2.5). The character of relaxation of the unstable mode, $\psi_{0,0}(\vec{x})$, depends on the dimensionality of the space, which is connected with qualitative differences in the dependence of the effective potential, V_{eff}, on the coordinate $\vec{x}$. If, at $d \geq 3$, the effective potential is positive at the center of the equilibrium configuration $\Psi_*(\vec{x})$ and changes the sign while passing through the minimum in the interface region (Fig. 5.2b), in the one-dimensional case $V_{\text{eff}} < 0$ at the center of the equilibrium configuration and becomes positive at $x \to \infty$ ($\Psi \to 0$) (Fig. 5.2a). As a result of such a behavior of V_{eff}, for $d \geq 3$ the eigenfunction of the critical configuration $\psi_{0,0}(\vec{x})$ differs from zero only in the spherical layer that coincides with the interface, and the evolution of the nucleus is connected with the increase (decrease) of the radius of curvature of localization $\psi_{0,0}(\vec{x})$ (effective radius of the nucleus, x_*). In the space with $d = 1$, a similar eigenfunction is concentrated at the center of a near-critical configuration, and the nucleus growth begins with an increase in the amplitude of the field of the order parameter. Only then an increase in the effective radius, x_*, is found [346, 347].

It is easy to verify that for the same initial conditions, (T, Ψ_o), the work of formation of a one-dimensional nucleus is smaller than that of a three-

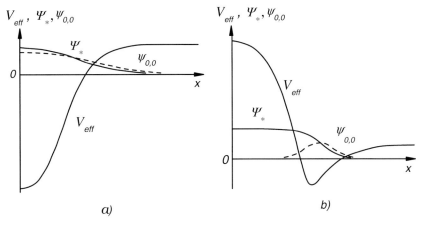

Fig. 5.2 Effective potential, V_{eff}, distribution of the order parameter in a critical nucleus, Ψ_*, and eigenfunction $\psi_{0,0}$ of the operator of stability, $\hat{L}$, for a one-dimensional (a) and a three-dimensional (b) system.

dimensional one. Even in an isotropic system a heterophase fluctuation is spherical only on average, and the field, $\Psi(\vec{x})$, will change differently in different directions from its center. Bursts of the field, $\Psi(\vec{x})$, along some randomly distinguished directions may be interpreted as quasi-low-size heterophase fluctuations. The appearance of quasi-low-size fluctuations is most probable at comparatively high degrees of metastability, when their effective size is sufficiently small, and shape fluctuations are more significant. If in a region of weak metastability the value of the order parameter, Ψ, at the center of a heterophase fluctuation is lower than the equilibrium value, such a three-dimensional fluctuation is practically always precritical and disperses. A one-dimensional fluctuation, on the contrary, may prove to be supercritical and begin to grow, at first at the expense of an increase in the amplitude of Ψ, and then in the effective radius, x_*. In the process of expansion, the convex sections of an inhomogeneity will have a smaller growth rate than the concave sections. The field $\Psi(\vec{x})$ spheroidizes, and the value of Ψ at its center tends to Ψ_+. A quasi-low-size nucleus becomes three dimensional, but when it happens, it may already prove to be supercritical [347, 348].

By definition, from Eqs. (5.19) and (5.21) for the work of formation of a critical nucleus, we have

$$W_* = \min \max \triangle F[\Psi] = \triangle F_0 Y_*[\Psi_*], \tag{5.27}$$

where

$$Y_* = 2 \int_0^\infty (\Psi_o + \Psi)(\Psi_o - \Psi)^3 d^d x. \tag{5.28}$$

The integral, Eq. (5.28), is temperature independent, increases infinitely with Ψ_0 tending to ± 1 and is equal to zero on the spinodal. According to Eqs. (5.2), (5.3), and (5.20), in the vicinity of the critical point the value of W_* is directly proportional to $t^{2-d/2}$.

The criterion of weak metastability, Eq. (5.22), already contains the condition of a small intensity of fluctuations, and considering Eqs. (5.8) and (5.20) it may be written as

$$G_* = \frac{W_*}{k_B T} = \frac{\kappa^{d/2}\varphi_s^{4-d}}{k_B T_c b^{d/2-1}} Y_* = \widetilde{Gi}^{-1} Y_* \gg 1. \tag{5.29}$$

Inequality (5.29) is always fulfilled if $r_0 \to \infty$ or $d > 4$. In these cases, the mean-field approximation becomes rigorous, and the region of strong fluctuations in the vicinity of the critical point is absent.

At $d = 3$, we have

$$G_* = \frac{\kappa^{3/2}\varphi_s Y_*}{k_B T_c b^{1/2}} = \frac{|a|\varphi_s^2 \xi^3 Y_*}{k_B T_c} = \left(\frac{t}{Gi}\right)^{1/2} Y_*. \tag{5.30}$$

Substitution of Eq. (5.25) into Eq. (5.28) and the transition to dimensionless quantities yield

$$Y_* = \frac{256\pi}{81}\left(\frac{|a|\varphi_s}{|h_0|}\right)^2 = \frac{16\pi}{9}\left(\frac{R_*}{\xi}\right)^2. \tag{5.31}$$

Using the formula for the surface tension, Eq. (3.119), and the distribution, Eq. (5.23), we get

$$\sigma = 2\int\kappa\left(\frac{d\varphi}{dz}\right)^2 dz = \frac{4}{3}\kappa^{1/2}\varphi_s^2|a|^{1/2} = \frac{4}{3}\varphi_s^2\xi|a|. \tag{5.32}$$

From Eqs. (5.27), (5.31), and (5.32) the Gibbs formula for the work of formation of a critical nucleus follows as

$$W_* = \frac{4}{3}\pi\sigma R_*^2 = \frac{4}{3}\frac{\pi\sigma^3}{\varphi_s^2|h_0|^2}. \tag{5.33}$$

According to Eqs. (5.29) and (5.30), the area of applicability of nucleation theory, which is based on the mean-field approximation, is determined both by the individual properties of a substance (Ginzburg number, Gi) and the parameters that characterize the thermodynamic state of the metastable phase (distance from the critical point t, critical-nucleus radius R_*). The ratio, R_*/ξ, is a scale invariant since the critical-nucleus radius has the same scaling behavior as the correlation radius. For regions of weak and strong fluctuations, Eqs. (5.31), (5.10), and (5.11) result in [349]

$$\frac{R_*}{\xi} = \frac{4}{3}\frac{|a|\varphi_s}{|h_0|} = \frac{4}{3}\frac{B_o\, t^{\gamma+\beta}}{\chi_0\, |h_0|} = \frac{4}{3}\frac{B_o}{\chi_0}S, \tag{5.34}$$

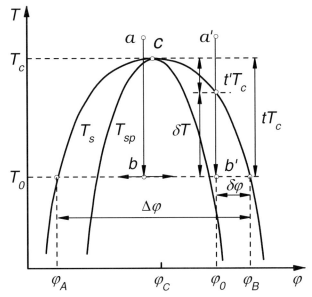

Fig. 5.3 Phase diagram in the range of first-order phase transitions. T_s: binodal; T_{sp}: spinodal; C: critical point; a, a': initial states; b, b': initial unstable and metastable states.

where S is the scaled dimensionless supersaturation. The Gibbs number, $G_* = G_*(S)$, is a universal function of the scaled dimensionless supersaturation. In experiments on nucleation kinetics, the quantity to be measured is usually not the parameter S, but the value of supercooling, $\delta T = T_s(\varphi_o) - T_o$, or supersaturation, $\delta \varphi = \varphi_s(T_o) - \varphi_o$ (Fig. 5.3). At a small depth of penetration into the metastable region, we have

$$h_o \simeq \left.\frac{\partial h}{\partial \varphi}\right|_s \delta\varphi \simeq \left.\frac{\partial h}{\partial \varphi}\right|_s \left(\frac{d\varphi_s}{dT}\right) \delta T = a\varphi_s \beta \frac{\delta T}{tT_c} \tag{5.35}$$

or

$$h_o \simeq \left.\frac{\partial h}{\partial \varphi}\right|_s \delta\varphi \simeq a\delta\varphi. \tag{5.36}$$

Substitution of Eq. (5.35) into Eqs. (5.30) and (5.31) gives

$$G_* = \frac{W_*}{k_B T_c} = \left(\frac{X_o}{X}\right)^2, \tag{5.37}$$

where

$$X = \frac{\delta T}{tT_c} \tag{5.38}$$

is the reduced supercooling

$$X_o = 2\left(\frac{\pi}{3}\right)^{1/2}\left(\frac{\sigma_o^3}{k_B T_c}\right)^{1/2}\frac{X_o}{\beta B_o^2}, \tag{5.39}$$

and σ_o is the critical amplitude of the surface tension. In writing Eqs. (5.37)–(5.39), we used the definitions of the surface tension, Eq. (5.32), and the correlation radius, Eq. (5.6).

After a similar substitution of Eq. (5.36) into Eqs. (5.30) and (5.31), we have

$$G_* = \frac{W_*}{k_B T_c} = \left(\frac{Y_o}{Y}\right)^2, \tag{5.40}$$

where

$$Y = \frac{2\delta\varphi}{\beta\varphi_s} \tag{5.41}$$

is the reduced supersaturation

$$Y_o = 4\left(\frac{\pi}{3}\right)^{1/2}\left(\frac{\sigma_o^3}{k_B T_c}\right)^{1/2}\frac{X_o}{\beta B_o^2}. \tag{5.42}$$

The factor $2/\beta$ is included in the definition of Y_o and Y for comparison with the results of Refs. [350, 351].

5.3
Steady-State Nucleation

A stationary solution of the problem of the random walk of a local inhomogeneity, $\delta\varphi(\vec{r})$, in the field of external forces given by the functional, Eq. (5.14), is obtained in Section 2.7 (Eq. (2.86)). This solution determines the resulting density of fluxes of the order-parameter field, $\varphi(\vec{r})$, via the saddle point of the functional $\Delta F_*[\varphi_*]$, i.e., the nucleation rate J, and may be written as Eq. (2.112). Since the thermodynamic potential, $F[\varphi]$, in the Landau theory simultaneously is an effective Hamiltonian, the equilibrium density of probability, P_{eq}, of appearance in a unit volume of a metastable system of a near-critical field, $\varphi(\vec{r})$, is determined by the expression [349, 352]

$$P_{eq} = \frac{\exp\left[-\frac{F_*}{k_B T}\right]\int \exp\left[-\frac{1}{2k_B T}\sum_{\vec{s}}' e_{\vec{s}} x_{\vec{s}}^2\right] D[x_{\vec{s}}]}{\exp\left[-\frac{F_0}{k_B T}\right]\int \exp\left[-\frac{1}{2k_B T}\sum_{i} e_{oi} x_{oi}^2\right] D[x_{oi}]}. \tag{5.43}$$

In the numerator, integration is performed with respect to every possible field $\varphi(\vec{r})$, in the denominator with respect to patterns corresponding to the homogeneous metastable state. The prime in the sum signifies exclusion from the summation of the unstable mode with a negative eigenvalue of e_0 ($n = 0, l = 0$). By virtue of the stability of the metastable phase with respect to continuous changes in the state variables, all eigenvalues e_{oi} are positive, and eigenfunctions are plane waves with wave vectors, $\vec{q}$. The spectrum of e_{oi} is continuous

$$e_{oi} = q^2 + \frac{1}{\kappa}\frac{\partial^2 f}{\partial \varphi^2}, \tag{5.44}$$

and in the long-wave limit $e_{oi} \simeq \xi^{-2}$ holds. The calculation of the Gaussian integrals in Eq. (5.43) [350] leads to

$$P_{eq} = \aleph \prod_{n \geq 0, l > 1} \left(\frac{2\pi k_B T}{e_{n,l}}\right)^{(2l+1)/2} \prod_i \left(\frac{e_{oi}}{2\pi k_B T}\right)^{1/2} \exp\left(-\frac{\Delta F}{k_B T}\right), \tag{5.45}$$

where

$$\aleph = \left(\frac{1}{3}\int (\nabla \varphi_*)^2 d\vec{r}\right)^{3/2} = \left(\frac{2\pi R^2 \sigma}{3\kappa}\right)^{3/2} \tag{5.46}$$

is the contribution of the translational mode.

If it is granted that large-scale inhomogeneities do not change the fluctuation spectrum of the initial phase, then from Eq. (5.45) for the average number of critical nuclei in a unit volume of the metastable phase at equilibrium conditions, we have

$$P_{eq} \simeq \left(\frac{2\pi R_*^2 \sigma}{3\kappa}\right)^{3/2} \left(\frac{\kappa}{\pi k_B T \xi^2}\right)^2 \exp\left(-\frac{\Delta F}{k_B T}\right). \tag{5.47}$$

At stationary conditions, according to Eq. (2.86), the average number of critical nuclei in a unit volume of the system is

$$P_{st} = \left(\frac{k_B T}{2\pi |e_0|}\right)^{1/2} P_{eq}(0). \tag{5.48}$$

The pre-exponential factors in the distributions P_{eq} and P_{st} are functions of the scaled dimensionless supersaturation. Using Eqs. (2.51), (5.37), (5.38), and (5.39), from Eqs. (5.47) and (5.48), we have

$$P_{st} = \frac{A}{\xi^3}\left(\frac{X_0}{X}\right)^{7/3} \exp\left[-\left(\frac{X_0}{X}\right)^2\right], \tag{5.49}$$

where $A \simeq X_0^3/(12 \cdot 3^{1/2})$. As has been shown by Gunther et al. [352], Eq. (5.49) does not agree with the theory of scale invariance at large values of X (small distances from the critical point, high supersaturations). For the elimination of this drawback, Langer and Schwartz [351] introduced into Eq. (5.49) an additional factor $(1 - X/X_0)^\phi$, where $\phi = 10/3 + 1/\delta \simeq 3.55$.

After finding the distribution function of heterophase fluctuations, P_{eq}, the calculation of the nucleation rate is reduced to determining the increase of the decrement of the critical configuration, λ_0. If, as expected, the system has one large-scale mode, an order parameter $\varphi(\vec{r})$, whose relaxation is governed by some single factor, the operator $\hat{\Gamma}$ in Eq. (2.64) contains only one kinetic coefficient. In particular, with the diffusive mechanism of the nucleus growth in the vicinity of the critical point of mixing $\Gamma_c = D_g/(\partial \mu/\partial c)_T$, where D_g is the diffusion coefficient, and from Eqs. (2.73) and (2.82) for λ_0, we have

$$\lambda_0 = \frac{2 D_g \sigma}{(\partial \mu/\partial c)_T R_*^3 (\Delta c)^2}. \tag{5.50}$$

Here, Δc is the concentration difference in the co-existing phases.

A thermodynamic system is described generally by a set of hydrodynamic modes, and the order parameter is defined as a linear combination of fields, $a_i(\vec{r}, \tau)$. The effective thermodynamic potential of the system, $E[a_i]$, is given as the Landau expansion [62] the coefficients of which are determined through the quantities being measured. The kinetic coefficients, Γ_{ij}, are found from the linearized hydrodynamic equations. This approach to describe the kinetics of nucleation in a liquid–vapor system and a binary solution separating into layers in the vicinity of the corresponding critical points was already suggested by Patashinsky and Shumilo [65, 349]. The state of a one-component system was described by the particle density fields, $\rho(\vec{r}, \tau)$, and the density of entropy, $s(\vec{r}, \tau)$. The functional $E[\rho, s]$ is the internal energy of the system. For the kinetic factor in Eq. (2.112), we get [65]

$$B = \frac{2 \Lambda k_B T^2 \rho''}{l^2 \rho'^2} \frac{1}{(R_* + L)} \left(\frac{\sigma}{k_B T}\right)^{1/2}, \tag{5.51}$$

where

$$L = \frac{1}{6\xi} \frac{\Lambda(\eta_v + \frac{4}{3}\eta)(\rho'/\rho'')^2}{dT}, \tag{5.52}$$

l is the heat of evaporation per molecule, $d = dp_s/dT$.

If $L \ll R_*$, the nucleus growth is limited by heat supply, and from Eq. (5.51), we have

$$B \equiv B_{c11} = \frac{2 \Lambda k_B T^2 \rho''}{l^2 \rho'^2 R_*} \left(\frac{\sigma}{k_B T}\right) = \frac{2 \Lambda k_B T}{l d R_*} \frac{\rho''^2}{\rho' \Delta \rho} \left(\frac{\sigma}{k_B T}\right). \tag{5.53}$$

This result differs from Eq. (3.59) by the factor $\rho''^2/\rho' \Delta \rho$.

If $\Lambda \gg R_*$, nucleation is controlled by the viscosity

$$B \equiv B_{c12} = \frac{12k_B T^3 d\xi}{l^2 \rho'^2 (\eta_v + \frac{4}{3}\eta)(\rho'/\rho'')^2} \left(\frac{\sigma}{k_B T}\right)^{1/2} \qquad (5.54)$$
$$= \frac{12k_B T \rho'' \rho'^2 \xi}{(\eta_v + \frac{4}{3}\eta)(\Delta\rho)^2} \left(\frac{\sigma}{k_B T}\right)^{1/2}.$$

Equation (3.55) in Section 3.2 corresponds to this case. When $\eta \gg \eta_v$ and $p'' = \rho'' k_B T$, Eq. (5.54) differs from Eq. (3.55) by the factor $18(\xi/R_*)(\rho'/\Delta\rho)^2$.

Langer and Turski [350] calculated the kinetic factor, B, for the cases of nondissipative nucleus growth by taking into account heat dissipation. According to these authors, viscous and inertial forces are weak in the vicinity of the liquid–vapor critical point and were neglected. When the determining factor of the nucleus growth is the thermal regime, we get

$$B \equiv B_{c21} = \frac{2}{3\pi\sqrt{3}} \left(\frac{\sigma}{k_B T}\right)^{3/2} \left(\frac{R_*}{\xi}\right)^4 \frac{\Lambda \sigma T}{l^2 R_*^3 \rho'^3}. \qquad (5.55)$$

(Note that a mistake was made in Ref. [350] in the deduction of Eq. (5.55). Kawasaki [353] was the first to mention it. He pointed out that the rate of the nucleus growth in the Langer–Turski formula did not satisfy the hypothesis of dynamic scaling. In Eq. (5.55), this inaccuracy is corrected, and a factor that fits it to the theory of scale invariance is introduced). This expression differs from both Eqs. (3.55) and (5.53) of Patashinsky and Shumilo. If energy dissipation is excluded, then

$$B \equiv B_{c22} = \frac{1}{3\pi\sqrt{3}} \left(\frac{\sigma}{k_B T}\right)^{3/2} \left(\frac{R_*}{\xi}\right)^4 \left(\frac{2\sigma\rho''}{m\rho'^2 R_*^3 \Delta\rho^2}\right)^{1/2}. \qquad (5.56)$$

The expression for the nucleation rate, Eq. (2.112), in the vicinity of the critical point may be presented in the scale-invariant form [354]

$$J = \hat{J} \cdot \tilde{J}(S) t^j, \qquad (5.57)$$

where $\hat{J}$ is the nonuniversal amplitude factor, j is the critical exponent, and $\tilde{J}$ is the universal function of scaled supersaturation. The value of j and the type of the function, $\tilde{J}(S)$, are determined by the properties of the relaxing field, i.e., whether the order parameter is conserved or not. The following equations have been obtained for a one-component liquid–vapor system in the thermodiffusion regime [354]

$$\tilde{J}(X/X_0) = \left(\frac{X}{X_0}\right)^{2/3} \left(1 + \frac{X}{X_0}\right)^\phi \exp\left[-\left(\frac{X_0}{X}\right)^2\right], \qquad (5.58)$$

$$\hat{J} = \frac{D_{T_0} X_0^6}{\zeta_0^5}, \qquad (5.59)$$

$$j = \nu(d+z). \qquad (5.60)$$

Here, z is the critical index and D_{T_0} is the critical amplitude of thermal conductivity.

5.4
Peculiarities of New Phase Formation in the Critical Region

As the critical point is approached, a region of strong fluctuations is entered where the conditions of applicability of the mean-field approximation, Eqs. (5.7) and (5.8), are violated. An expression for the density of free energy of a homogeneous system, $f(\varphi, T)$, in a region of strong fluctuations is found by the renormalization group methods [63]. However, in a metastable region this approach cannot be fully realized as a successive statistical averaging over fluctuations on all length scales inevitably excludes from consideration not fully stable (metastable) states. At low supersaturations, when the probability of appearance of a competing phase is low, one can divide fluctuations into homophase and heterophase fluctuations and perform averaging only over the first type of fluctuations. In contrast, at $R_* \simeq \xi$ such a division becomes impossible. As has been shown by Sarkies and Frenkel [355], the application of Widom's scaled equation of state [356] in calculating the work of formation of a critical nucleus yields lower values for the activation barrier than the Landau expansion Eqs. (5.15), (5.16), and (5.17). A decrease in the value of W_* is also observed when the ideas of finite-size scaling [357] are utilized in the nucleation problem. According to Ref. [357], in systems that contain interfaces of finite thickness the surface tension does not become zero at the critical point of a macroscopic system (parent phase). In this case, the Gibbs number is written as $G_* = G_{cl*} + \delta G_{fs}$, where $\delta G_{fs} < 0$.

A different approach to construct a theory of nucleation in a region of large radii of correlation of fluctuations is suggested by Binder and Stauffer [350]. These authors proceed from the cluster model of Fisher [358]. In this model, the equilibrium distribution function of clusters with respect to the number of molecules contained in them is written as

$$P_{eq}(n) = q_0 n^{-(2+1/\delta)} \exp\left[-hn - b_0 t n^{1/\beta\delta}\right], \qquad (5.61)$$

where q_0, b_0 are the dimensional individual constants. The exponent contains terms proportional to the volume and the surface of a cluster. The pre-exponential term (geometric factor) gives the most compact form of a cluster, corresponding to the minimum of its surface entropy. Clusters are believed to be noninteracting. The constants q_0 and b_0 are found from comparing experimental (p, ρ, T)-data with the equation of state of a cluster gas via $p = k_B T \sum_{n=1}^{\infty} P_{eq}(n)$.

According to Binder and Stauffer [354], the theory based on the mean-field approximation is applicable if $G_* > 100$ ("geometric region"). The droplet model should be used when $1 < G_* \leq 100$. This region is called "fluctuation-dominated." Here, the scale function of the nucleation rate has the form

$$\tilde{J} \simeq A \left(\frac{\delta T}{tT_c}\right)^{(j-1/2)(\gamma+\beta-1)} \exp\left[-X_\phi \left(\frac{\delta T}{tT_c}\right)^{-1/(\beta\delta-1)}\right], \quad (5.62)$$

where A, X_ϕ are the individual constants and $j = 2 - \alpha + \gamma + 2\nu$. At the boundary of the "geometric" and the "fluctuation-dominated" region ($G_* \simeq 50 - 70$) it is advisable to take for the Gibbs number, Eq. (5.37), retaining the pre-exponential factor of Eq. (5.62). In the vicinity of the critical point this procedure leads to higher superheats (supercoolings) than those given by the classical approach.

In the case of an unconserved order parameter, for the coefficient of diffusion of nuclei on the axis of their sizes, one can write [65]

$$D_R \sim \Gamma_n \kappa \left(\frac{Gi}{t}\right)^{1/2} \left(\frac{\xi}{R_*}\right)^2. \quad (5.63)$$

At low supersaturations ($R_* \gg \xi$), in the region of weak (Gi < t) and strong (Gi > t) fluctuations the coefficient of generalized diffusion is small. The small value of D_R and the concentration of large nuclei in the metastable phase justify the assumption of ignoring the probability of their coalescence. When $G_* \simeq 1$, the density of heterophase fluctuations is high, and the activation barrier does not limit their increase. Here, evidently, it is necessary to take into account the interaction of precritical formations even at the stage of nucleation.

McGraw and Reiss [359], McGraw [360] believe that the interaction of nuclei (effect of excluded volume) in the vicinity of the critical point has also an effect in the range of weak metastability. Regarding precritical aggregates as a gas of hard spheres, the authors [359] have shown that taking into account the effect of excluded volume leads to a renormalization of the surface tension in the expression for the work of formation of a critical nucleus, Eq. (5.33). The effective surface tension is written as

$$\sigma_{\text{eff}} = \sigma_\infty + \sigma_{\text{ts}}, \quad (5.64)$$

where σ_{ts} is the surface tension of a hard sphere system. The dependence of σ_{eff} on the curvature of the separating surface is not taken into account. The value of σ_{ts} is calculated in the framework of the Lebowitz–Helfand theory [361]. In the vicinity of the critical point, one can observe good agreement between σ_{eff} and the results of computation of the surface tension from classical homogeneous nucleation theory by data on the attainable superheating of binary solutions [362]. Heermann [363] took into account the dependence $\sigma(R_*)$ employing the Tolman formula and also obtained limiting superheatings close to those observed by experiment [362].

Owing to the inhibition of diffusional processes, a new phase nucleus growing in the vicinity of the critical point has time to interact only with its nearest surroundings, so that the actual volume, Ω, of a nucleus-parent phase system is much smaller than the total volume, V. The value of Ω depends on the time, τ, of formation of a nucleus, and the factors that determine the rate of establishment of equilibrium. In a one-component system, in the vicinity of the critical point the limiting factor is heat conduction, and for Ω one can write [364]

$$\Omega = C_\omega (D_T \tau)^{3/2}, \tag{5.65}$$

where C_ω is a coefficient of proportionality.

The work of formation of a nucleus in a system of finite volume is written as

$$W = W(R) + \Delta W(R, \Omega). \tag{5.66}$$

Here, $W(R)$ is the work of formation of a new phase in a macroscopic system $(V \to \infty)$, which is determined by Eq. (3.1), and $\Delta W(R, \Omega)$ is a correction depending on the dimensions of the nucleus of a new phase and the volume, Ω.

The character of the dependence of W on R for different values of Ω is shown in Fig 5.4. At $\tau \to \infty$, the value of Ω increases without limit, and $\Delta W \to 0$. In the vicinity of the critical point, the same interval of time τ determines a relatively small effective volume, Ω. As it is evident from Fig 5.4, if Ω is smaller than a certain lower value Ω_0, there is no maximum on the curve of the dependence $W = W(R)$. Nucleation is suppressed, and a certain time, τ_0, is required for its beginning. The relation between the expectation time, τ_0, and the value of scaled supercooling, $\delta T/tT_c$, is determined by the conditions $\delta W = 0$ and $\delta^2 W = 0$. For a one-component system, we have [364]

$$\frac{\delta T}{tT_c} = A_0 t^{-\varphi} \tau_0^{-3/8}, \tag{5.67}$$

where

$$A_0 = 2^{11/4} 3^{-3/4} (d_0 \beta)^{-1/2} B_0^{-1/2} \chi_0^{1/4} \Omega_0^{-1/4} (2D_{T_0})^{-3/8}, \tag{5.68}$$

$$\varphi = \frac{\beta}{2} + \frac{\gamma}{4} + 3\frac{\nu}{8} \simeq 0.7.$$

Here, d_0 is a universal constant of the order of unity. Taking into account the delay effect at the stage of formation of a critical nucleus leads to an increase of the limiting superheat of the liquid. However, as shown by the results of numerical evaluations, the value of such a correction is small. At a sufficiently large distance from the critical point of the classical liquid $t = 5 \cdot 10^{-4}$, the characteristic times of thermodiffusion on a length scale of the order of the critical-nucleus radius, $R_* \simeq 1.5 \cdot 10^{-7}$ m, do not exceed $8 \cdot 10^{-6}$ s.

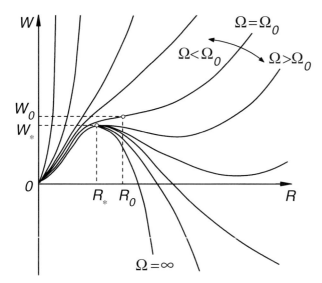

Fig. 5.4 Work of formation of a new-phase nucleus in a system of finite volume, Ω.

A decrease in the coefficient of generalized nuclei diffusion, D_R, at the approach of the critical point results in an increase of the time of establishment of the stationary nuclei size distribution. When the characteristic time of an experiment becomes comparable with the time lag, τ_l, the nucleation process is nonstationary. The importance of taking into account the phenomenon of nonsteady state effects in nucleation in the vicinity of the critical point was first noted in Ref. [365]. As is shown in Refs. [366, 367], the value of τ_l depends on the rate of transfer of a system into the metastable state. The rate of penetration into the metastable region determines the characteristic length of relaxation, l_r. Perturbations with a linear size, $R < l_r$, are in the state of equilibrium with the phase surrounding them and may be described by the Gibbs distribution. At $l_r \simeq R_*$ (slow penetration), equilibrium in a system of precritical heterophase fluctuations is rapidly achieved, and the stationary nucleation regime sets in after the time, τ_l, determined by Eq. (2.198). If $l_r \simeq \xi$ (rapid penetration), then in the time of the transfer of the system into the metastable region only heterophase fluctuations with $R < l_r$ are capable to reach equilibrium, and the formation of a stationary flow requires a certain additional time, in which a stationary distribution of heterophase fluctuations from l_r to R_* in sizes is achieved. Since the coefficient of generalized diffusion, Eq. (5.63), is inversely proportional to the square of the nucleus radius, for large nuclei this time interval may be sufficiently large. In Refs. [366, 367], when solving the nonstationary problem for the case $l_r \simeq \xi$ the authors obtained the result

$$\tau'_l = \wp_0 \tau_l, \tag{5.69}$$

where

$$\wp_0 = 2\Gamma\left(\frac{3}{4}\right)\left(\frac{k_B T}{4\pi\sigma R_*^2}\right)^{3/4} \exp\left(\frac{\pi\sigma R_*^2}{6 k_B T}\right). \tag{5.70}$$

Here, $\Gamma(x)$ is the Euler Gamma function.

For liquid xenon at $t = 5 \cdot 10^{-3}$, the surface tension equals $\sigma \simeq 10^{-5}\,\text{Nm}^{-1}$. A superheating, at which $R_* \simeq 10^{-7}$ m holds, corresponds to a stationary nucleation rate, $J = 10^7\,\text{m}^{-3}\text{s}^{-1}$. According to Eq. (2.198), the time of establishment of a stationary flux of nuclei at a slow penetration into the metastable region is $\tau_l \simeq 10^{-5}$ s. At rapid penetration, it is $\wp_0 \simeq 4 \cdot 10^6$, and from Eq. (5.69), we have $\tau'_l \simeq 40$ s.

The decay of the metastable state leads to the formation of a stable phase and, besides nucleation, includes the stage of nuclei growth. At high superheatings of a liquid away from the critical point, the growth of the vapor phase is of explosive character. The characteristic time of decay of 1 cm^3 of liquid xenon superheated at atmospheric pressure to the temperature T_n, at which the mean expectation time of the first viable nucleus corresponds to $\bar{\tau} = 1$ s, is $\tau_g \simeq 10^{-6}$ s. In the vicinity of the critical point, when $t = 5 \cdot 10^{-4}$ s, and $\bar{\tau} = 1$ s, $\tau_g \simeq 10$ s holds. The volume fraction, $\eta(\tau)$, of the vapor phase in the liquid at time, τ, is determined by Eq. (3.84). At the initial stage of decay ($\eta(\tau) \ll 1$), the interactions of growing centers can be neglected. Expanding the exponential function in Eq. (3.84) into a series, we have

$$\eta(\tau) = \int_0^\tau J(\tau')V(\tau - \tau')d\tau'. \tag{5.71}$$

Equation (5.71) is known as the free-growth approximation [117]. When the growth of the vapor phase is limited by heat conduction, we instead get

$$[R(\tau - \tau')]^2 = R_*^2 + D_T \frac{\varphi_0 - \varphi''}{\varphi_s}(\tau - \tau'). \tag{5.72}$$

A similar law of growth is observed in the case of a stratifying solution, where diffusional processes are limiting the growth. At short distances from the critical point and low supercoolings, we arrive at

$$\varphi_0 - \varphi'' = \varphi_s \frac{d\varphi_s}{dT}\delta T = \varphi_s \beta \frac{\delta T}{t T_c}. \tag{5.73}$$

Neglecting the effects of nonstationarity, from Eqs. (5.71)–(5.73), we have

$$\eta(\tau) = \frac{8\pi}{15}\left(D_T\frac{\beta\delta T}{tT_c}\right)^{3/2} J\tau^{5/2}. \tag{5.74}$$

Equation (5.74) determines the time, τ_g, required for the formation of a prescribed fraction of the vapor phase. If $\eta(\tau_g) = 1/2$, then

$$\tau_g = \left[\frac{16\pi}{15}\left(D_T\frac{\beta\delta T}{tT_c}\right)^{3/2} J\right]^{2/5}$$

$$= t^{3\nu}\left[\frac{16\pi}{15}\left(D_{T_0}\frac{\delta T}{tT_c}\right)^{3/2}\widetilde{\widetilde{J}}\left(\frac{\delta T}{tT_c}\right)\right]^{2/5}. \tag{5.75}$$

In writing Eq. (5.75), the scaling equation for the nucleation rate, Eq. (5.57) ($d = z = 3$), and the asymptotic law, $D_T = D_{T_0}t^\nu$, were used.

Taking into account in Eq. (5.71) the nonstationarity of the process of nucleation, according to Eq. (2.197), we arrive at

$$\eta(\tau) = \int_0^\tau J \exp\left(-\frac{\tau_l}{\tau'}\right) V(\tau - \tau')d\tau' = \frac{8\pi}{15}\left(D_T\frac{\beta\delta T}{tT_c}\right)^{3/2} J\tau^{5/2}\varsigma(\tau_l, \tau), \tag{5.76}$$

where

$$\varsigma(\tau_l, \tau) = \frac{5}{2}\int_0^1 \exp\left(-\frac{\tau_l}{\tau y}\right)(1-y)^{3/2}dy. \tag{5.77}$$

In the case of "rapid" ($\tau < \tau_l$) penetration into the metastable region, we have

$$\eta(\tau) \simeq \left(D_T\frac{\beta\delta T}{tT_c}\right)^{3/2} J\left(\frac{\tau^{7/2}}{\tau_l}\right)\exp\left(-\frac{\tau_l}{\tau}\right). \tag{5.78}$$

In the case of "slow" penetration, we instead obtain

$$\eta(\tau) \simeq \left(D_T\frac{\beta\delta T}{tT_c}\right)^{3/2} J\exp\left(-\frac{\tau_l}{\tau}\right). \tag{5.79}$$

Equations (5.74) and (5.76) do not take into account the change of supersaturation in the process of nuclei growth. Here, fluctuational formation of new nuclei is practically excluded, and the increase of size of large nuclei takes place at the expense of dissolution of smaller ones (coalescence process) [368]. A simplified version of the theory accounting for both the processes of formation of nuclei and their growth in the vicinity of the critical point of mixing of a binary solution was suggested by Langer and Schwartz [351]. The authors proceeded from the equation of motion for the nuclei size distribution function, $P(R, \tau)$, which is written as

$$\frac{\partial P}{\partial \tau} = -\frac{\partial}{\partial R}\left[\frac{dR}{d\tau}P\right] + j. \qquad (5.80)$$

The first term in the right-hand side of Eq. (5.80) describes nuclei growth and the second term their generation. The function $j(R)$ is given by the expression

$$J = \int_{R_*}^{\infty} j(R)dR. \qquad (5.81)$$

The integral

$$N(\tau) = \int_{R_*}^{\infty} P(R, \tau)dR \qquad (5.82)$$

determines the number of nuclei in a unit volume. Only supercritical aggregates are considered to be a part of a new phase. Taking the first-order derivative of Eq. (5.82) with respect to time yields

$$\frac{dN}{d\tau} = J - P(R_*)\frac{dR_*}{d\tau}. \qquad (5.83)$$

Multiplication of Eq. (5.80) by R with subsequent integration results in

$$\frac{d\bar{R}}{d\tau} = \left\langle \frac{dR}{d\tau} \right\rangle + \frac{1}{N}\int_{R_*}^{\infty}(R - \bar{R})j(R)dR + (\bar{R} - R_*)\frac{P(R_*)}{N}\frac{dR_*}{d\tau}, \qquad (5.84)$$

where $\bar{R}$ is the average value of the nucleus radius at the time, τ. The first term in the right-hand side of Eq. (5.84) is the average rate of deterministic growth of a nucleus. The integral in the second term may be taken by the passage method and gives

$$\frac{1}{N}\int_{R_*}^{\infty}(R - \bar{R})j(R)dR \cong \frac{J}{N}(R_* - \bar{R} + \frac{1}{2}\delta R_*). \qquad (5.85)$$

Here, δR_* is the width of the activation barrier at the height $W_* - k_B T$. The function, $P(R)$, has a sharp peak close to R_*, and is equal, in a first approximation, to

$$P(R_*) \cong \frac{Nb_0}{\bar{R} - R_*}, \qquad (5.86)$$

where the constant b_0 is determined at the coalescence stage ($\tau \to \infty$) in the framework of the Lifshitz–Slezov theory [368].

Equations (5.83) and (5.84), supplemented by the conservation law of dissolved substance (for the notations see Fig 5.3),

$$\frac{\delta\varphi_0 - \delta\varphi}{\Delta\varphi - \delta\varphi} \simeq \frac{4}{3}\pi \bar{R}^3 N, \tag{5.87}$$

form a complete system of equations for the problem under consideration. The initial conditions for this system of equations are

$$\delta\varphi = \delta\varphi_0, \quad N = 0 \quad \text{at} \quad \tau = 0. \tag{5.88}$$

After a transition to dimensionless variables

$$s = \frac{Y}{Y_+}, \quad \theta = \frac{DY_+^3 \tau}{24\zeta^2}, \quad q = \frac{\bar{R}Y_+}{2\zeta}, \quad q_* = \frac{R_* Y_+}{2\zeta} \simeq \frac{1}{s}, \quad u = \frac{64\pi \zeta^3 N}{Y_+}, \tag{5.89}$$

the temperature-dependent quantities in Eqs. (5.83), (5.84), and (5.87) disappear and one obtains

$$Y_+ - Y = uq^3, \tag{5.90}$$

$$\frac{du}{d\theta} = \tilde{J}(s) - \frac{ub_0}{q - q_*}\frac{dq_*}{d\theta}, \tag{5.91}$$

$$\frac{dq}{d\theta} = \frac{1}{q^2}\left(\frac{q}{q_*} - 1\right) + \frac{\tilde{J}(s)}{u}(a - q - q_*) + b_0\frac{dq_*}{d\theta}, \tag{5.92}$$

where $\tilde{J}(s)$ is given by Eq. (5.58) and $a \cong 0.2$ holds.

With $\tilde{J}(s) = 0$, the system of equations, Eqs. (5.90), (5.91), and (5.92), is reduced to the system of equations of the Lifshitz–Slezov coalescence theory [368], from which the asymptotic laws follow for the average nucleus radius, $\bar{R}(\tau)$ (the law of 1/3), the total number of nuclei, $N(\tau)$, and the supersaturation, $\delta c(\tau)$. Excluding from Eqs. (5.90)–(5.92) the terms responsible for coalescence, we obtain the law of free growth, Eq. (5.71). The result of solving the system of equations, Eqs. (5.90)–(5.92), depends on the value of the initial supersaturation, Y_+. For low supersaturations with a height of the activation barrier, $G_* \geq 20$, the decay of the metastable phase proceeds at a practically constant initial level of metastability as a result of growth of a small number of new-phase centers. Here, the free-growth approximation, Eq. (5.71), can be used. At high supersaturations ($G_* \leq 10$), nucleation proceeds rapidly, and the forming nuclei of a new phase immediately grow in the Lifshitz–Slezov regime ($\bar{R} \sim \tau^{1/3}$). As the two-phase state is approached, nuclei appear and disappear at a practically equal rate.

5.5
Experimental Investigations of Nucleation in the Vicinity of Critical and Tricritical Points

A metastable system in the vicinity of singular points of a thermodynamic surface of states is a convenient object for an experimental study of homogeneous nucleation kinetics. As a result of the correlation radius being large, many of the inhomogeneities and the defects on the walls of a measuring cell, and foreign inclusions whose size is smaller than ζ have no effects on the formation of a new phase. At the same time, the small width of the metastable region imposes stringent requirements on both the accuracy of the parameters measured in experiments and the data on thermodynamical properties used in calculations by formulae of the kinetic nucleation theory. Here, it is desirable to possess the totality of physical properties of the substance under investigation obtained on portions from one batch with the use of the same measuring systems. Experimental data [369, 370] testify that even in soldered ampoules a "withdrawal" of the critical temperature and some other properties of solutions can be observed in the course of time.

As objects of investigation of the kinetics of phase separation in the vicinity of critical points both one-component liquids and binary solutions are chosen. The first evidence of a discrepancy between the homogeneous nucleation theory and experiment in the critical region of a one-component liquid was obtained by Dahl and Moldover [50] in experiments on isochoric heat capacity. Systematic investigations of this phenomenon were continued in Refs. [365] and [371], respectively, for carbon dioxide and xenon. As the critical point is approached, owing to a decrease in the difference of densities of coexisting phases and the inhibition of diffusional processes, the boiling-up of a metastable liquid loses its explosive character. This feature requires different methods of registration of the moment of phase transition than with regard to the interval $T < T_c$. In Refs. [365, 371], the decay of the metastable state was registered by the thermal effect of a phase transition. As distinct from carbon dioxide, xenon possesses an enhanced radiation stability, which made it possible to realize the conditions of homogeneous nucleation in relatively large liquid volumes. The measuring system was a modified calorimeter [365] (Fig 5.5). Liquid xenon was superheated in a glass ampoule (5) ($V \simeq 0.34 \, \text{cm}^3$), which was in contact with a copper sleeve (6). The sleeve was secured inside a system of thermostating screens (1, 3). A standard platinum resistance thermometer was mounted on the inner screen (3). The temperature of the tracking screen (3) was maintained equal to the ampoule temperature with an error that did not exceed ± 1 mK. At ampoule heating rates of $(0.1–1.0) \, \text{mK} \cdot \text{s}^{-1}$, the temperature lag was less than 1 mK and reached 3 mK at $\dot{T} \simeq 5 \, \text{mK·s}^{-1}$. The temperature regime was controlled by differential copper–constantan thermocouples (shown by arrows). All the wires inserted into the measuring system had been previously brought into contact with a thermostated ring (4). The surfaces of the screens and the ring were silver plated. With the help of a copper capillary (8), the ampoule (5) is connected to a thermocompressor, which ensured the creation of the required pressure in the system and its maintenance with an error of ±50 Pa. The value of the excess pressure was measured by a standard dead-weight pressure-gauge tester. The water hammer caused by liquid boiling-up in the ampoule was reduced by a receiver (7) ($V \simeq 0.5\,\text{cm}^3$). The measuring system was mounted in a vacuum chamber, which was immersed in a Dewar flask with petroleum ether. The cooling was realized by pumping nitrogen vapors through a coil located in the Dewar flask.

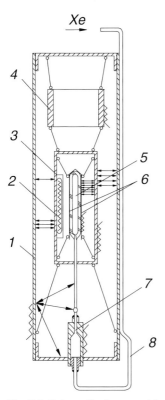

Fig. 5.5 Schematic diagram of the chamber of a setup for determining the temperature of liquid superheating in the vicinity of the critical point.

The boiling-up temperature for superheated liquid xenon was measured in the pressure range from 4.8 to 5.8 MPa, which corresponds to distances from the critical point along the saturation line $t' = 1 - T_s(p)/T_c = (0.1\text{–}5.0) \cdot 10^{-2}$. The heating rates were $(0.05\text{–}5.0)\,\text{mK}\cdot\text{s}^{-1}$. The error in measuring the boiling-up temperature did not exceed ±3 mK. From 3 to 10 boiling-up events were registered at given T and p, and the most probable boiling-up temperature of a liquid was found as the arithmetic mean. The results of determining T_b at rates $\dot{T} = 1$ mK·s^{-1} are presented in Fig 5.6. Ibidem one can see T_b as a function of the xenon heating rate. Along with experiments on isobaric liquid-xenon heating [365], experiments have been performed [372] in which the most probable boiling-up temperature, T_b, was determined in the process of an isochoric liquid stretch. Xenon condensed into glass ampoules of volume $V \simeq 0.7$ cm^3, which, after being filled, were soldered and placed in the measuring cell of a Calve calorimeter [51]. The boiling-up temperature was measured at cooling rates from 0.1 to 2.0 mK·s^{-1}. Similar experiments have earlier been made with carbon dioxide [371].

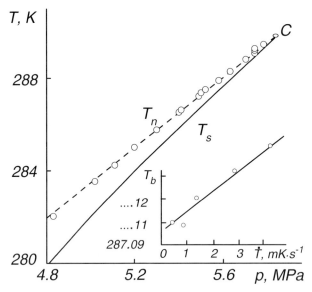

Fig. 5.6 Boiling-up temperature for liquid xenon in the vicinity of the critical point ($\dot{T} = 1$ mK·s^{-1}). T_s: binodal; T_n: temperature of attainable superheating (theory, $J = 10^7$ m^{-3}·s^{-1}). In the inset, the heating-rate dependence of the boiling-up temperature is shown ($p = 5.511$ MPa).

In the vicinity of the critical mixing point, nucleation was investigated in binary solutions methylcyclohexane-perfluoromethyl-cyclohexane (C_7H_{14}–C_7F_{14}) [362,369,373], methanol-cyclohexane [374], isobutylic acid–water (IBA–H_2O) [369,375], 2.6-lutidin–water [376–378]. A phase transition in such systems was usually registered using optical methods to detect the abrupt change in the intensity of the light passing through the measuring cell. The source of optical radiation was a He–Ne-laser, whose beam, for separating the effects of heterogeneous nucleation on the walls, was focused onto the center of an optical dish. The initial state was the two-phase state of a system or the homogeneous homophase state. As was already mentioned in Ref. [369], if cooling was realized from a one-phase region, in experiments higher supercoolings could be observed as compared with those from a two-phase region. When Havland et al. [369] took the samples of C_7H_{14}–C_7F_{14} of Heady and Cahn [362], they obtained a good agreement in phase-separation temperatures with their results in entering the metastable region from a one-phase state. Reproducibility was absent if a two-phase state was chosen as the initial state.

Depending on the type of phase diagram, a solution was transferred into a metastable state by lowering the pressure, which shifted the line of phase equilibrium, or temperature (systems with an upper critical point: IBA–H_2O, C_7H_{14}–C_7F_{14}), or by heating (systems with a lower critical point: 2.6-lutidin–water). Cooling was realized by decreasing the temperature of the thermostated bath [369], heating, by passing a current pulse through the solution [378] or via microwave radiation [374]. An optical cell with a sample was cooled (heated) continuously or employing the step method till the appearance of a phase separation signal. In a number of experiments, a combined way of entering a metastable region was chosen, i.e., pressure release to a value which was in compliance with a supercooling making up about 80% of the limiting value, with a subsequent step-by-step heating of a sample by 1–5 mK with an intermediate time lag from 0.5 to 2 min. In experiments with binary solutions, the interval from the line of phase equilibrium to the temperature of phase transition, T_b, was usually passed within several minutes, though in some experiments [369] this time was equal to hours. Cooling rates varied in the range from 1 to 100 mK·min^{-1}. In special experiments [378], studying the dependence of T_b on the speed of entering the metastable region, the value of $\dot{T}$ was as large as 2–30 mK·s^{-1}. All authors, beginning with the pioneering papers by Sundquist and Oriani [373], Heady, and Cahn [362], established that in the vicinity of the critical mixing point experimental values of $\delta T_b = T_s - T_b$ exceed theoretical values $\delta T_n = T_s - T_n$.

An object for investigating the kinetics of spontaneous nucleation, unique in its own way, are superfluid solutions of ^{4}He–^{3}He. Liquid ^{4}He changes from the normal into the superfluid state at a temperature $T_\lambda = 2.172$ K (λ-transition). The solution in ^{4}He of the light isotope ^{3}He moves the λ-point into a region of lower temperatures. Down to $T_t = 0.872$ K, the state of the solution changes continuously. A further decrease in the temperature results in a phase stratification (Fig 5.7). The lower (superfluid) phase of the solution is enriched with the isotope ^{4}He, the upper (normal) with the isotope ^{3}He. The point at which the λ-line is replaced by the line of phase transition of first order is a critical point of higher order (tricritical point).

The first investigations of the kinetics of phase transitions in metastable solutions of ^{4}He–^{3}He in the vicinity of the tricritical point were made by groups

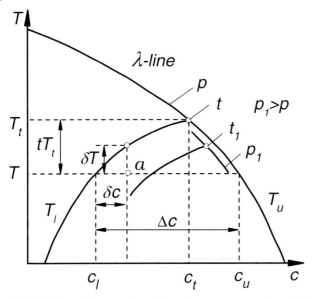

Fig. 5.7 Phase diagram of a solution at pressures p and p_1. T_l, T_u are branches of the stratification curve for the superfluid and the normal phases, respectively; t: tricritical point; a: given metastable state.

of Leiderer [379, 380] and Hoffer [381, 382]. In all experiments, the pressure dependence of the phase-separation curve (Fig 5.7) was employed to transfer the system into a metastable state. The decay of the metastable state was registered by the method of light scattering. A peculiar feature of ^{4}He–^{3}He solutions consists in the small difference in the refractive indices of the superfluid and the normal phase and, as a consequence, a small value of repeated light scattering. This feature makes it possible to use the method of light scattering not only to register the moment of phase separation, but also for studying the dynamics of growth of new-phase centers [380]. In Ref. [379], supercoolings are investigated of both the superfluid and the normal phase of a solution. The data from Refs. [381, 382] refer only to the superfluid phase. Despite the asymmetry of the curve of co-existence of the phases, the values of relative supercooling $(T_b - T_s)/(T_t - T_s)$ for the two phases of a solution coincide within the experimental error. The effective cooling rates are 150 mK·s^{-1} [379] and 13 80 mK·s^{-1} [382].

A good agreement between experimental data and the classical homogeneous nucleation theory has been observed at $t = (T_t - T_s)/T_t > 3 \cdot 10^{-2}$, as well as a considerably higher disagreement between theory and experiment in the value of supercooling compared to solutions of organic liquids when $t \to 0$.

5.6
Comparison of Theory and Experiment

The attainable superheating of liquefied gases and their solutions at pressures $p \leq 0.7 p_c$ is the subject of consideration in Chapters 3 and 4. We will now discuss the results of experiments on the kinetics of phase separation in the vicinity of both critical and tricritical points. As it is shown in Section 5.3, the general form of the equation for the stationary nucleation rate, Eq. (2.112), in the vicinity of singular points of the thermodynamic surface of states remains unchanged. As before, the dominant role is played by the exponential factor.

One or another indication of a phase transition is registered in experiments on nucleation kinetics. It may be the boiling-up of a superheated liquid, the appearance of a certain quantity of a new phase in phase-separating solutions registered by a measuring system, etc. At given values of the state variables, the characteristic sign of a phase transition may be correlated to a certain expectation time, τ_b. In the regime of continuous changes, the value of any state parameter, e.g., temperature, is registered at the moment of manifestation of the phase-transition, and it is denoted by the subscript, b. The parameters τ_b, T_b, p_b, etc. are quantities measured directly. On their basis, information can be obtained on J, T_n, p_n and so on.

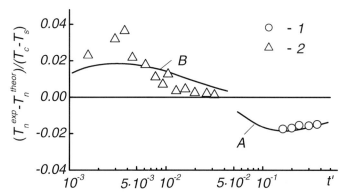

Fig. 5.8 Deviations of experimental values of the superheating temperature of liquid xenon from those calculated by the Döring–Volmer theory in a macroscopic approximation ($J = 10^7$ m$^{-3} \cdot$ s^{-1}). 1: [145]; 2: [365]; A: calculation by Eqs. (2.112) and (3.15) with $z_0 = 1$; B: calculation from Eq. (5.75) for $\tau_g = 1$ s.

In Fig 5.8, experimental values of the boiling-up temperature, T_b, for liquid xenon obtained by the method of continuous isobaric heating ($\dot{T} = 0.2$ mK·s^{-1}) are compared with the results of calculating the temperature of attainable superheating, T_n, by the Döring–Volmer theory (Eqs. (2.112) and (3.15), $z_0 = 1$) [365]. The work of formation of a critical bubble has been determined in a macroscopic approximation, $\sigma = \sigma_\infty$, with the use of the values of σ_∞ and p_s from Refs. [155, 383, 384]. The values of the reduced temperature along the boundary curve, $t' = 1 - T_s(p)/T_c$, have been assigned to the abscissa. The values of t', $t = 1 - T_n(p)/T_c$ and the relative liquid superheating, $\Delta T/tT_c = [T_n(p) - T_s(p)]/tT_c$, are connected by the relation $\Delta T/tT_c = t'/t - 1$. Parallel to the data from Ref. [365], Fig. 5.8 presents the values of the temperature of attainable superheating for xenon, T_n, obtained in experiments determining the mean lifetime [145]. The systematic underheating of liquid xenon to theoretical values of T_n at $t' > 10^{-1}$ is mainly connected (see Section 3.9) with the neglecting of the dependence of the critical bubble surface tension on the curvature of the separating surface.

In the asymptotic vicinity of the critical point, the expression for the nucleation rate, Eq. (2.112), is written as

$$J = \rho z_0 B \exp\left[-G_0\left(\frac{t^{(3/2)\mu-\beta}T_c}{\Delta T}\right)^2\right], \quad (5.93)$$

where for xenon

$$G_0 = \frac{16\pi}{3k_B T_c} \frac{\sigma_0^3}{l_0^2 \rho_c^2} \simeq 54.05. \quad (5.94)$$

The kinetic factor in the Döring–Volmer theory does not depend on superheating. For an isobaric process and $J - 10^7$ m$^{-3} \cdot$ s^{-1}, we have

$$\frac{\Delta T}{t^{(3/2)\mu-\beta}T_c} = \frac{G_0^{1/2}}{[\ln(\rho_c z_0 B/J)]^{1/2}} \simeq 1.5. \quad (5.95)$$

At $t' < 10^{-2}$, boiling-up temperatures, T_b, for liquid xenon registered in experiments are much higher than the temperature of attainable superheating, T_n, calculated from Eq. (5.75). For $t' \simeq 10^{-3}$, the value of $T_b - T_s$ is approximately two times as large as the theoretical value of $T_n - T_s$.

The relative supercooling, $\delta T/tT_c$, in an isochoric process ($J = 10^7$ m$^{-3} \cdot$ s^{-1}) is

$$\frac{\delta T}{tT_c} = \frac{G_0^{1/2}}{[\ln(\rho_c z B/J)]^{1/2}} \simeq 0.1425. \quad (5.96)$$

The results of experiments regarding the measurement of boiling-up temperatures for liquid xenon in soldered glass ampoules are given in Fig 5.9 [372]. Ibidem experimental data can be found on the supercooling of carbon dioxide [371], helium-3 [50], 2.6-lutidin–water solutions [376], methylcyclohexane–perfluoromethylcyclohexane [362] and isobutylic acid–water [369]. Systems with essentially different properties in the vicinity of critical points demonstrate an amazing agreement in the deviations of the measured values of T_b

from the temperatures of limiting supersaturation, T_n, calculated by homogeneous nucleation theory. On the logarithmic scale, these data are grouped along one line, which points to the asymptotic divergence of the ratio $(T_s - T_b)/(T_c - T_s)$ at the critical point.

As distinct from the Döring–Volmer theory, the kinetic factors B of other versions of the homogeneous nucleation theory (Sections 3.2 and 5.3) depend on the supersaturation. The most significant dependence of B on G_* is given by the Binder-Stauffer theory [354] constructed on the basis of Fisher's droplet model. The decrease of the kinetic factor with an approach to the critical point results in an increasing relative supercooling, but this fact does not eliminate the anomalous disagreement between theory and experiment.

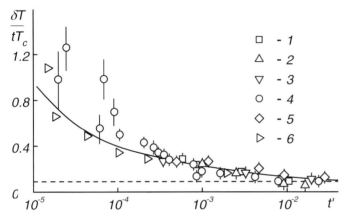

Fig. 5.9 Reduced supercoolings of liquid xenon (1), carbon dioxide (2) [371], helium-3 (3) [50], solutions 2.6-lutidin–water (4), methylcyclohexane-perfluoromethylcyclohexane (5) [362] and isobutylic acid–water (6) [369]. The *solid line* shows the calculation from Eq. (5.75) for $\tau_g = 1$ s; the *dashed line*, the Döring–Volmer theory.

If at a sufficiently large distance from the critical point liquid boiling-up takes place, due to a high rate of bubble growth the nucleation process may be identified with the appearance in it of the first viable bubble, and T_b with T_n. At $T \to T_c$, the mean expectation time of appearance of the first viable bubble, $\bar{\tau}$, is much shorter than the time of its subsequent growth, τ_g, and $T_b \neq T_n$. A characteristic parameter of phase transition in the vicinity of the critical point is the volume fraction of the vapor phase η, which in isothermal conditions is determined by Eq. (5.74). In the regime of continuous heating (cooling) at a constant rate $\dot{T} = dT/d\tau$, for the volume of a nucleus at temperature, T', one can write (evaluation from above)

$$V(T - T') = \frac{4}{3}\pi \left\langle \left(D_T \frac{\beta \delta T}{t T_c}\right)^{3/2} \right\rangle \left[\frac{T - T'}{\dot{T}}\right]^{3/2}, \qquad (5.97)$$

where the angle brackets signify the average value on the interval (T, T') [117]. In a free-growth approximation, Eq. (5.71) (isochoric process of penetration into a metastable region), we have

$$\eta(t) = \frac{4}{3}\pi \left(\frac{T_c}{|\dot{T}|}\right)^{5/2} \left\langle \left(D_T \frac{\beta\delta T}{tT_c}\right)^{3/2} \right\rangle z_0 \rho_c B \varsigma(t, t'). \tag{5.98}$$

Here

$$\varsigma(t, t') = \int_t^{t'} \exp\left[-G_0 \left(\frac{x}{x - t'}\right)^2\right] |x - t|^{3/2} dx. \tag{5.99}$$

Given the fraction of a new phase at a certain time τ_g, Eqs. (5.74) and (5.98) determine the stationary nucleation rate. In order to find J, it is necessary to know what fraction of the incipient phase corresponds to the signal of its appearance registered in the experiment. Owing to a very strong dependence of J on T, Eqs. (5.74) and (5.98) are noncritical to the value of η. The results of calculating the boiling-up temperatures of liquid xenon for $\tau_g \simeq 1$ s and $\eta \simeq 0.2$ are shown in Fig 5.8 by the line B. The moment of liquid boiling-up in an ampoule registered in an experiment approximately corresponds to such a vapor-phase fraction (determined by photomicrography of a sample). The good agreement of theory and experiment makes it possible to assert that the considered model adequately depicts the main features of the phenomenon being observed, at least to $t' \simeq 10^{-3}$. At $t' > (10^{-3}$–$10^{-4})$ and $J < 10^{17}$ m^{-3}. s^{-1}, processes of diffusional growth of new-phase centers and their coalescence are separated in time, which justifies the use of a free-growth approximation, Eq. (5.71). In the nearest vicinity of the critical point, the condition of independence of nuclei growth is violated. Experimental data for one-component liquids are not available here. In experiments with a minimum of the dissolution curve of binary solutions, distances from the critical mixing point reached $t' = 2 \cdot 10^{-5}$. Here, it is already necessary to use the Langer–Schwartz approach [351], which makes it possible to determine the value of the initial supersaturation, $Y(0) = Y_+$, given the amount of the new phase, η, formed in the time, τ_g.

The time of observation scaled according to Eq. (5.89) is a universal function of the initial supersaturation, Y_+

$$\theta_g(Y_+) = \frac{D_g Y_0^3 \tau_g t^{3\nu}}{24\tilde{\varsigma}_0^2}, \tag{5.100}$$

$$Y_+ = \left[1 - \left(1 + \frac{\delta T}{T_c - T_s}\right)^{-\beta}\right] \frac{1}{\beta Y_0}, \tag{5.101}$$

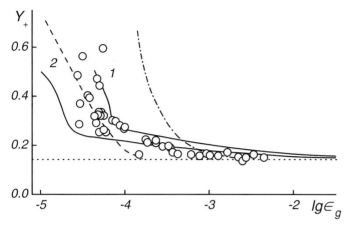

Fig. 5.10 Reduced initial supersaturation, Y_+, as a function of the reduced temperature, ϵ_g. *Dots* show experimental data for the isobutylic acid–water system [369]; 1, 2: calculation by the Langer–Schwartz theory [351] for $\tau_g = 10^2$ and 10^3 s, respectively; *dashed line*: Furukawa theory [364]; *dashed-dotted line*: theory of McGraw and Reiss [359]; *dotted line*: Döring–Volmer theory.

where Y_0 is determined by Eq. (5.42). The value of τ_g is prescribed by experimental conditions. If a new reduced temperature is introduced

$$\epsilon_g = \left(\frac{D_g Y_0^3}{24 \zeta_0^2}\right)^{1/3\nu} t, \qquad (5.102)$$

Eq. (5.100) establishes a relation between the scaled temperature, ϵ_g, and the supersaturation, Y_+, for a given time, τ_g

$$\epsilon_g^{3\nu} = \frac{\theta_g(Y_+)}{\tau_g}. \qquad (5.103)$$

In Fig 5.10, the results of calculating the dependence of Y_+ on ϵ_g in the isobutylic acid–water system for two values of time τ_g are compared with experimental data [369]. According to the evaluations of the authors of Ref. [369], at the chosen experimental conditions the characteristic times, τ_g, are 200–400 s.

The Furukawa theory [364] is also in satisfactory agreement with the experimental data [369]. Figure 5.10 shows the results of calculating Y_+ by Eq. (5.67) for $\tau_0 = 20$ s and $A_0 = 10^{-3.07}$. By considering the increase in the isothermal compressibility of a solution with supersaturation it is necessary to increase the time τ_0. If the compressibility increases by two or three times, the value of τ_0 reaches 80–1800 s, which is already close to the characteristic time intervals of Ref. [369].

Taking into account the effect of excluded volume at the nucleation stage also leads to an increase in the limiting supersaturation. In the theory of Lebowitz and Helfand [361], the effective value for the surface tension of a system of hard spheres, σ_{ts} (Eq. (5.64)), depends on the wave vector of cutting, q_*. McGraw [360] has determined the value of q_* for the isobutylic acid–water system by data on the shear viscosity ($q_*^{-1} = 4.24$ nm) and obtained considerable discrepancies between theory and experiment. These discrepancies are eliminated with $q_*^{-1} = 12.0$ nm.

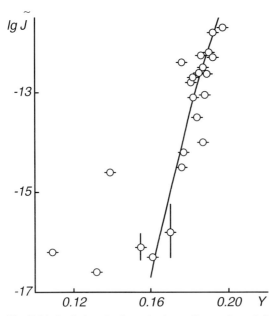

Fig. 5.11 Scaled nucleation rate depending on the relative supersaturation of an isobutylic acid–water solution [375]. The line shows results of calculation by the classical homogeneous nucleation theory.

Using the technology of two-level supercooling, which is widely applied in thermodiffusion chambers for studying the condensation of supersaturated vapors, Siebert and Knobler [375] in experiments with isobutylic acid–water solutions separated the nucleation stage from the stage of nuclei growth. A solution was supercooled down to a certain temperature, T_0 ($\delta T = 10 - 25$ mK) and held at that level for a certain time τ ($\tau = 5$–300 s), within which the formation and the accumulation of nuclei took place. The initial supersaturation in this case practically did not change ($Y \simeq Y_+$). The number of new-phase nuclei forming in a unit volume in the time τ is determined by

$$N = J(Y)\tau. \tag{5.104}$$

On expiry of the time τ, the solution temperature increased to $T_1 < T_s$ (in this case the origination of new centers practically did not stop) and only centers of size $\bar{R} > R_*(T_1)$ proved to be viable. Within 5–10 min they grew up to sizes at which they could be detected and counted with the help of a microscope ($N = 5$–1000). The dependence of the nucleation rate on the value of relative supersaturation obtained from Eq. (5.104) is presented in Fig. 5.11. Ibidem one can see the results of calculating J by homogeneous nucleation theory (Langer–Turski version). In the investigated range of nucleation rates the Gibbs number has values of the order $G_* = (31$–$50)$. For the temperatures T_n obtained from the homogeneous nucleation theory to coincide with T_b, it is necessary to divide the Gibbs number by 15!

The growth of new-phase centers in the vicinity of evaporation and mixing critical points was investigated in Refs. [377, 385]. The experiments consisted in determining the average radii of nuclei at the last stages of their growth, finding the average density of new-phase centers and some other parameters which could be evaluated from the theories of Langer–Schwartz [351] and Lifshitz–Slezov [368]. The data obtained point to the fact that, in the vicinity of

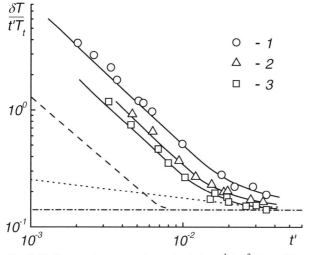

Fig. 5.12 Reduced supercoolings of solutions ^{4}He–^{3}He at different cooling rates [382]. 1: $\dot{T} = 80$ mK·s^{-1}; 2: 21; 3: 13. A *dashed line* shows data for the 2.6-lutidin–water system [378], $\dot{T} = 30$ mK·s^{-1}; a *dotted line* shows the same dependence [376] with $\dot{T} = 1$ mK·s^{-1}; a *dashed-dotted line* represents results of the classical homogeneous nucleation theory.

the critical point, the discrepancies between the temperature of homogeneous nucleation, T_n, and the temperature of phase transition, T_b, are not linked with the peculiarities of nucleation, but are consequence of the critical inhibition of the growth rate of the nuclei.

Peculiarities in the kinetics of near-critical phase separation similar to those of classical liquids and solutions are also demonstrated by quantum solutions of ^{4}He–^{3}He in the vicinity of the tricritical point (Fig 5.12). At $t' = [T_t(p) - T_s]/T_t(p) < 0.02$, relative supercoolings of the superfluid solution phase are described by the simple power law $\delta T/t'T_t = at^b$, where $t = [T_t(p) - T]/T_t(p)$ [382]. The value of the coefficient, a, depends on the cooling rate increasing with it. The superscript, b, in a first approximation does not depend on $\dot{T}$ and is equal to $\simeq 1.28$. For $t' > 0.012$, the value of $\delta T/t'T_t$ at all cooling rates is approximately constant and close to that calculated from homogeneous nucleation theory. As distinct from classical solutions, quantum solutions demonstrate a stronger disagreement in the values of T_b and T_n as the tricritical point is approached, which may be a specific feature of the tricritical behavior. Being a higher order critical point, the tricritical point has a boundary dimension $d_t = 3$. The latter statement means that, with an accuracy of logarithmic corrections, the tricritical phenomenon may be described even in the framework of the mean-field theory, i.e., the tricritical behavior is almost classical.

In the vicinity of the tricritical point, the value of $\delta c/\Delta c$ is related to the supercooling, δT, and the distance from T_t by the expression

$$\frac{\delta c}{\Delta c} = 0.818 \frac{\delta T}{\delta T + t'T_t}. \tag{5.105}$$

According to the Furukawa theory [364], which takes into account the effect of crowded state, the value of supercooling δT, and consequently of $\delta c/\Delta c$, is determined by Eq. (5.67), where the index ϕ is equal to 1.125 (tricritical indices $\beta = \gamma = \nu = 1$), whereas in solutions of organic liquids, $\phi \simeq 0.70$. The tricritical index, $\phi = 1.125$, is close to the value $b = 1.28$ obtained from experimental data. The results of experiments on the kinetics of phase separation in solutions of ^{4}He–^{3}He [381, 382] are in good agreement with the Langer–Schwartz theory [351] if $\tau_g = (0.5 \pm 0.1)$ s, which is close to the characteristic times of the experiment.

Bodensohn et al. [380] investigated the dependence of the number density of normal-component centers in the superfluid phase of solutions ^{4}He–^{3}He on the rate of the transfer of the system into the metastable state. Experiments were performed in the vicinity of the tricritical point ($t' = 0.02$) at cooling rates $dt/d\tau$ ranging from $2 \cdot 10^{-2}$ to $3 \cdot 10^{-1}$ s^{-1}. An increase in the cooling rate results in an increasing number of new-phase centers, N. According to the measurements made, $N \sim (dt/d\tau)^{1.55 \pm 0.1}$. Numerical integration of the system of equations, Eqs. (5.90), (5.91), and (5.92). yields a similar power law for the number of new-phase centers in a unit volume. Here, the exponent is equal to $\simeq 1.5$.

5.7
Nucleation in the Vicinity of a Spinodal Curve

As a result of an unlimited increase of the susceptibility with an approach to the boundary of essential instability, the spinodal curve, here, similar to states in the vicinity of the critical point, the intensity and the linear dimension of fluctuations of the thermodynamic parameters of the system are anomalously large. Let us assume that nonequilibrium states of a system in the vicinity of the spinodal are fully determined by only one hydrodynamic mode, the order-parameter field, $\varphi(\vec{r},\tau)$. For definiteness we suppose that $\varphi(\vec{r},\tau)$ is the density $\rho(\vec{r},\tau)$, thus the initial metastable phase is a one-component liquid. The relaxation of the field, $\rho(\vec{r},\tau)$, to the stable state is connected with the formation and growth of nuclei. In the vicinity of the spinodal, the energy barrier separating the metastable from the stable phase is small, and to describe nucleation it will be sufficient to consider only density fluctuations for which $|\rho - \rho_0| \sim |\rho_{sp} - \rho_0|$ (ρ_0 is the density of a homogeneous metastable phase). The change of the Helmholtz free energy, $\Delta F[\rho]$, connected with such a fluctuation is determined by Eq. (2.36), where we can limit ourselves to cubic terms of the expansion of Δf into a series in powers of $\rho - \rho_0$ [8, 58]

$$\Delta f = -\frac{1}{6} f''' \left[3(\rho_{sp} - \rho_0)(\rho - \rho_0)^2 - (\rho - \rho_0)^3 \right]. \tag{5.106}$$

Now moving to dimensionless variables

$$x = \frac{\rho - \rho_0}{2(\rho_{sp} - \rho_0)}, \tag{5.107}$$

we obtain

$$x = \left(\frac{|f'''||\rho_{sp} - \rho_0|}{8\kappa} \right)^{1/2}, \quad r = \frac{r}{R_0}, \tag{5.108}$$

where R_0 is a certain characteristic linear scale of a heterophase fluctuation.

In a space of dimensionality, d, the nonequilibrium functional, Eq. (2.36), with allowance for Eqs. (5.106)–(5.108), is written as

$$\Delta F[\rho] = \Delta F_0 \int_0^\infty \left[4x^2 - \frac{8}{3}x^3 + (\nabla x)^2 \right] d^d x = \Delta F_0 Y[x], \tag{5.109}$$

where

$$\Delta F_0 = 2^{\frac{3}{2}d-1} \kappa^{d/2} |-f'''|^{1-d/2} |\rho_{sp} - \rho_0|^{3-d/2}. \tag{5.110}$$

If pressure p is presented as a series in powers of density differences, $\rho_{sp} - \rho_0$, i.e., as

$$p - p_{sp} = \frac{1}{2} \rho_{sp} |-f'''| (\rho_{sp} - \rho_0)^2, \tag{5.111}$$

then from Eq. (5.110) for ΔF_0, we have

$$\Delta F_0 = 2^{1/2+5d/4}\kappa^{d/2}|-f'''|^{-1/2-d/4}\rho_{sp}^{-3/2+d/4}|p - p_{sp}|^{3/2-d/4}. \qquad (5.112)$$

The problem of finding the critical configuration of the field, $\rho(\vec{r})$, is reduced to finding the "lowest," and consequently the most accessible for the system, passage point of the hypersurface, $\Delta F[\rho]$. The Euler equation of the variational problem, $\delta\Delta F[\chi]/\delta\chi|_{\chi_*} = 0$, for spherically symmetric nuclei is written as

$$\chi'' + \frac{d-1}{x}\chi' - 4\chi + 4\chi^2 = 0, \qquad (5.113)$$

$$\chi' = 0 \quad \text{at} \quad x \to 0 \quad \text{and} \quad x \to \infty; \qquad \chi \to 0 \quad \text{at} \quad x \to \infty. \qquad (5.114)$$

Since Eq. (5.113) does not contain any additional variables except $\chi(x)$, at fixed d its solution is a certain universal function. $\chi_*(x)$ (Fig. 5.13).

The work of formation of a critical nucleus is

$$W_* = \Delta F_* = \min\max \Delta F[\chi] = \Delta F_0 \min\max Y\{\chi\} = \Delta F_0 Y_*, \qquad (5.115)$$

where

$$Y_* = \frac{4}{3}\int \chi^3 d^d x \qquad (5.116)$$

is a constant depending only on the space dimensionality. In the one-dimensional case, Eq. (5.113) has an analytic solution [346, 386]. For $d = 3$, numerical methods were used to obtain $I_* = 43.66$; $\chi_*(x = 0) = 4.19$; $x_{1/2}(\chi = \chi_*/2) = 0.61$ (Refs. [8, 58, 386] give results that differ from this one). The distribution of the reduced density in a critical heterophase fluctuation,

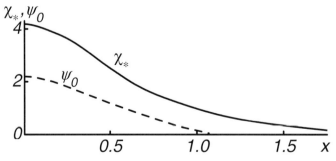

Fig. 5.13 Distribution of the reduced density in a critical heterophase fluctuation, χ_*, and the eigenfunction, ψ_0, of the stability operator, $\hat{L}$, in the vicinity of the spinodal curve.

χ_*, and the eigenfunction, ψ_0, of the stability operator, $\hat{L}$, in the vicinity of the spinodal curve are shown in Fig. 5.13.

From Eqs. (5.107), (5.108), (5.110), (5.112), and (5.115) it follows that, while the characteristic linear dimension of the critical nucleus, $R_0 = r_{1/2}$, increases without limit ($R_0 \sim |\rho_{sp} - \rho_0|^{-1/2} \sim |p_0 - p_{sp}|^{-1/4}$), the work of formation of a nucleus, W_*, and the density at its center, $\rho(r = 0)$, decrease, respectively, as $|\rho_{sp} - \rho_0|^{3-d/2} \sim |p_0 - p_{sp}|^{3/2-d/4}$ and $|\rho_{sp} - \rho_0| \sim |p_0 - p_{sp}|^{1/2}$ [8, 58].

The consideration presented here assumes that fluctuations in the metastable phase are small, i.e., the amplitude of density fluctuations, $\delta\rho = \rho - \rho_0$, in a region with a linear scale of the order of the correlation radius, ζ, is small compared with the value of $|\rho_{sp} - \rho_0|$. The criterion of smallness of fluctuations (Ginzburg criterion), Eq. (5.8), will be written here as [386, 387]

$$\frac{(\rho_{sp} - \rho_0)^2}{r_0^{-2}\zeta^{2-d}v_m^{d-4}} \gg 1. \tag{5.117}$$

In accordance with the determination of the correlation length, Eq. (5.6), in the vicinity of the spinodal, we have

$$\zeta^2 \simeq \frac{\kappa}{|-f'''||\rho_{sp} - \rho_0|}. \tag{5.118}$$

Substitution of Eq. (5.118) into Eq. (5.117) gives

$$\frac{\kappa^{d/2}|-f'''|^{1-d/2}|\rho_{sp} - \rho_0|^{3-d/2}}{k_B T} \gg 1. \tag{5.119}$$

The numerator in Eq. (5.119) coincides with an accuracy of the numerical coefficient with the expression for ΔF_0 (see Eq. (5.110)).

Criterion (5.119) is always fulfilled if $d > 6$, or if the interaction radius $r_0 \to \infty$. In these cases the mean-field approximation becomes true. When fluctuations are suppressed, the work of formation of a critical nucleus and, consequently, the lifetime of the metastable phase are infinitely large [24]. For the spinodal and for the critical point there exists a threshold dimension, $d_{sp} = 6$ ($d_{sp} > d_c = 4$), above which the mean-field approximation is always correct.

According to Fig. 5.13, in the vicinity of the spinodal the density distribution in a critical nucleus is different from a similar distribution in a region of weak metastability. If in the latter case a critical nucleus is a certain compact formation (bubble, drop) the volume V of which is connected with its linear dimension, R_*, by the relation $V \sim R_*^d$, and the density at the center is close to the equilibrium value determined from the condition of equilibrium of macroscopic phases, then in the vicinity of the spinodal a critical nucleus, even at the center, contains no stable phase and wholly consists of a transition layer. The continuums approach examined here describes a certain averaged structure of a heterophase fluctuation. Investigations of nucleation in three-dimensional Ising models by the Monte–Carlo method have shown that at a small amplitude and to a large extent the critical configuration has a ramified structure of a percolation cluster, whose fractal dimension d_f is smaller than the space dimension d, i.e., $V \sim R_*^{d_f}$, where $d_f < d$ [388–391].

Let us now examine the evolution of a near-critical configuration [386]. The corresponding relaxation process is described by an equation of the type of Eq. (2.64), in which $a_i = \rho(\vec{r}, \tau)$. We assume that the relaxing parameter is not conserved. We introduce the dimensionless time

$$\theta = \frac{\tau}{(4/\Gamma_n| - f'''||\rho_{sp} - \rho_0|)} = \frac{\tau}{\tau_0}. \tag{5.120}$$

Then from Eqs. (2.64), (5.107), (5.108), (5.109), and (5.120) it follows that

$$\frac{\partial \chi}{\partial \theta} = -\frac{\delta Y[\chi]}{\delta \chi} + g(\vec{x}, \theta), \tag{5.121}$$

where $g(\vec{x}, \theta)$ is a dimensionless random force imitating thermal fluctuations.

By linearizing Eq. (5.121) with respect to small deviations $\phi(\vec{x}, \tau) = \chi(\vec{x}, \theta) - \chi_*(\vec{x})$, we get

$$\frac{\partial \phi}{\partial \theta} = \Delta \phi - 4\phi + 8\chi_* \phi + g(\vec{x}, \theta). \tag{5.122}$$

We write $\phi(\vec{x}, \theta)$ as

$$\phi(\vec{x}, \theta) = \sum_n a_n(\theta) \psi_n(\vec{x}). \tag{5.123}$$

Substitution of Eq. (5.123) into Eq. (5.122) gives the equations for the expansion amplitudes

$$\frac{da_n}{d\theta} = \lambda_n a_n + g_n(\theta) \tag{5.124}$$

and eigenfunctions

$$\left[\frac{d^2}{dx^2} + \frac{2}{x} \frac{d}{dx} - 4 + 8\chi_*(x) \right] \psi_n(x) = \lambda_n \psi_n(x). \tag{5.125}$$

Here, $g_n(\theta)$ is the corresponding harmonics of $g(\vec{x}, \theta)$, λ_n are the eigenvalues of the linear operator

$$\hat{L} = [\Delta - 4 + 8\chi_*(x)]. \tag{5.126}$$

The operator $\hat{L}$ has a finite number of discrete eigenvalues. The eigenfunctions, $\psi_n(\vec{x})$, corresponding to $\lambda_n > 0$, describe perturbations of the extreme configuration, $\chi_*(x)$, which respect to which it is stable. The harmonics $\psi_1(\vec{x})$ with $\lambda_1 = 0$ represents a translational mode. The instability of the extreme configuration is connected with the harmonics $\psi_0(\vec{x})$, which is the solution of Eq. (5.125) at the only possible negative eigenvalue of $\lambda_0 < 0$. By means of a numerical solution of Eq. (5.125) with the boundary conditions $\chi_0 \to 0$ at $x \to \infty$ and $\chi_0 x = 0$ at $x \to 0$, we have obtained $\lambda_0 = -9.392$. The eigenfunction $\psi_0(x)$ is shown in Fig. 5.13. As distinct from the case of weak metastability, where in a space of dimension $d \geq 3$ the eigenfunction with $\lambda_0 < 0$ is equal to zero at the nucleus center and is practically localized only in the spherical layer coinciding with its interface (see Fig. 5.2b), a similar eigenfunction in the vicinity of the spinodal is mainly concentrated at the nucleus center, as it is in the one-dimensional case (see Fig. 5.2a). Thus, the evolution of near-critical configurations in the vicinity of the spinodal is reduced to the change of their amplitudes, and not of the effective radii.

The fluctuational growth of precritical configurations of the field, $\chi(x)$, may be regarded as a diffusional process along the phase axis of amplitude c_0 of the unstable mode [386]

$$\frac{da_0}{d\theta} = \lambda_0 a_0 + g_0(\theta). \tag{5.127}$$

Here, the first term on the right-hand side determines the rate of deterministic growth, and the second one accounts for the contribution of fluctuations.

Just as it was done in describing the relaxation of weakly metastable states, a Fokker–Planck equation can be formulated for the distribution function of configurations of the field, $\chi(\vec{x}, \theta)$, by the amplitude, a_0

$$\frac{\partial P}{\partial \theta} = -\frac{\partial}{\partial a_0}(\lambda_0 a_0 P) + D\frac{\partial^2 P}{\partial a_0^2}. \tag{5.128}$$

The stationary flux of nuclei through the saddle point of the functional, Eq. (5.109), along the axis of a_0 into the region $a_0 > 0$ is

$$J = \lambda_0 a_0 P - D\frac{dP}{da_0}. \tag{5.129}$$

At $c_0 \to +\infty$, we have $P(a_0) \to 0$. The values of $a_0 < 0$ correspond to configurations with precritical amplitudes. If we demand that for such values of a_0 the distribution $P(a_0)$ should coincide with the equilibrium distribution, $P_{eq}(a_0)$, then from Eq. (5.129) for the nucleation rate we get

$$J = C_0 \left(\frac{|\lambda_0|k_B T}{2\pi \Delta F_0}\right)^{1/2} \exp\left(-\frac{W_*}{k_B T}\right) \tag{5.130}$$

or in dimensional units ($d = 3$)

$$J = C_0 2^{-7/4} \Gamma_n \kappa^{-9/4} |-f'''|^3 (\rho_{sp} - \rho_0)^{7/4} \left(\frac{|\lambda_0|k_B T}{2^{14}\pi}\right)^{1/2} \exp\left(-\frac{W_*}{k_B T}\right). \tag{5.131}$$

A metastable phase can be considered as a statistically defined state only if the characteristic time of its existence, $\bar{\tau} = (JV)^{-1}$, is much larger than the time of establishment of a local equilibrium, $\tau_0 \simeq (\Gamma_n| - f'''||\rho_{sp} - \rho_0|)^{-1}$. According to Eq. (5.131), the time $\bar{\tau}$ is exponentially large compared to τ_0 at $W_*/k_B T \gg 1$. Since the value of W_* differs from that of ΔF determining the range of applicability of the approximation of weak fluctuations, Eq. (5.119), by a sufficiently large factor Y_*, the criterion of applicability of the mean-field theory in the vicinity of the spinodal for the description should be written as

$$\frac{\Delta F_0}{k_B T} Y_* = \frac{W_*}{k_B T} \gg 1. \qquad (5.132)$$

In the nearest vicinity of the spinodal, the theory under consideration is not applicable. The reason is not the strengthening of the interactions between fluctuations of scale ξ, but the short time of formation of a critical configuration. When the height of the activation barrier, W_*, is comparable with the average energy of thermal motion per degree of freedom, it makes no sense to speak about a nucleation process. The authors of Ref. [386] determine the boundary of stability of a fluctuating system employing the condition $W_*/k_B T = 1$. They call this boundary the physical spinodal. Along this line, the relation

$$|\rho_{sp} - \rho_{sp}^{\phi}| = \left(\frac{k_B T}{2^{3d/2-1} \kappa^{d/2} |-f'''|^{1-d/2} Y_*} \right)^{\frac{1}{3-d/2}} \qquad (5.133)$$

holds. In the region $\rho_{sp}^{\phi} > \rho > \rho_{sp}$, the initial homogeneous state of a liquid, stable with respect to long wave-length fluctuations, transforms into a heterophase state under the action of fluctuations of the order of ξ in a time comparable to the time of establishment of local equilibrium.

The approach considered presupposes that the resultant flux of nuclei goes through the saddle point of the surface, $F[\rho]$. With an approach to the spinodal, both the characteristic size of the critical configuration of the field, $\rho(\vec{r}, \tau)$, and the time of its formation increase. It is quite possible that under these conditions a system leaves the state of metastable equilibrium not through the configuration that is the most advantageous energetically, but by means of formation of the configuration to which the minimum time of its formation corresponds. Such a configuration may be more compact, and correspond to a point along the ridge of the thermodynamic potential surface, $\Delta F[\rho]$, with a higher value of the activation energy than at the saddle point.

5.8
Theory of Spinodal Decomposition

A first-order phase transition implies the existence of a region on the thermodynamic surface where the homogeneous state of a substance is absolutely unstable. This region is bounded by a spinodal (see Fig. 5.3). In passing through the spinodal of a binary (one-component) system the susceptibility, $\chi_T \sim (\partial c/\partial \mu)_T$ (isobaric heat capacity, $c_p \sim (\partial S/\partial T)_p$), becomes negative, and consequently also the diffusion coefficient, $D_g \sim \chi_T^{-1}$ (thermal diffusivity, $D_T \sim c_p^{-1}$) behaves similarly. Formally it is equivalent to changing the sign of time in the equation of diffusion (thermal diffusivity) and leads to the appearance of solutions with exponentially increasing concentration (temperature) differences. A substance that is brought into the unstable state when relaxing, loses space homogeneity rapidly. At the initial stage of relaxation, a peculiar grain-cell (modulated) structure forms. Owing to thermodynamic instability, such structurization is called spinodal decomposition.

The initial stage of spinodal decomposition in a one-component system was first considered by Zeldovich and Todes [69]. A phenomenological theory of formation of space-periodic structures in binary melts was developed by Cahn [392–394]. Three stages can be distinguished regarding the kinetics of evolution of unstable states: the initial, intermediate, and completing stages. The relaxation rate at the last two stages, when fragments of phases form from a continuous structure and their growth begins, depends on many factors [395–399].

We shall examine the kinetics of the process of restoration of equilibrium in a system which at a time, $\tau = 0$, is instantaneously transferred from the stable state (see Fig. 5.3, point a) into a region where the homogeneous state of the initial phase is absolutely unstable (point b). Hereafter, unless otherwise stated, we assume the existence in the system of only one large-scale mode, the order-parameter field, for which for sake of definiteness the concentration field, $c(\vec{r}, \tau)$, will be chosen. The instantaneous penetration into the region of unstable states means that no essential system rearrangement takes place in the time of the transfer, variations in the order parameter remain at the level of thermal noise.

The relaxation of the unstable state to equilibrium is described by Eq. (2.64). The realization of the spinodal decay is possible only in systems with a conserved order parameter. In the vicinity of the critical point, the thermodynamic potential, $F[c]$, is written as given by Eq. (2.33). Substitution of Eq. (2.33) into Eq. (2.64) yields

$$\frac{\partial c(\vec{r},\tau)}{\partial \tau} = \Gamma_c \nabla^2 \left(\frac{\partial \Delta f}{\partial c} - 2\kappa \nabla^2 c + \right) + \zeta(\vec{r},\tau). \tag{5.134}$$

Expanding the derivative, $\partial\Delta f/\partial c$, into a Taylor series in terms of deviations $\delta c(\vec{r},\tau) = c(\vec{r},\tau) - c_0$, where c_0 is the concentration of the homogeneous unstable system at the time $\tau = 0$, we have

$$\frac{\partial\Delta f}{\partial c} = \left(\frac{\partial\Delta f}{\partial c}\right)_{c_0} + \left(\frac{\partial^2\Delta f}{\partial c^2}\right)_{c_0}\delta c + \sum_{n=3}^{\infty}\frac{1}{(n-1)!}\left(\frac{\partial^n\Delta f}{\partial c^n}\right)_{c_0}(\delta c)^{n-1}. \quad (5.135)$$

Limiting ourselves to terms linear in $\delta c(\vec{r},\tau)$, after substitution of Eq. (5.135) into Eq. (5.134) and transition to the Fourier transform of the correlation function of fluctuations of the order parameter

$$S(\vec{q},\tau) \equiv \int G(\vec{r},\tau)\exp(i\vec{q}\vec{r})d^d r, \quad (5.136)$$

where

$$G(\vec{r},\tau) \equiv \langle\delta c(\vec{r}'+\vec{r},\tau)\delta c(\vec{r}',\tau)\rangle, \quad (5.137)$$

we get

$$\frac{\partial S(\vec{q},\tau)}{\partial\tau} = 2\omega(\vec{q})S(\vec{q},\tau) + 2k_B T\Gamma_c q^2. \quad (5.138)$$

Here, the amplification factor is given by

$$\omega(\vec{q}) = -\Gamma_c q^2\left[\left(\frac{\partial^2\Delta f}{\partial c^2}\right)_{c_0} + 2\kappa q^2\right]. \quad (5.139)$$

The integration of Eq. (5.138) results in

$$S(q,\tau) = S(q,0)\exp[2\omega(q)\tau] + S(q,\infty)\{1 - \exp[2\omega(q)\tau]\}. \quad (5.140)$$

The value of $S(q,0)$ is given by the initial conditions. At $\tau \to \infty$, we have

$$S(q,\infty) = \frac{-k_B T\Gamma_c q^2}{\omega(q)}. \quad (5.141)$$

Equations (5.138) and (5.139) are the final result of the linear theory of spinodal decomposition of Cahn and Cook [394, 400, 401].

In the stable and the metastable regions, $(\partial^2\Delta f/\partial^2)_T > 0$, and for all wavelengths, $\lambda = 2\pi/q$, the amplification factor in Eq. (5.139) is negative (Fig. 5.14). Inside the spinodal, the value of $\omega(q)$ is positive if

$$q < q_c = \sqrt{-\frac{1}{2\kappa}\left(\frac{\partial^2\Delta f}{\partial c^2}\right)_{c_0}}. \quad (5.142)$$

The homogeneous state proves to be unstable with respect to perturbations with wavelengths, $\lambda > \lambda_c = 2\pi/q_c$. Equation (5.139) reflects the competition between the forces leading at small q to phase separation (the term in square brackets) and the factor q^2 responsible for the diffusion and impeding the increase of long-wavelength perturbations. As a result, perturbations with $q_m = q_c/2^{1/2}$ grow most rapidly (see Fig. 5.14), which leads to the formation of a cell structure with a characteristic linear scale, $l \simeq 2\pi/q_m$.

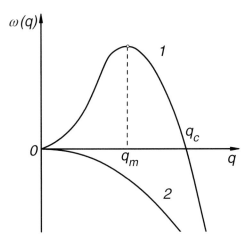

Fig. 5.14 Amplification factor in the unstable (1) and the metastable (2) regions.

In the first papers dealing with the theory of spinodal decomposition [392–394], spontaneous fluctuations were not taken into account (the Cahn linear theory). Then

$$S(\vec{q}, \tau) = S(\vec{q}, 0) \exp\left[2\omega(\vec{q})\tau\right]. \tag{5.143}$$

From Eq. (5.143) it follows that in the region of unstable states fluctuations die out completely if $q > q_c$ and increase exponentially if $q < q_c$. If the fluctuation noise is taken into account, the increase of $S(\vec{q}, \tau)$ (Eq. (5.140)) will take place not only at $\omega(q) > 0$, but also at $\omega(q) < 0$ if $S(q, 0) < S(q, \infty)$. At $\tau = 0$, the critical value of the wave number determined by the condition $\partial S(q, \tau)/\partial \tau = 0$ is

$$q_c^* = \left\{\frac{1}{2\kappa S(q,0)} \left[k_B T - \left(\frac{\partial^2 \Delta f}{\partial c^2}\right)_{c_0} S(q,0)\right]\right\}^{1/2}. \tag{5.144}$$

From Eq. (5.144) it follows that $q_c^* > q_c$. The maximum of $S(q_c^*, \tau)$ is shifted toward small q, and at $\tau \to \infty$ the value of $q_c(\tau)$ tends to q_c [402]. The time at which $q_c^*(\tau)$ reaches the value of $q = q_c$ is equal to

$$\tau_c \simeq 2\kappa S(q,0) \left\{\Gamma_c \left(\frac{\partial^2 \Delta f}{\partial c^2}\right)_{c_0} \left[k_B T - \left(\frac{\partial^2 \Delta f}{\partial c^2}\right)_{c_0} S(q,0)\right]\right\}^{-1}. \tag{5.145}$$

For $\tau < \tau_c$, the Cahn–Cook linear theory presents the dependence of $S(q,\tau)$ on q and τ qualitatively and quantitatively correctly, predicting, however, an unlimited growth of fluctuations for $q < q_c$ and $\tau \to \infty$. When the deviations $(c(\vec{r},\tau) - c_0)$ cease to be small, in the expansion of Eq. (5.135) one cannot restrict oneself only to terms that are linear in $\delta c(\vec{r},\tau)$. Fluctuations corresponding to different q's become statistically dependent, and the problem turns essentially nonlinear. The linear theory of spinodal decomposition is based on the mean-field approximation. The criterion of applicability of the linear theory by Cahn [387, 402]

$$\langle [\delta(\vec{r},\tau)]^2 \rangle_l \ll (c - c_{sp})^2 \tag{5.146}$$

is similar to that used in describing nucleation in the vicinity of the spinodal, Eq. (5.117). Averaging in Eq. (5.146) is made over volumes $v_m = l^d \sim q_m^{-d}$. At the approach to the spinodal curve, the quantity q_m^{-1} grows as the correlation length ξ. Moreover, the expression on the left hand side of Eq. (5.146) may be expressed as

$$\langle [\delta c(\vec{r},\tau)]^2 \rangle_l \simeq \langle [\delta c(\vec{r},0)]^2 \rangle_l \exp[2\omega(q_m)\tau]$$
$$\simeq r_0^{-2} q_m^{d-2} v_m^{d-4} \exp[2\omega(q_m)\tau], \tag{5.147}$$

Taking into account both mentioned circumstances, the limiting time for the applicability of the linear may be estimated from Eq. (5.146) as [402]

$$\tau_* \simeq \omega^{-1}(q_m) \ln \left(r_0 | c - c_{sp} | v_m^{2-d/2} q_m^{1-d/2} \right). \tag{5.148}$$

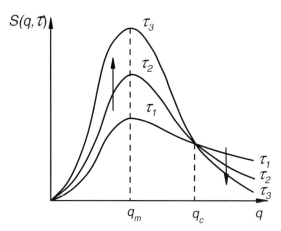

Fig. 5.15 Evolution of the structure factor according to the linear theory of spinodal decomposition.

At $\tau < \tau_*$, the intensity of density fluctuations with a wavelength, $\lambda = \lambda_m$, will increase, but λ_m in this case will not change (Fig. 5.15). Fluctuations "condense," but do not "roughen." To a certain degree this process is similar to the process of nucleation on the metastable side in the vicinity of the spinodal (see Section 5.7), where heterophase fluctuations first increase their amplitudes, and only then the growth of their linear dimensions begins.

The limitation of the Cahn–Cook theory also manifests itself in the fact that by substituting Eq. (5.139) into Eq. (5.141) a formula of the Ornstein–Zernike type is obtained, where the component corresponding to the correlation length is negative. Since this component is proportional to the coefficient at the linear term of the expansion of Eq. (5.135), in order to arrive at the correct sign it is necessary to take into account the nonlinear corrections of higher orders.

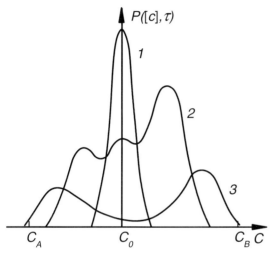

Fig. 5.16 Evolution of the distribution $P([c], \tau)$. Curves 1–3 correspond to times $\tau_3 > \tau_2 > \tau_1$.

The first attempt to include nonlinear terms in the equation of relaxation, Eq. (5.134), was made by Langer [395] and Langer et al. [396]. As shown in Ref. [396], to describe the evolution of an unstable system, it is necessary to supplement Eq. (5.134) by an equation for the functional of the distribution $P([c], \tau)$ of configurations of the field, $c(\vec{r}, \tau)$. If we take into account the interaction of fluctuations in a mean-field approximation [64] and assume that at any time the distribution $P([c], \tau)$ coincides with the Gaussian distribution centered at c_0, then for $S(\vec{q}, \tau)$ we obtain Eq. (5.138), where

$$\omega(q, \tau) = -\Gamma_c q^2 \left[\left(\frac{\partial^2 \Delta f}{\partial c^2} \right)_{c_0} + \frac{1}{2} \left(\frac{\partial^4 \Delta f}{\partial c^4} \right)_{c_0} \langle \delta c^2(\tau) \rangle + 2\kappa q^2 \right]. \quad (5.149)$$

Since the correlator, $\langle \delta c^2(\tau) \rangle$, is a positive increasing function of time, and $(\partial^4 \Delta f / \partial c^4)_{c_0} > 0$, the critical wave number, q_m, will decrease in the process of decay. This property will lead to a decrease in the rate of build-up of unstable Fourier components of concentration. However, since the second component in the right-hand side of Eq. (5.149) is always smaller than the first one, in the limit, $\tau \to \infty$, such a modification of the theory does not give the Ornstein–Zernike formula for $S(q, \infty)$. The conversion of sign in the component preceding $2\kappa q^2$ in Eq. (5.149) can be only obtained by assuming that the distribution $P([c], \tau)$ changes its shape in time as shown in Fig. 5.16. The binodal character of $P([c], \tau)$ signifies the appearance at intermediate stages of the decay of a bound modulated structure, in which the concentration in cells is determined by the position of the peaks of $P([c], \tau)$, and the characteristic linear dimension of cells by their halfwidth. The equation for $S(q, \tau)$ will still look like Eq. (5.138), where

$$w(q, \tau) = -\Gamma_c q^2 [A(\tau) + 2\kappa q^2]. \tag{5.150}$$

For the function $A(\tau)$, Langer et al. [396] suggested the approximation

$$A(\tau) S(\vec{q}, \tau) = \sum_{n=2}^{\infty} \frac{1}{(n-1)!} \left(\frac{\partial^n f}{\partial c^n} \right)_{c_0} S_n(\vec{q}, \tau). \tag{5.151}$$

Here, $S_n(\vec{q}, \tau)$ are the Fourier transforms of the correlation functions of higher orders

$$G_n(\vec{r}, \tau) = \int \delta c^{n-1}(\vec{r} + \vec{r}', \tau) \delta c(\vec{r}', \tau) d\vec{r}'. \tag{5.152}$$

As distinct from the linear theory, the nonlinear theory of spinodal decomposition [396] gives a nonexponential growth of $S(q_m, \tau)$ at all values of q and a shift of its maximum toward small q's in time. Thus, nonlinear effects slow down the structurization of a system in composition and lead to a time dependence of the characteristic size of the cell structure. Numerical calculations have yielded [403]

$$S(q_m, \tau) \sim \tau^{0.81}, \qquad q_m \sim \tau^{-0.212}. \tag{5.153}$$

The solution of Eq. (5.138) in the approximation of Eq. (5.151) gives the correct Ornstein–Zernike asymptotics with $A(\tau \to \infty) > 0$. However, ignoring the interaction of the order parameter $c(\vec{r}, \tau)$ with the other large-scale modes will cause the theory of Langer, Bar-on, and Miller not to describe the approach of the system to equilibrium.

The different theories of spinodal decomposition, discussed here briefly, are strictly applicable only to media with an infinitely high viscosity (solid bodies). In solutions and one-component liquids flows caused by pressure, concentration and temperature gradients may have a considerable effect on the decomposition process. Such flows lead to hydrodynamic interactions of fluctuations of the order parameter and the appearance of a nonlocal term in the equation for the functional of the distribution $P([c], \tau)$ [404, 405]. In order to account for the hydrodynamic interactions, Kawasaki [404], Kawasaki and Ohta [405] used the approximation of Langer, Bar-on, and Miller, Eq. (5.151). The equations obtained only allow one a numerical solution and are applicable at the intermediate stages of the spinodal decomposition process.

At the end of the intermediate stage of relaxation, stable phase boundaries are formed in the system, and heterogeneities develop in the form of embedments of one phase into the other. Further relaxation is then connected with the coarsening of grains of the competing phases and is in many respects similar to the process of nuclei growth in a metastable system. At $\tau \to \infty$, the linear dimension of the separating phase increases by the law

$$l \simeq q_m^{-1} \sim \tau^p, \tag{5.154}$$

and the dissipation intensity as

$$S(q_m, \tau) \sim \tau^n. \tag{5.155}$$

At the intermediate stage of the decay $n > 3p$, and at the completing one, $n = 3p$ [406]. The exponents in the growth laws, Eqs. (5.154) and (5.155), also depend on the dimensionality of space, d [407]. If the determining mechanism at the completing stage of the decay is coagulation of grains, then $p = 1/3$ [406, 407]. The same value of the exponent in Eq. (5.154) will be found in the case of the coalescence of grains [368].

In solutions and one-component liquids, the growth rate of grains at the last stages of spinodal decomposition may be limited by viscosity, inertial forces, and heat supply. Depending on the relations between the factors mentioned, the exponent p in the growth law, Eq. (5.154), may take the values ($d = 3$): 2/5 (the determining factor is heat supply and inertial forces) [397], 1/3 (heat supply and viscosity) [398], 2/3 (surface tension and inertial forces) [399]. As it is shown by Siggia [408], the grains of separating phases in the vicinity of the critical point form a bounded structure. This property leads to the growth of separating phases at the expense of a substance flow from some regions to others. Such a growth mechanism is limited by surface tension, viscosity and described by Eq. (5.154) with $p = 1$.

Using the ideas of the dynamic theory of scale invariance, Binder and Stauffer [354] wrote the structural factor as

$$S(q, \tau) = L^d(\tau)\widetilde{S}[qL(\tau)], \tag{5.156}$$

where the scale-invariant function, $\widetilde{S}(x)$, has no explicit dependence on time. The form of this function was found by Furukawa [409, 410]

$$\widetilde{S}(x) = \frac{(1+\gamma/2)x^2}{\gamma/2 + x^{2+\gamma}}, \quad x = qL(t), \tag{5.157}$$

where $\gamma = 2d$ if the entry into the unstable region is performed passing the critical point, and $\gamma = d+1$ in all other cases. Equation (5.156) predicts that the values of $S(q,\tau)$ obtained at various times at different depths of penetration into the unstable region, in variables $S(q,\tau)q_m^d(\tau)$, q/q_m should be described by a single curve.

5.9
Experimental Studies of Spinodal Decomposition

Observation of spinodal decomposition is possible if the time of "preparation" of the unstable state, τ_m is shorter than the characteristic time, τ_d, of phase separation. Owing to a very small value of τ_d in a one-component system, which is $(10^{-10}$–$10^{-9})$ s [89,411,412], experimental investigations of spinodal decomposition are confronted with considerable difficulties. Binary liquid mixtures are more convenient objects for studying spinodal decomposition as the diffusion coefficients in such systems are approximately two orders of magnitude lower than the thermal diffusivities of pure liquids. Owing to the asymptotic divergence of the diffusion coefficient at $T \to T_c$, in the vicinity of the critical point values of τ_d might be expected which are not too low. Besides, being the point of tangency of the spinodal and the binodal, the critical point allows a system transfer from a stable to the unstable region bypassing metastable states.

Penetration into an unstable region may be realized either by a rapid decrease (increase for systems with a lower critical point) of temperature or by an abrupt change of pressure, which shifts the co-existence curve [382]. The pressure release is a less inertial method of realizing an unstable state than the temperature shift. However, it may be accompanied by adiabatic heating or cooling of the liquid. A drawback of the methods of heating (by Joule heat [412], microwave radiation [413], intensive stirring [414]) is the appearance of a temperature difference between the liquid and the heat bath that surrounds it. The change of concentration, as a means of the transfer of the system into an unstable state, is a rather complicated method. Since the theory of spinodal decay is formulated in the reciprocal space, its most direct proof is given by diffraction methods, first of all by small-angle light scattering. The quantity that is usually observed in experiments is the intensity of scattered light $I(\vec{q},\tau)$, which is proportional to the structure factor, $S(\vec{q},\tau)$. The modulus of the wave vector of scattering, $\vec{q}$, is related to the length of the radiation wave, λ_0, and the angle between the incident and the scattered beam by the equation $|\vec{q}| = 4\pi \sin(\Theta/2)/\lambda_0$. In order to observe fluctuations of composition (density) with a wavelength $\lambda \simeq 10^{-4}$ cm at $\lambda_0 \simeq 632.8$ nm, it is necessary to have a scattering angle $\Theta \simeq 0.05$ rad.

In experiments on spinodal decomposition, the angular light distribution is registered by photomultipliers or photodiodes [382], a photodiode in combination with a rotable mirror, or by rapid filming [415]. Each of the methods listed has its advantages and drawbacks. Thus, the method of a rotable mirror is applicable only to systems where the time constant of the spinodal-decay process is larger than 100 s. The drawbacks of rapid filming include a narrow dynamic range and the complexity of data analysis, though the procedure of conducting measurements is very convenient.

Up to now, spinodal decomposition has been investigated only in one cryogenic system, in a solution of ^{4}He–^{3}He [381, 382, 415]. In all other papers, the objects of study are organic liquids with critical temperatures close to room temperature. In liquid solutions of ^{4}He–^{3}He, the unstable region contacts the region of stable states at the tricritical point. In Ref. [415], the transfer of a solution into the unstable region was realized by a pressure release at a rate of 10 MPa s^{-1}, which is equivalent to a rate of temperature decrease of ~ 4 K s^{-1}. Pressure and temperature were registered by a capacity pressure transducer and a germanium resistance thermometer, respectively. The optical setup included a laser, a linear grid of 512 photodiodes with automatic scanning and a digital oscillograph. The operation of the systems was controlled by a minicomputer.

Spinodal decomposition is connected with the generation of inhomogeneities in the systems of a certain characteristic scale. In an isotropic system, the angular distribution of the intensity of scattered light has the form of a ring. The ring radius is determined by the most rapidly growing Fourier-component with a wavelength, $\lambda_m = 2\pi/q_m$. According to the linear theory of spinodal decomposition, the brightness of the ring increases exponentially, and its radius will not change in time. Experiments on solutions of ^{4}He–^{3}He [381, 382, 415] and solutions of organic liquids [412–414, 416–418] have not revealed such a relaxation regime. Almost immediately after the appearance of scattered light a ring was formed, which was continuously collapsing. For solutions of ^{4}He–^{3}He, the characteristic collapse time is ~ 0.3 s, which is approximately 1000 times less than the time registered for disintegrating mixtures of organic liquids [417–424]. The latter property is connected with a considerably smaller value of the diffusion coefficient in solutions like isobutylic acid–water, 2.6-lutidine–water than in solutions of helium isotopes. The scattering ring in a solution of ^{4}He–^{3}He formed in the process of pressure release (with a time constant of the hydraulic system ~ 0.2 s), whereas in solutions of organic liquids a sharp image of the ring appeared only after approximately 30 s of observations. The total time of observation was ten times longer than that in solutions of helium isotopes. In this time, the characteristic size of inhomogeneities in the isobutylic acid–water system increased from 1.6 to 3.3 μm (the linear regime of relaxation was registered in one of the first papers on spinodal decomposition in binary liquid mixtures [416]; however, as shown by subsequent investigations [417], it was the result of experimental errors).

Some peculiarities of spinodal decomposition of solutions of ^{4}He–^{3}He may be connected with the specific features of the tricritical system, which is characterized by two order parameters, the macroscopic wave function Ψ, which describes the superfluid phase and is a two-component vector, and the concentration, c, of the isotope ^{3}He in ^{4}He. The interaction of the vector and the scalar order parameter leads to a considerable asymmetry of the curve of co-existence of the normal and the superfluid phase and, consequently, to different values of such properties as the diffusion coefficient, the correlation length, etc., on the left and the right branches of the co-existence curve. The peculiarities of the linear and the intermediate stages of spinodal decay in tricritical systems are examined in Refs. [419, 420], Ref. [421] is devoted to the late stage of decay. The results of these papers show that for systems of tricritical composition the relaxation of the unstable state obeys the laws for critical systems. Thus, in the vicinity of the critical and the tricritical point of solutions of simple liquids and liquids with similar properties no linear stage of spinodal decay is observed. According to Eq. (5.148), this result is due to the small values of the radius of intermolecular interaction, the large values of the diffusion coefficient, and also to the proximity of the spinodal. In solutions of high-molecular compounds, in polymeric mixtures, where the values of r_0 are large, the viscosity is high, and the diffusion coefficient is small, the linear stage is observed [422, 423].

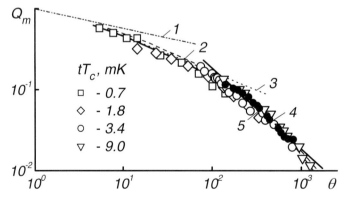

Fig. 5.17 Reduced wave number, Q_m, as a function of the reduced time, θ, for the isobutylic acid–water system (*light dots* [418] and a *bold solid line* [417]) and a solution of helium isotopes (*dark dots* [382]). 1: theory of Langer, Bar-on, and Miller [396]; 2: Kawasaki, Ohta [405]; 3: Lifshitz, Slezov [368] and Binder and Stauffer [354]; 4: Furukawa [410]; 5: Siggia [408].

As the characteristic linear scale for the spinodal decay, the correlation radius ξ may be chosen, which is determined as $\xi^{-1} = \lim_{\tau \to 0} q_m(\tau) = q_m(\tau = 0)$. One may distinguish here the time interval $\tau_c = \xi^2/D$, where $D = \lim_{\tau \to 0}[2\omega(q_m)/q_m^2]$. In the linear regime, $D = \omega(q)/q^2|_{q=0}$ holds. The scales introduced make it possible to go over to dimensionless variables

$$\theta = \frac{\tau}{\tau_0}, \qquad Q_m = \frac{q_m(\theta)}{q_m(0)} \tag{5.158}$$

and write the laws (5.154) and (5.155) in reduced form as

$$S_m(\theta, Q_m) \sim \theta^n, \qquad Q_m(\theta) \sim \theta^{-p}. \tag{5.159}$$

Figure 5.17 shows Q_m as a function of θ in isobutylic acid–water mixtures [417, 418] and in a solution of ^{4}He–^{3}He [382]. The depth of penetration into the region of unstable states for the isobutylic acid–water system is indicated in the figure. All the data refer to mixtures of critical or tricritical (system ^{4}He–^{3}He) composition. In Ref. [418], the transfer of the solution into a region of unstable states is realized by lowering the temperature, in Refs. [382, 417], by the method of pressure release. The excellent agreement between data for the isobutylic acid–water system obtained in two different experiments [417, 418] is a strong indication that the results are independent from the methods of obtaining an unstable state. In constructing the function $Q_m(\theta)$ for mixtures of organic liquids, values of ξ and D from Refs. [414, 424] were used.

In the system ^{4}He–^{3}He, the values of ξ and D are different in the normal and the superfluid phases [425, 426]. In reduced coordinates, the (Q_m, θ)-data for solutions of helium isotopes agree with the results for mixtures of organic liquids if $\xi = 10^{-5}$ cm, $D/\xi^2 \simeq 5 \cdot 10^3$ s^{-1}. The fact that the results, which have been obtained at different values of the depth of penetration into the unstable region and pertaining to systems of different chemical nature, form a single curve in coordinates $Q_m(\theta)$ is a convincing corroboration of the hypothesis of scale invariance for the spinodal decay. Even at the lowest values of θ experimental data disagree with the results of the nonlinear theory of spinodal decay by Langer, Bar-on, and Miller [396], and up to $\theta \simeq 100$ are in good agreement with the theory by Kawasaki and Ohta [405]. The latter is indicative of the important role of hydrodynamic effects, which are partially accounted for in Ref. [405].

Though the results presented in Fig. 5.17 cannot be approximated in the whole region of variations of θ by the power law Eq. (5.159), two section on the dependence $Q_m(\theta)$ can be distinguished: the first corresponds to the interval $0.3 \leq Q_m \leq 0.6$ ($\theta \leq 80$), where the exponent $p \simeq 0.3$, the second to $0.08 \leq Q_m \leq 0.1$ ($\theta \geq 80$), and here $p \simeq 1$. At small values of θ, experimental data agree with the conclusions of the coalescence theory by Lifshitz and Slezov [368] and the theory of cluster ripening by Binder and Stauffer [354]. The

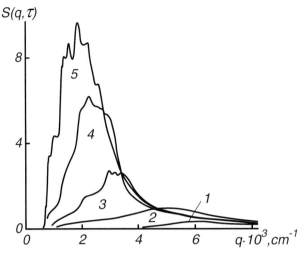

Fig. 5.18 Evolution of the structure factor of a solution of ^{4}He–^{3}He. 1: $\tau = 24$ ms; 2: 48; 3: 72; 4: 96; 5: 120.

linear law of Q_m decrease holds true at the last stages of the decay, which corroborates the Siggia theory [408] taking into account the effect of the surface tension on the dynamics of decomposition of the solution. The linear growth law is observed as long as $q_m \leq q_g = \xi(g\Delta\rho/\sigma)^{1/2}$, where g is the free fall gravity acceleration, $\Delta\rho$ is the density difference of the separating phases. In mixtures of organic liquids at a depth of penetration into the unstable region $tT_c > 1$ mK, we have $q_g < 7 \cdot 10^{-3}$ and the values of $q_m < q_g$ are located outside the region accessible for measurements. Owing to the small value of σ, in solutions of ^{4}He–^{3}He the value of q_g is about eight times as large, and gravitational effects may influence the pattern of light scattering [382]. Leiderer et al. [427] investigated the last stages of spinodal decay in solutions of helium isotopes ($0.5 < \tau < 1.5$ s) and discovered the asymmetry of the scattering ring, which had the shape of an ellipse. The asymmetry of scattering leads to different values of the vector, q_m, in the vertical ($q_{m,h}$) and the horizontal ($q_{m,l}$) directions, $q_{m,h}$ being a weaker function of time than $q_{m,l}$.

Figure 5.18 presents normalized angular distributions of the intensity of scattered light in a mixture of ^{4}He–^{3}He cooled down to temperatures 0.01 K lower than the tricritical point [382]. The curves are numbered according to the moments of observations following the transfer of the system into the unstable region. The time interval between each of the presented curves is 24 ms. If measurements of the diameter of the scattering ring in mixtures of organic liquids at the intermediate stage of the decay corroborate the theory by Kawasaki, Ohta [405], then only a qualitative agreement is observed for data on the scattering intensity.

The calculation of $S(q,\tau)$ at the last stages of the decay is a rather complicated mathematical problem. According to Refs. [354, 409], the similarity

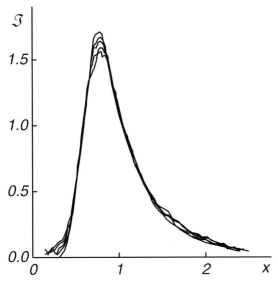

Fig. 5.19 Scaling function of the structure factor of a solution of ^{4}He–^{3}He (normalized and averaged versions of the functions shown in Fig. 5.18).

relation, Eq. (5.156), holds true here. To check this relation, the authors of Ref. [382] presented it in a somewhat different form as

$$\bar{S}(q,\tau)q_1^3(\tau) = \Im\left(\frac{q}{q_1(\tau)}\right), \tag{5.160}$$

where

$$\bar{S}(q,\tau) = \frac{S(q,\tau)}{\int\limits_0^\infty q^2 S(q,\tau)dq}, \tag{5.161}$$

and

$$q_1(\tau) = \frac{\int\limits_0^\infty q S(q,\tau)dq}{\int\limits_0^\infty S(q,\tau)dq} \tag{5.162}$$

is the characteristic size of formations. As can be seen in Fig. 5.19, at the same depth of penetration into the unstable region the function $\Im(x)$ has the same form for different times. The form of $\Im(x)$ does not change while passing from the diffusional regime of growth ($p = 1/3$) to the regime where the surface tension is the limiting factor ($p = 1$). The function $\Im(x)$ retains its form under variations of the depth of penetration into the unstable region [428] as well as while passing to other solutions [429].

The peak of the function $\Im(x)$ is located within the region of values $0.7 \leq x \leq 0.9$. Beyond this region, the simple power law [382]

$$\Im(x) \sim x^{\phi} \tag{5.163}$$

holds, where $\phi \simeq -4$ if $x \geq 0.9$, and $\phi = 4$ when $x \leq 0.7$. The asymptotic behavior of $\Im(x)$ at $x \gg 1$ and $x \ll 1$ was examined by Furukawa [409, 410]. The scaling function suggested by him, Eq. (5.157), does not agree with experimental data on binary liquid mixtures. This failure may be connected with the neglecting of hydrodynamic effects in the model.

The existence of a particular case of the general similarity relationship, Eq. (5.160), for the particular value of $q = q_m$ results from the following simple considerations [418]. If we assume that the "grained" structure forming in the intermediate and in the final stages of spinodal decay is close packed and has a characteristic linear size L, and that the intensity of scattering from every cell of such a structure is proportional to L^6, which corresponds to the limiting case of Rayleigh–Ganz in scattering theory, the structure factor $S(q, \tau)$ is proportional to the number of cells in a unit volume multiplied by L^6. Thus, $S(q, \tau) \sim L^{-3}L^6 = L^3 \sim q_m^{-3}$, and the product $q_m^3 S(q_m, \tau)$ is independent of time.

6
Nucleation Kinetics Near the Absolute Zero of Temperature

6.1
Quantum Tunneling of Nuclei

Thermo-fluctuational nucleation is impossible near zero absolute temperatures. Here nuclei of the new phase may appear as a result of quantum tunneling of heterophase fluctuations through the energetic barrier. Lifshitz and Kagan were the first to examine the kinetics of quantum nucleation at first-order phase transitions [8]. In their analysis they supposed incompressibility of the metastable liquid and that no dissipation occurs in the process of evolution of the critical nuclei and in further growth of the supercritical nuclei. Neglecting the dissipation of energy means that new excitations do not appear at the quantum underbarrier decomposition of the metastable state, and all parameters of the system adapt themselves adiabatically to the value of a single distinguished parameter. Here the radius of a nucleus of the new phase is chosen as such a parameter.

The process of quantum decomposition of a metastable phase differs significantly from the mechanism of decomposition proceeding via thermal fluctuations. The growth dynamics of a heterophase fluctuation at thermo-fluctuational nucleation determines only the value of the kinetic factor (see Eq. (2.112)) and does not influence the work of critical cluster formation. However, in the case of quantum tunneling the kinetic energy of a heterophase fluctuation is directly part of the exponent in Eq. (2.112).

In treating quantum nucleation, an approach is commonly employed which is based on the transition to an imaginary time τ, in order to describe the quantum tunneling of nuclei through the barrier [430–434]. The probability of nucleation in a quasiclassical approximation is defined with exponential accuracy by the following expression:

$$P \sim \exp\left(-\frac{I_*}{\hbar}\right), \qquad (6.1)$$

where I_* is the value of the effective action of I on the extremum trajectory,

Explosive Boiling of Superheated Cryogenic Liquids. Vladimir G. Baidakov
Copyright © 2007 WILEY-VCH Verlag GmbH & Co. KGaA, Weinheim
ISBN: 978-3-527-40575-6

$R(\tau)$. An extremal has a period, $\iota = (\hbar/k_B T)$. The action of I is calculated per period. A nucleus is a macroscopic formation at small-scale metastability, and in the absence of dissipation of energy the Lagrangian of the system is a function of the radius, R, and the speed, $\dot{R}$, of the interphase boundary

$$I[R(\tau)] = \int_{-\frac{1}{2}\iota}^{\frac{1}{2}\iota} \left[\frac{1}{2}M(R)\dot{R}^2 + W(R)\right] d\tau, \tag{6.2}$$

$$W(R) = 4\pi R^2 \sigma - \frac{4}{3}\pi R^3 \rho'' \Delta\mu. \tag{6.3}$$

Here, $M(R) = 4\pi R^3 \rho_{\text{eff}}$ is the excess mass of a nucleus, $\rho_{eff} = (\Delta\rho)^2/\rho'$ is the effective density of the excess mass, $\Delta\rho = \rho' - \rho'', \rho', \rho''$ are the densities of the metastable and the ambient phases, respectively, and $\Delta\mu = \mu' - \mu''$ is the difference of chemical potentials. The first term in the integrand is the kinetic energy and the second term is the potential energy of the nucleus. The integration is performed with respect to the imaginary time from the entry point to the exit point of the barrier.

There are two types of trajectories which lead to an extremum of the effective action, I. The first one is the classical trajectory that does not depend on time and which passes through the maximum of the potential energy, $W(R)$ (Fig. 6.1). This trajectory implies zero values of the kinetic energy and leads to the classical thermo-fluctuational regime of nucleation

$$I_* = \frac{\hbar W_*}{k_B T}, \quad W_* = \frac{4}{3}\pi R_*^2 \sigma. \tag{6.4}$$

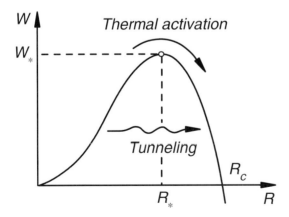

Fig. 6.1 Two mechanisms of overcoming the potential barrier in nucleation.

The second trajectory evidently depends on time and describes quantum tunneling through the potential barrier. In this case, the appearance of a nucleus of critical dimension, R_*, in a metastable system is not sufficient for the decomposition of the homogeneous state. The removal of supersaturation begins only after the appearance of a nucleus of the dimension $R_r (R_r > R_*)$, which meets the exit point from the potential barrier. The value of R_* depends on the energy of the metastable phase where tunneling through the barrier proceeds. If the tunneling process starts from a zero energetic level, which is realized at $T = 0$, then

$$R_r = R_c = \frac{3}{2} R_*. \tag{6.5}$$

The extremum trajectory at zero temperature is found from the condition

$$\frac{M(R)\dot{R}^2}{2} - W(R) = 0, \tag{6.6}$$

the solution of which is [431]

$$\frac{|\tau(x)|}{\tau_c} = \frac{\pi}{2} - \mathrm{arctg}\left[\left(\frac{x}{1-x}\right)^{1/2}\right] + [x(1-x)]^{1/2},$$
$$x(\tau) = 0, \qquad |\tau(x)| > \frac{\pi \tau_c}{2}. \tag{6.7}$$

Here, $x = (R/R_c)$ and $\tau_c = (\rho_{\mathrm{eff}}/\Delta\rho)(\rho' R_c^3/2\sigma)^{1/2}$. In the absence of dissipation, τ is the real time. The substitution of $R(\tau)$ from Eq. (6.7) into Eq. (6.2) gives the value of the extremum action

$$I_* = \frac{5\pi}{64} (8\pi\sigma M R_c^4)^{1/2} = \frac{5\sqrt{2}\pi^2}{16} \left(\frac{\rho_{\mathrm{eff}}}{\Delta\rho}\right) (\rho'\sigma)^{1/2} R_c^{7/2}. \tag{6.8}$$

Since the time at $T = 0$ on the extremal is finite and the action (Eq. (6.2)) is local, it is same at $T > 0$ [8], so that P evidently does not depend on the temperature right up to the transition to the activation regime. The temperature of this transition is determined by equality of the energy of activation at classical and quantum nucleation

$$T_* = \frac{128\hbar}{135\pi k \tau_c} = \frac{128\sqrt{2}\hbar}{135\pi k} \frac{\Delta\rho}{\rho_{\mathrm{eff}}} \left(\frac{\sigma}{\rho' R_c^3}\right)^{1/2}. \tag{6.9}$$

By analogy with thermo-fluctuational nucleation we get

$$J = C_0 B \exp\left(-\frac{I_*}{\hbar}\right), \tag{6.10}$$

where C_0 is the number of virtual centers of nucleation in a unit volume of the metastable phase, $C_0 \simeq \rho'$, B is the kinetic factor which is proportional to zero temperature vibration frequency. At $T = 0$, we obtain [8]

$$B_{T=0} = \left[\frac{12\pi^{3/2}}{\Gamma(1/4)}\right]^{4/7} \frac{(4\pi\sigma)^{5/7} R_c^{6/7}}{\hbar^{3/7} M^{2/7}}. \qquad (6.11)$$

Here, $\Gamma(x)$ is the Gamma function. At the order of magnitude, the kinetic factor at quantum tunneling coincides with the value of B at thermo-fluctuational nucleation (for liquid helium $B \simeq 2.8 \cdot 10^{12}$ s^{-1} [8]).

If $T \neq 0$, a viable nucleus appears as a result of the optimal combination of the thermal activation and tunneling. According to Lifshitz and Kagan [8], nuclei appear in the systems described by Eq. (6.2) at $0 < T < T_*$ only as a result of the quantum mechanism (tunneling from zero level), and nucleation is exclusively thermo activated at $T > T_*$. The frequency of nucleation is again determined by Eq. (6.10), where the exponent for $T < T_*$ is calculated by Eq. (6.8), and the kinetic coefficient has an additional factor

$$B_{(T<T_*)} = B_{(T=0)} \left(1 - \frac{T}{T_0}\right)^{-1}, \quad T_0 = \frac{135}{128} T_*. \qquad (6.12)$$

It is necessary to use the classical theory of nucleation in the range of temperatures, $T > T_*$.

Burmistrov and Dubovskii [430] examined the influence of viscous friction on the expectance of the underbarrier tunneling of nuclei. The approach developed in Refs. [431–433] leads to the determination of the effective action. This quantity is not local and an imaginary time was used to account for dissipation. In the case of an incompressible quantum liquid, one obtains

$$I[R(\tau)] = I_{\text{rev}} + \Delta I_{\text{diss}}, \qquad (6.13)$$

where

$$\Delta I_{\text{diss}} = \frac{\eta}{4\pi} \int_{-1/2t}^{1/2t} \left\{ \int_{1/2t}^{1/2t} \left\{ \frac{\pi T [\gamma(R_\tau) - \gamma(R_{\tau'})]}{\sin \pi T} \right\}^2 \frac{d\tau'}{\tau - \tau'} \right\} d\tau, \qquad (6.14)$$

and I_{rev} is determined by Eq. (6.2). We obtain the value of $\gamma(R)$ through the friction coefficient $\alpha(R)$ as

$$\gamma(R) = \int \left[\frac{\alpha(R')}{\eta}\right]^{\frac{1}{2}} dR'. \qquad (6.15)$$

According to Eqs. (6.1) and (6.13), the internal friction decreases the possibility of quantum nucleation and leads to an evident dependence of the exponent in Eq. (6.1) on temperature.

In the case of weak dissipation, we can consider the term ΔI_{diss} in Eq. (6.13) a perturbation. The integral, Eq. (6.14), can be calculated in this case along the trajectory, Eq. (6.7). In comparison with the dissipation-free kinetics, the temperature of transition to the activation regime, T_*, decreases by ΔT_*, which is proportional to $\Delta I_{\text{diss}*}/I_*$. The dissipation term in Eq. (6.13) is a result of the interaction of the variable $R(\tau)$ with the excitations of the metastable phase. If the free length of excitations is $l(T) \gg R_c$ (Knudsen limit), then

$$\Delta I_{\text{diss}*} = 4\pi \rho_{\text{eff}}(\rho')^{-1} \frac{\eta R_c^4}{l} \left[\frac{1.49}{\pi} + \frac{13}{768} \left(\frac{\pi k_B T \tau_c}{2\hbar} \right)^2 \right]. \tag{6.16}$$

At $l(T) \ll R_c$ (the hydrodynamic limit), we have instead

$$\Delta I_{\text{diss}*} = \frac{64\pi}{9} \rho_{\text{eff}}(\rho')^{-1} \eta R_c^3 \left[\frac{3.99}{\pi} + 0.035 \left(\frac{\pi k_B T \tau_c}{\hbar} \right)^2 \right]. \tag{6.17}$$

In the case of a strong dissipation the underbarrier tunneling is determined entirely by the internal friction, and we can neglect the kinetic energy in $I_{\text{rev}*}$ (see Eq. (6.2)). As a result we have [430] for the extreme value of the effective action

$$I_* = \eta \gamma^2 (R_c) \Im(t), \tag{6.18}$$

where $\Im(t)$ is a universal function of the reduced temperature, $t = 4k_B T \eta \gamma^2 R_c (27\hbar W_*)^{-1}$. Consequently, dissipation leads to the emergence of a new limiting temperature T_l, at which the free length, $l(T_l)$, is equal to R_c. It influences not only the exponent in Eq. (6.10), but also the kinetic factor and the shear viscosity is now a constituent part of it. These features lead to a strong temperature dependence of $B(T)$.

The account of the compressibility of a metastable phase also leads to a nonlocal property of the effective action. By restricting his consideration to the main temperature-dependent term in developing I as a series in powers of T and inverse powers of the speed of sound C, Korshunov [433] obtained a correction to the effective action in dependence on the compressibility of the liquid as

$$\Delta I_{\text{comp}*} = -490 \left(\frac{3\pi}{4} \right)^6 \frac{(\sigma \rho_{\text{eff}})^2}{C(\rho'' \Delta \mu)^3} \left(\frac{k_B T}{\hbar} \right)^4. \tag{6.19}$$

The negative value, $\Delta I_{\text{comp}*}$, is caused by the fact that, when a nucleus appears, only a part of the metastable liquid has enough time to start moving and the total kinetic energy is smaller than in the absence of compressibility.

Quantum nucleation is a result of the manifestation of quantum phenomena at the macroscopic level. The theory of quantum nucleation was used in the research of the solidification of isotopes ^{3}He and ^{4}He [435], the kinetics of stratification of liquid solutions of ^{4}He–^{3}He [436], the initiation of a B-phase in a A-phase of superfluid ^{3}He [437], the stability of supercooled liquid hydrogen [438], the processes of the faceting of quantum helium crystals [439], the kinetics of appearance of molecular hydrogen in metastable metal hydrogen [440], the decomposition of the current states of Josephson contacts [441], and the initiation of quantum vortices in superfluid ^{4}He [442]. Recently, quantum effects at nucleation have been intensively analyzed experimentally.

6.2
Limiting Supersaturations of ^{4}He–^{3}He-Solutions

Supersaturated superfluid solutions of ^{4}He–^{3}He are one of the few metastable systems where we can expect the appearance of macroscopic quantum tunneling of nuclei in the liquid phase. Such solutions remain liquid up to zero temperature, dissolution of ^{3}He in ^{4}He being limited here. Stratification begins when the concentration of a light isotope reaches 0.064 mole fraction (see Fig. 5.7). According to Lifshitz's [435] estimates of the temperature of quantum crossover for superfluid solutions of helium isotopes, we can expect here $T_* \simeq 14$ mK.

In contrast to solutions of classical liquids, the presence of a liquid–vapor boundary in the quantum system ^{4}He–^{3}He eliminates nonequilibrium with respect to phase stratification. The phase enriched with ^{3}He develops continuously at insignificant supersaturations from a film of foreign atoms of ^{3}He existing at such a boundary [443]. The opposite effect takes place for a liquid–solid boundary: due to van der Waals forces the vessel walls and particles are covered with a film of ^{4}He [444], these surfaces cannot act as centers of nucleation any more. Thus, in the absence of a vapor phase, nucleation in supersaturated superfluid solutions ^{4}He–^{3}He is most likely to develop according to a homogenous mechanism.

Significant supersaturations of solutions of ^{4}He–^{3}He were first distinguished in experiments in the determination of their physicochemical properties [445, 446]. Brubaker and Moldover [447] examined limiting supercoolings of superfluid solutions of helium isotopes and obtained values of δT_n, which are by orders of magnitude lower than predicted by the classical theory of homogeneous nucleation. Both the temperature and the concentration range of the examined state parameters were significantly extended in the following papers [448–453].

The methods of continuous change of concentration [448–450], of pressure release [449] and of simultaneous change of concentration and pressure [451–453] were used in the study of the kinetics of spontaneous nucleation. Limiting supersaturations were registered according to the change of the first sound velocity [448–450], of NMR-signal amplitudes [449], of a liquid inductive capacity [448–453]. The experiments included the interval of temperatures from 400 µK to 250 mK and the interval of pressures from 0.05 to 0.5 MPa.

Phenomena peculiar to superfluid solutions like thermo-osmosis and the spouting effect [448] are used in the method of continuous change of concentration. The chamber of the setup consists of two cells connected with each other by a capillary. One of them is a measuring cell ($V \simeq 6.17\,\text{cm}^3$), which has a capacitance-type transducer to measure dielectric permittivity and a piezosensor to determine the first sound velocity. The other cell is a controlling cell ($V \simeq 53.04\,\text{cm}^3$). Its temperature can vary within rather wide limits. The heating of the controlling cell leads to the flow of a light isotope into the measuring cell. The composition of the mixture in the measuring cell was determined simultaneously by acoustic and capacitive methods. The velocity of concentration change was setup by the temperature of the controlling cell.

The pressure release method is based on the fact that the phase equilibrium curve (the left branch of the phase diagram in Fig. 5.7) shifts in the direction of higher concentrations when pressure increases up to ~ 1 MPa. Therefore, the initial state of a homogeneous solution is close to the phase equilibrium curve at increased pressure, but a solution will be in a metastable state after pressure release. The measuring cell ($V \simeq 5.2\,\text{cm}^3$) has a small-volume space in the top part, which is covered by a NMR coil [449]. Since the amplitude, h, of a NMR echo-signal is proportional to the number of nuclei of ^{3}He, the dependence $h(\tau)$ carries the information about the change of concentration (supersaturation) of a solution. In Ref. [449], a solution has been beforehand transferred by pressure decrease ($\dot{p} \simeq 1.5 \cdot 10^{-4}\,\text{MPa·s}^{-1}$) into a metastable phase, where its lifetime was equal to hours, and then nucleation was initiated by heating with a rate ($\dot{T} \simeq 15 - 30\,\text{µK·s}^{-1}$). The results of Refs. [448–450] at the pressure $p = 0.05$ MPa are shown in Fig. 6.2. When the temperature is higher than 50 mK, the data obtained according to the methods of pressure release and of continuous change of concentration agree well with each other. At $T < 50$ mK, the method of pressure release gives higher supersaturations.

Before going over to the consideration of the kinetics of phase stratification of ^{4}He–^{3}He solutions at ultralow temperatures, we compare the data of Refs. [448–450] with the theory of homogeneous nucleation. Lifshitz [436] generalized the theory to superfluid solutions of ^{4}He–^{3}He. At $c < 0.6$, the metastable phase is a solution of normal Fermi liquid in a superfluid Bose liquid. The motion in such a system is described by the two-speed hydrody-

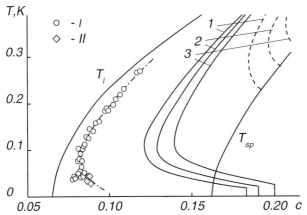

Fig. 6.2 Limiting supersaturations of the superfluid phase of solutions of ^{4}He–^{3}He. Experiment: I is the method of continuous change of concentration [448]; II is the method of pressure release [449]. Theory: solid lines refer to calculations according to Eqs. (2.112), (6.21), (6.22), and (6.24); dashed lines to solutions of Eqs. (2.112), (6.21), (6.22), and (6.23) for $J = 10^{-10}$ m$^{-3}\cdot$s^{-1}, 10^0 (2), and 10^{10} (3). T_l is the phase equilibrium line; T_{sp} is the spinodal curve.

namics [454]. The effective mass density of a nucleus is determined by

$$\rho_{\text{eff}} = \rho'_n + \frac{(\rho'' - \rho'_s)^2}{\rho'_s}, \tag{6.20}$$

where ρ'_n and ρ'_s are the normal density and the density of the superfluid component, respectively, in a supersaturated solution, and ρ'' is the density in a nucleus. In the regime of thermo-activation nucleation, without taking into account energy dissipation, we obtain [436] for the kinetic factor

$$B_{(T>T_*)} = B_{(T=0)} \left[\frac{2}{3} \left(\frac{\pi M}{\sigma} \right)^{1/2} \frac{k_B T}{\hbar} - 1 \right]^{-1}. \tag{6.21}$$

Here, $B_{(T=0)}$ is given by Eq. (6.11).

The work of the formation of a critical nucleus in a solution is given by

$$W_* = \frac{16\pi}{3} \frac{\sigma^3}{(\Delta\mu\rho'')^2}, \tag{6.22}$$

where $\Delta\mu$ is the chemical potential difference of ^{3}He in the metastable phase and in a nucleus at the given temperature and pressure. $\Delta\mu\rho''$ determines the supersaturation of the parent phase and is not directly measured in the experiment. Its value and also the composition of a critical nucleus are calculated from the conditions of equilibrium of the critical nucleus with the environment. At small supersaturations [455]

$$\Delta\mu \simeq (\mu' - \mu_l)\left(1 - \frac{c_l}{c_u}\right), \tag{6.23}$$

where the indices l and u specify the quantity referring to the lower and the upper phase, which are in equilibrium at a planar interface. This approximation is often used in literature

$$\Delta\mu \simeq \left(\frac{\partial\mu}{\partial c}\right)\bigg|_{c=c_l} (c - c_l). \tag{6.24}$$

At $T = 100$ mK and $\Delta c = c - c_l = 0.05$, the discrepancies between the data of Eqs. (6.23) and (6.24) reach 42%. The results of calculation of limiting supersaturations of Δc_n from Eq. (2.112) using Eqs. (6.21)–(6.24), and $C = \rho', z_0 = 1$, for three values of the nucleation frequency are shown in Fig. 6.2.

Considering the weak dependence of $B_{(T>T_*)}$ and ρ' on supersaturation and the temperature independence of σ at $T < 300$ mK, we get from Eqs. (2.112), (6.22), and (6.24)

$$G_* = \frac{W_*}{k_B T} \sim (\Delta c_n^2 T)^{-1} = \text{const}. \tag{6.25}$$

Experimental data [448–450] for $T > 50$ mK are qualitatively in agreement with Eq. (6.25). This result proves the thermo-fluctuational mechanism of nucleation in this temperature range.

In the regime of continuous increase of concentration the effective frequency is determined by

$$J(c_n) = \frac{\dot{c}}{V}\left(\frac{\partial G}{\partial c}\right)_{c=c_n}. \tag{6.26}$$

From the bar diagram of stratification event distribution at $T = 90$ mK, $\dot{c} = 4 \cdot 10^{-6}$ s^{-1} and $\Delta c_n = 1.1 \cdot 10^{-2}$ we obtain $\delta c_{1/2} \simeq 1.5 \cdot 10^{-3}$ [450]. Then, according to Eqs. (6.25) and (6.26), we have $(\partial G/\partial c)_{c=c_n} = 1.63 \cdot 10^3$, and $J = 1.1 \cdot 10^3$ m^{-3}s^{-1}. Figure 6.2 shows the significant difference of the theoretical and the experimental results for the given nucleation frequency. The refusal of the approximation (6.24) and the use of Eq. (6.23) in the calculations increase the difference. Such a discrepancy may be connected both with the approximations used in calculations according to the formulae of the homogeneous nucleation theory and with the influence of heterogeneous nucleation centers and external fields. The anomalous increase of limiting supersaturation at $T \to 0$ according to Eq. (6.25) requires to take into account the dependence $\sigma(R)$ at rather small frequencies of nucleation [455].

Satoh [451–453] used the method of simultaneous change of concentration and pressure for the search of limiting supersaturations of solutions of ^{4}He–^{3}He in the range of temperatures from 400 µK to 160 mK. The measuring cell ($V \simeq 77\,\text{cm}^3$) had two capacitance concentration sensors in the top and the bottom part. Initially, ^{4}He was condensed in the cell. There, it neutralized the centers of nucleation by damping the interior surfaces of the cell. Then, a predetermined quantity of ^{3}He, which did not change during the experiment, was filled into the cell. The conveying into the cell of HeII was performed through a special filter which can only be passed by the superfluid component. As a result of the increase of solubility of ^{3}He with the rise of pressure and the ratio ^{4}He/^{3}He the concentrated phase turned into the weak phase. After preparing a compressed one-phase state, the process of dilatation began due to the outflow of HeII, which was accompanied by the rise of concentration of ^{3}He. The moment of stratification was registered as a leap in the indications of concentration transducers.

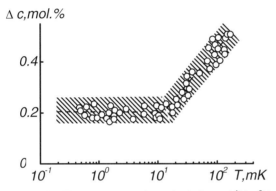

Fig. 6.3 Limiting supersaturations of solutions of ^{4}He–^{3}He close to the absolute zero of temperature [453].

The results of measurements are shown in Fig. 6.3. For temperatures lower than 10 mK, the value of Δc_n does not depend on temperature. This effect can indicate the appearance of quantum tunneling of nuclei. The increase of Δc_n at $T > 10$ mK is connected with the pressure reduction. In order to compare the theory of homogeneous nucleation and the experimental data of Refs. [448–450] and [451–453], the latter should be corrected to a single value of pressure and nucleation frequency or the data should be converted to some degree of metastability "isomorphous" to a one-component system. Burmistrov et al. [456] suggested to use in this respect the value of $\Delta\mu\rho''$, which directly defines the work of formation of a critical nucleus, Eq. (6.22), as the reference supersaturation. Considering the practical independence of the interfacial tension in the investigated state parameter region from temperature, the choice of $\Delta\mu\rho''$ as the measure of metastability was equivalent to the admission of the radius of a critical nucleus. To convert the experimentally registered values of Δc_n and p_n, the following formula [456] was introduced:

$$(\Delta \mu \rho'')_n \simeq \left\{ \rho'' \left[v_u - v_l - (x_u - x_l) \frac{\partial v}{\partial x} \right] \left(\frac{\partial p}{\partial c} \right)_T \right\} \Delta c_n, \tag{6.27}$$

where x_l, x_u and v_l, v_u are the mass concentrations of ^{3}He and the specific volume in the weak and the concentrated phase at phase equilibrium at a planar interphase boundary.

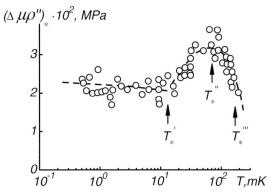

Fig. 6.4 The value of $(\Delta \mu \rho'')_*$ as a function of temperature.

The value of $\Delta \mu \rho''$ at temperatures lower than $T'_* \simeq 10$ mK does not depend within the experimental error on temperature and has a maximum at $T''_* \simeq 70$ mK (Fig. 6.4). Further temperature increase is connected with a decreasing degree of metastability of the solution, which is in agreement with the classical theory of nucleation. The temperature independence of $\Delta \mu \rho''$ at $T < T'_*$ is a sound argument for the quantum tunneling of nuclei. In the range $T < T'_* (\Delta \mu \rho'')_{n,I} \simeq 0.024$ MPa and from Eq. (6.22) for $\sigma = 0.24$ μN $\cdot$ m^{-1}, we get $W_*/(k_B) \simeq 17$ K. The radius of a critical nucleus under such conditions is $R_* \simeq 2.5$ nm, the number of atoms of ^{3}He contained in it equals $n''_* \simeq 10^3$.

The effective action in a hydrodynamic approximation is determined by Eq. (6.8), where the effective density of the additional mass is given by Eq. (6.20). In the absence of dissipation of energy, the temperature of the change of the nucleation mode is $T_* = \hbar W_*/k_B I_* \simeq 5$ mK [453]. This temperature is close to T'_* and is more than an order of magnitude lower than the value of T''_* to which it should be compared. Two reasons for such a great discrepancy can be supposed. First, calculations of $(\Delta \mu \rho'')_{n,I}$ may be inaccurate so that only the order of the value of T_* can be determined reliably. Second, the hydrodynamic approximation may be incorrect at ultralow temperatures, where the length of the track of quasiparticles of ^{3}He exceeds significantly the radius of a critical nucleus. The temperature at which the value of $(\Delta \mu \rho'')$ in the regime of thermo-fluctuational nucleation is equal to its value at the quantum tunneling stage is denoted in Fig. 6.4 as T_*'''. Substitution of $(\Delta\mu\rho'')_{n,I}$ into Eqs. (2.112), (6.22), (6.23), and (6.24) gives $T_*''' \simeq 200$ mK, which is in good agreement with the experimental value of $T_*''' \simeq 150$ mK.

In the case of liquid solutions of ^{4}He–^{3}He, mass diffusion will be the key mechanism of energy dissipation. The coefficient of diffusion of ^{3}He atoms rises with temperature decrease and below T_*' dissipative processes in the kinetics of nucleation may be neglected [456, 457]. However, their contribution to the kinetics of nucleation at $T > T_*'$ is significant. The authors of Ref. [453] link the rise of $(\Delta\mu\rho'')_n$ on the interval T_*', T_*'' with the demonstration of dissipation in quantum tunneling of nuclei.

6.3
Formation of Quantum Vortices in Superfluid Helium

Phase transitions of second-order (λ-transitions) take place in liquid ^{4}He at a pressure of the saturated vapor and a temperature, $T_\lambda = 2.172$ K. The low-temperature phase (HeII), which evolves as a result of the phase transition, is superfluid. Superfluid helium is a mixture of two components: a normal one, which behaves as a common classical liquid, and a superfluid one with zero viscosity. The hydrodynamic velocity, $(\vec{v}_n, \vec{v}_s)$, and the density, (ρ_n, ρ_s) [454], are determined by the respective properties of both components. The flow of the superfluid component is always potential, i.e., it obeys the property rot $\vec{v}_s = 0$.

If in a volume of HeII we create a relative motion of the components (for example, a rotation by moving some external object through the liquid which is initially at rest), the superfluid component will flow without friction in contact with the normal component, on the vessel walls or on the surface of a moving object. Such a state of motion is metastable since a state without relative motion but with the same total momentum is more probable from a thermodynamic point of view. The velocity corresponding to the boundary of essential instability of the metastable state of motion is determined by (Landau number) [454]

$$|\vec{v}_s - \vec{v}_n| = \min \left[\frac{\epsilon(p)}{p}\right] = v_{sp}, \qquad (6.28)$$

where $\epsilon(p)$ and p are the energy and the momentum of elementary excitations in liquid helium, respectively. According to Eq. (6.28), the critical (spinodal) velocity is approximately equal to 60 mm s^{-1}. Experimental data [458, 459] show the disruption of a superfluid motion in HeII at significantly lower values of the relative velocity.

Similarly to the common types of first-order phase transitions, the metastability of a superfluid motion cannot be removed simultaneously in the whole volume of a liquid because of its potential character. However, the mutual friction between the normal and the superfluid component at the condition rot $\vec{v}_s = 0$ can appear on some special vortex filaments [460]. These vortex filaments have a thickness of the order of the size of the atoms of the superfluid liquid. Their quantum nature is manifested in the fact that the circulation of the velocity of superfluid motion, $\check{\kappa}$, is quantized along this filament, i.e., $\check{\kappa} = n\hbar/m$, where n are integer numbers. Obviously, only the filaments with the quantum number, $n = 1$ [454], are really stable in liquid HeII.

At a predetermined vortex length, the work of formation of a closed ring is less than that of a filament. As a consequence, in the case of a flow of an undefined configuration the appearance of vortex rings will be most probable. Iordansky [461], and later Langer and Fisher [462], supposed that vortex rings in superfluid helium are formed as a result of thermal fluctuations and play a role which is similar to the role of Gibbs nuclei in the liquid–vapor phase transition. The critical size of a vortex ring, R_*, is found from the condition of equilibrium between the force of friction and the hydrodynamic forces which affect the ring. In a quasiclassical approximation, the work of formation of a circular vortex ring is determined by its energy, $E(R)$, and momentum, $\vec{p}(R)$, in liquid at rest [460]

$$W = E(R) - \vec{p}(R)\vec{v}_s, \tag{6.29}$$

where

$$E(R) = \frac{\check{\kappa}^2}{2} R\rho_s \left(\eta - \frac{7}{4}\right), \quad \eta = \ln\frac{8R}{a_0}, \tag{6.30}$$

$$p(R) = \pi\check{\kappa}R^2\rho_s. \tag{6.31}$$

Here, a_0 is the radius of a vortex nucleus (a cortex). From Eq. (6.29) for the vortex critical radius and for the work of its formation (accurate to terms which are proportional to $(a_0/R)^2$) we have

$$R_* = \frac{\check{\kappa}}{4\pi v_s}\left(\eta - \frac{3}{4}\right), \tag{6.32}$$

$$W_* \simeq \frac{\check{\kappa}^3}{16\pi v_s}\rho_s \left(\eta - \frac{1}{4}\right)\left(\eta - \frac{11}{4}\right). \tag{6.33}$$

The fluctuational increase of vortex rings is treated as a diffusion process in the multidimensional space of their quantum numbers [460].

In order to describe such processes quantitatively, Iordansky took an unstable variable as the main parameter for the description which in this case was the radius of the ring, integrated it with respect to all stable variables and then obtained a one-dimensional Fokker–Planck equation [61], in which $P(R,\tau)$ is the distribution function of vortex rings with respect to their radii. The increment of growth of an unstable mode is found due to the solution of a linearized hydrodynamic problem. For further analysis the exact value of the pre-exponential factor in a stationary solution of Eq. (2.112) does not play a significant role. Therefore, following the general ideas of Kramers' theory [83], the equation for the frequency of vortex formation can be written as

$$J = C_0 \frac{\omega_0}{2\pi} \exp\left(-\frac{W_*}{k_B T}\right), \tag{6.34}$$

where $\omega_0 \simeq 5 \cdot 10^{12}$ s^{-1} is some characteristic frequency of processes on the atom-size scale, C_0, as before, is the normalization constant of the distribution function of vortex rings. In a first approximation, we may write $C_0 \simeq \rho'_s$. From Eqs. (6.33) and (6.34), for the limiting velocity of a superfluid motion we obtain

$$v_{s,n} \simeq \frac{\hbar^3 \rho'_s \left(\eta - \frac{1}{4}\right)\left(\eta - \frac{11}{4}\right)}{16\pi m^3 k_B T \ln\left(\frac{\rho'_s \omega_0}{2\pi J}\right)}. \tag{6.35}$$

In the vicinity of the λ-point, the temperature dependence of $v_{s,n}$ is mainly determined by the temperature dependence of the density of the superfluid component, ρ'_s. It can be described by the power law [454]

$$\rho'_s(T) \simeq \rho'_{s0}\left(1 - \frac{T}{T_\lambda}\right)^{2/3}. \tag{6.36}$$

Substitution of Eq. (6.36) into Eq. (6.35) yields

$$v_{s,n}(T) \simeq v_{s,n0}\left(1 - \frac{T}{T_\lambda}\right)^{2/3}, \tag{6.37}$$

where

$$v_{s,n0} \simeq \frac{\hbar^3 \rho'_{s0}\left(\eta - \frac{1}{4}\right)\left(\eta - \frac{11}{4}\right)}{16\pi m^3 k_B T \ln\left(\frac{\rho'_s \omega_0}{2\pi J}\right)} \tag{6.38}$$

is the value of the velocity of superfluid motion at $T = 0$.

The dependence of the kind given with Eq. (6.37) was obtained in experiments on the flow of superfluid ^{4}He through filter materials with pore diameters of the order of 0.2 μm [458]. From the experimental data obtained, the

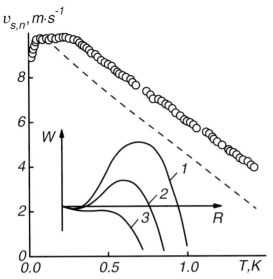

Fig. 6.5 The temperature dependence of the limiting velocity of superfluid motion in ^{4}He. *Dots*: experimental data [442]; *dashed line*: qualitative results of calculation according to Eqs. (6.34) and (6.29). In the inset, the dependence of the work of vortex ring formation on its radius is shown at velocities $v_s(1) < v_s(2) < v_s(3)$.

result $v_{s,n0} \simeq 3.8\,\text{m·s}^{-1}$ is found. According to Eq. (6.38), the value of $v_{s,n0}$ is approximately four times larger. According to Refs. [458,462], this compliance can be regarded as the confirmation of the thermo-fluctuational mechanism of vortex formation in a superfluid liquid because of the complexity of the pore configuration in filter materials. With decreasing temperature, the critical radius of a vortex ring decreases and Eq. (6.33) obtained by the expansion with respect to the small parameter, a_0/R, becomes invalid. The results of the numerical computation of $v_{s,n}$ from Eqs. (6.29) and (6.34) supplemented by the criterion $(dW/dR)_* = 0$ are shown in Fig. 6.5. At $T < 1$ K, the function $v_{s,n}(T)$ is close to the linear dependence $v_{s,n} = v_{s,n0}(1 - T/T_0)$, where $T_0 \simeq 2.45$ K and $v_{s,n0} \simeq 10\,\text{m·s}^{-1}$. The linear character of the dependence $v_{s,n}(T)$ at low temperatures is confirmed by experiments on the flow of HeII through channels with an inner diameter less than one micron [459,463].

Near absolute zero, where the sizes of the characteristic vortex rings are of the order of nanometers, vortex formation is realized as a result of quantum sub-barrier tunneling. If this motion proceeds without dissipation of energy, the frequency of vortex formation does not depend on temperature and is described by Eq. (6.10). The constancy of J at temperatures lower than T_* implies the independence of the limiting value of the velocity of superfluid motion of $v_{s,n}$ from temperature.

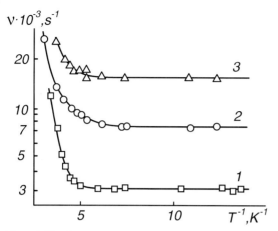

Fig. 6.6 The frequency of quantum vortex formation in superfluid helium at the passage of negative ions through it for different electric field intensities [464]. 1: $E = 8.85$ kV·m^{-1}; 2: 17.7; 3: 44.0.

The first experimental confirmations of quantum vortex tunneling near absolute zero of temperature were received from experiments on the motion of negative ions through liquid HeII [464] and on the flow of helium through gaps of submicron sizes [442, 463]. In Ref. [464], negative ions in liquid helium were accelerated to velocities of 50–65 m·s^{-1}. The frequency of quantum vortex formation was registered by a special inductive transducer. The experiments were carried out in the interval of temperatures $50 \leq T \leq 500$ mK at pressures less than 1.2 MPa. As it follows from Fig. 6.6, the frequency of vortex formation does not depend on temperature when it is less than 0.2 K.

In experiments on the outflow of helium through submicron gaps and canals, the limiting velocity of superfluid motion is a measured parameter. According to the data of Refs. [442, 463] the value of $v_{s,n}$ at $T < (150-200)$ mK is temperature independent (Fig. 6.5). The temperature of the change of the regime of vortex formation T_* (see Eq. (6.9)) is given by

$$T_* = \frac{\hbar \omega_0}{2\pi k_B}. \tag{6.39}$$

If in Eq. (6.39) we use the experimental value of $T_* \simeq 150$ mK, then $\omega_0 \simeq 1.9 \cdot 10^{10}$ s^{-1}, which coincides by the order of magnitude with the frequency of vortex rotation [465].

The temperature dependences of the limiting velocity and the frequency of vortex formation in superfluid helium obtained in experiments with charged particles and the flow of HeII through submicron canals are conform with each other and with the theory of thermo-fluctuational formation of the rings. We cannot see this agreement in the absolute values of $v_{s,n}$. Near to $T = 0$, experiments with negative ions give $v_{s,n} \simeq 55$ m·s^{-1}. According to the data on the flow of HeII through the gaps, $v_{s,n} \simeq 10\,\text{m·s}^{-1}$. In the theory of thermo-fluctuational vortex formation we can obtain both the results, since it includes such directly immeasurable parameters as the vortex nucleus radius, a_0, and the frequency, ω_0. Each of the described experiments is characterized by its own frequency of vortex formation. This fact has not yet been taken into account as a tool of comparing theory and experiment. Vortex formation can be initiated by external actions and by easily activated centers which are some defects on the inner surface of a measuring camera, of canals and gaps, and admixtures of a light isotope of ^{3}He [459, 466]. Thus, atoms of ^{3}He "condensed" in the center of a vortex ring decrease in this way the magnitude of the activational barrier. Vortex formation can be initiated by cosmic radiation and the so-called remnant vortices. Latter are caused by the absence of viscous energy losses in the superfluid component, which leads to a long-term existence of vortex rings of undercritical size in liquid helium. It seems that to create "pure" conditions for vortex formation is as difficult as to create them for nucleation.

6.4
Quantum Nucleation Near the Boundary of Essential Instability

The classical theory of steady-state nucleation in a macroscopic approximation predicts an unlimited increase of the degree of metastability of a system at $T \to 0$. The account of the dependence of nucleus properties on supersaturation (its size) leads to the conclusion that the lines of constant value of nucleation frequency tend, as the temperature decreases, to the spinodal, where $W_* = 0$ holds. Thus, the transition from the thermo-activational mechanism of nucleation to the quantum mechanism of nucleation near absolute zero of temperature takes place in a range of strong metastability. Here, the potential energy of heterophase fluctuations is determined by Eq. (5.109), and the nucleus formed in passing the barrier does not possess any stable bulk phase properties even in the center of the nucleus.

Lifshitz and Kagan [8] developed the theory of dissipationless quantum nucleation near the boundary of instability. Considering a small change of density at the formation of heterophase fluctuations, a liquid is considered as an elastic solid body with a modulus of rigidity equal to zero. In this case, the kinetic energy of the fluctuation is determined by the field $\vec{\varepsilon}(\vec{r})$ and for the effective action near the spinodal from Eqs. (6.2) and (5.109) ($d = 3$) we have [8]

$$I(\theta) = \Delta F_0 \tau_0 \int_0^\theta \left\{ \int \left[\frac{1}{2} \left(\frac{\partial \vec{u}}{\partial \theta} \right)^2 - 4\chi^2 + \frac{8}{3}\chi^3 - (\nabla \chi)^2 \right] d^3x \right\} d\theta'$$

$$= \Delta F_0 \tau_0 \int_0^\theta \aleph\{\vec{u}\} d\theta'. \tag{6.40}$$

The dimensionless vector field

$$\vec{u} = \left(\frac{|f'''|\rho_0^2}{8\kappa|\rho_{sp} - \rho_0|} \right)^{1/2} \vec{\varepsilon}, \tag{6.41}$$

and the dimensionless time are introduced here

$$\theta = \frac{\tau}{\tau_0}, \quad \tau_0^2 = \frac{16m\kappa}{|f'''|^2(\rho_{sp} - \rho_0)^2\rho_0}, \tag{6.42}$$

and certain dimensionless quantities determined earlier, Eqs. (5.107) and (5.108), are also used.

The extremum value of the effective action is

$$Y_* = \Delta F_0 \tau_0 Y_0 = \frac{2^{13/4} m^{1/2} \kappa^2 |\rho_{sp} - \rho_0|^{1/2}}{|f'''|^{3/2} \rho^{1/2}} I_0, \tag{6.43}$$

where

$$Y_0 = \min \left| \text{Im} \int_0^\theta \aleph\{\vec{u}\} d\theta' \right| \tag{6.44}$$

is a numerical constant equal to $Y_0 \simeq 10^2$. The integral in Eq. (6.44) is taken with respect to the imaginary time along the optimal trajectory from the entry to the exit point of the barrier. The pre-exponential factor in Eq. (6.10) is by the order of magnitude equal to [8]

$$C_0 B \sim (R_0^3 \tau_0)^{-1} = \frac{|f'''|^{5/2}|\rho_{sp} - \rho_0|^{5/2} \rho_0^{1/2}}{2^{13/2} m^{1/2} \kappa^2}. \tag{6.45}$$

In the case of quantum tunneling of nuclei, the exponent in Eq. (6.10) does not depend on temperature, and the dependence of the effective action on the degree of metastability, which is given by the value of $\rho_{sp} - \rho_0$, is stronger than its effect on the Gibbs number at thermo-fluctuational nucleation. From the equality of exponents in the equations for thermo-activational and quantum nucleation, we have

$$T_* = \frac{\hbar}{4k} \frac{|f'''||\rho_{sp} - \rho_0|\rho_0^{1/2}}{m^{1/2}\kappa^{1/2}}. \tag{6.46}$$

Near the spinodal, the domain of applicability of the considered approach is limited by the applicability of the quasiclassical approximation, $I_*/\hbar \gg 1$. This criterion is similar to the inequality (5.132) for thermo-activated nucleation.

The effect of the finiteness of the relaxation time on the rate of quantum nucleation in the vicinity of a liquid–vapor spinodal has been considered by Burmistrov and Dubovskii [467]. Small density oscillations are sound waves. The character of the sound depends on the value of the product of the oscillation frequency with the relaxation time. The finiteness of the relaxation time results in the frequency dispersion of the sound velocity and its absorption. This effect is particularly important in quantum liquids as the relaxation time is strongly temperature dependent.

In the approximation of one relaxation time, the relation between the wave number, k, and the frequency, ω (dispersion equation), is written as

$$C_=^2 k^2 = \omega^2 \frac{1 - i\omega\tau_r}{1 - i\omega\tau_r (C_\infty/C_=)^2}. \tag{6.47}$$

Here, $\tau_r = \tau_r(T)$ is the relaxation time depending on the temperature T, $C_=$ and C_∞ are the low-frequency and the high-frequency limits of the sound velocity, respectively. The low-frequency limit, $\omega\tau_r \ll 1$, corresponds to an ordinary hydrodynamic sound, the high-frequency limit, $\omega\tau_r \gg 1$, to a collision-free sound propagation.

Examining the process of quantum growth of a new-phase nucleus with the help of an effective action on an imaginary time, Eq. (6.40), the authors of Ref. [467] found the nucleation rate and evaluated the temperature of the thermo-quantum crossover depending on the relaxation time, τ_r. Between the extreme limits $\tau_r = 0$ and $\tau_r = \infty$ the following relationship is true:

$$I_\infty(T) = \frac{C_=}{C_\infty} I_0 \left(\frac{C_=}{C_\infty} T \right). \tag{6.48}$$

At $T = 0$ and a zero value of the relaxation time ($\tau_r = 0$), the action is evaluated as

$$I_0 \approx 2560 k^2 \frac{\rho(\rho - \rho_{sp})^2}{C_=^3(\rho)}. \tag{6.49}$$

The quantum nucleation rate increases monotonically with increasing relaxation time, τ_r, with a saturation in the limit, $\tau_r = \infty$.

Depending on the ratio, $C_\infty/C_=$, one can distinguish several regimes of quantum nucleation. If $C_\infty/C_= \geq 1$, then, depending on the ratio between the relaxation time and the proximity to the spinodal, the low-frequency or the high-frequency regimes of nucleation can be observed, respectively. In the case of a strong inequality, $C_\infty/C_= \gg 1$, a crossover from the low-frequency regime proceeds through a transition regime, which corresponds to the viscous overdamped nucleation regime. If the relaxation time, $\tau_r(T)$, increases without limit at $T \to 0$, the temperature dependence of the nucleation rate, $J = J(T)$, may have a nonmonotonic behavior with a minimum in the region of temperature, T_*, of transition from classical to quantum nucleation. The lines of limiting supersaturation of the metastable phase (J = const) in this case have a maximum near T_*. The larger the ratio $C_\infty/C_=$, the higher the relative value of the effect. When $\tau_r^{-1}(T \to 0) \to 0$, then T_* decreases in proportion to the product $C_\infty(\rho) \cdot C_=(\rho)$ becoming zero on the spinodal simultaneously with the Gibbs number. Therefore, the nucleation process in the vicinity of the spinodal has a thermo-activation character.

Nucleation in the vicinity of the spinodal proves to be qualitatively different in Bose and Fermi liquids, respectively. In the Bose liquid (^{4}He), the relaxation time, τ_r, rapidly increases with decreasing temperature. Therefore, quantum nucleation corresponds to the high-frequency limit ($\omega \tau_r \gg 1$), and the sub-barrier increase of the density fluctuation takes place in the collision-free regime, when the length of the free path of excitations exceeds the characteristic size of a nucleus. Owing to the small value of the ratio $(C_\infty - C_=)/C_= \sim \rho_n(T)/\rho \ll 1$ at low temperatures, quantitative distinctions between the low-frequency and the high-frequency nucleation regime, and also temperature effects connected with the difference in the rates, C_∞ and $C_=$, are small. Therefore, a noticeable temperature change in the rate of quantum nucleation in ^{4}He is not to be expected.

In the normal Fermi liquid, the low-frequency and the high-frequency modes of sound propagation are connected with different physical mechanisms. Here, the sound velocities $C_=$ and C_∞ are different even at $T = 0$. The time of collisions between quasiparticles has a weaker temperature dependence than in the Bose liquid. Therefore, in liquid ^{3}He a noticeable minimum in the temperature behavior of the nucleation rate in the region of transition from the quantum to the thermo-fluctuational nucleation regime can be expected. This conclusion implies with regard to the lines J = const that the value of $\delta \rho_n = \rho_n - \rho_{sp}$ passes through a minimum in going over from the thermo fluctuation to the quantum nucleation regime with a subsequent, almost temperature-independent behavior at sufficiently low temperatures.

The kinetics of quantum nucleation near the diffusion spinodal of superfluid solutions of ^{4}He–^{3}He is examined in Ref. [455]. The spinodal is approximated by

$$c_{sp} = c_0 \left(1 + \beta T^{2,33}\right). \tag{6.50}$$

The parameters of this equation, $c_0 = 0.1646$ and $\beta = 4.217$, are determined at $T < 0.3$ K from experimental data on the chemical potential of ^{3}He in a solution. Equation (6.50) represents in a qualitatively correct way the configuration of the diffusion spinodal in the whole temperature interval from $T = 0$ to the tricritical point (Fig. 6.2). The influence coefficient, κ, is found from the data on interfacial tension on the planar surface of discontinuity, according to Eq. (3.119). Within the framework of the model of a regular solution we have

$$\sigma_\infty(T \simeq 0) = 2(\kappa e_0)^{1/2} \rho_0^{3/2}$$
$$\times \int_0^1 \left\{ [c(1-c)]^{1/2} + \frac{k_B T}{e_0} \frac{c \ln c + (1-c) \ln(1-c)}{[c(1-c)]^{1/2}} \right\} dc, \quad (6.51)$$

where e_0 is the energy of mixing. The integral of the first term in Eq. (6.51) is taken analytically, of the second term, numerically. The result of the integration is

$$\sigma_\infty(T \simeq 0) = 2(\kappa e_0)^{1/2} \rho_0^{3/2} \left[\frac{\pi}{8} - 1.205 \frac{k_B T}{e_0} \right]. \quad (6.52)$$

From Eq. (6.52) and the experimental data on σ_∞ [57] we get $\kappa = 2.5 \cdot 10^{-14}$ Jm^5 and $e_0 = 3.09 \cdot 10^{-23}$ J.

The asymptotic behavior of the lines of limiting supersaturation near the diffusion spinodal for the cases of thermo-fluctuational (Eq. (5.131)) and quantum (Eqs. (6.10), (6.43), (6.45)) nucleation is shown in Fig. 6.7. Nowadays, there is no theory which allows one a calculation of the effective action in that region of state parameters where the macroscopic description of nuclei is no longer applicable. Therefore, we estimate the temperature of change of the nucleation regime as follows. At low supersaturations and states near to the spinodal, the concentration dependence of T_* is determined by Eqs. (6.9) and (6.46). In the intermediate region of concentrations, we describe the behavior by assuming some kind of crossover between these limiting behaviors. Determining the crossover function as

$$\varphi(c) = b(1 + d\Delta c), \quad (6.53)$$

where $b = -23.8, d = -32.2$, we can write

$$T_* = \varphi(c)\Delta c^{3/2}(c_{sp} - c). \quad (6.54)$$

The dependence $T_*(c)$ reaches its maximum in the range of about 20 mK. According to Eq. (6.54), for the case of homogeneous formation of a new phase the transfer from the thermo activational to the quantum mechanism takes place in a rather narrow interval of temperatures and concentrations (see Fig. 6.7). This condition results in strict requirements concerning the methods of carrying out experiments on the discovery of quantum tunneling of nuclei in the system ^{4}He–^{3}He. We have to mention that estimations given above are only valid by the order of magnitude.

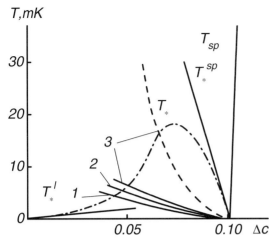

Fig. 6.7 Temperature of change of the nucleation mechanism in solutions of ^{4}He–^{3}He. T_*^l: near the line of phase equilibrium (Eq. (6.9)); T_*^{sp}: near the spinodal (Eq. (6.46)); T_*: result of Eq. (6.54). 1, 2, 3: asymptotic behavior of the lines of limiting supersaturation near the spinodal (Eq. (5.131)) for $J = 10^{-10}$ m^{3}s^{-1} (1), 10^0 (2), 10^{10} (3); *dashed line*: the expected behavior of the line of limiting supersaturation; T_{sp}: spinodal curve.

6.5
Quantum Cavitation in Helium

Helium-4 and helium-3 remain liquid as the temperature decreases to absolute zero (if the static pressure does not exceed $\sim$ 2.5 MPa for ^{4}He and 3.5 MPa for ^{3}He). This behavior is due to the fact that at a certain temperature these liquids become degenerate, and the energy of zero-point oscillations of the atoms exceeds the energy of interaction. For liquid ^{4}He, for which molecular-kinetic phenomena are described by the Bose–Einstein statistics, the temperature of quantum degeneration coincides with the temperature of the λ-transition, the transition from the normal (HeI) into the superfluid (HeII) phase. The Fermi liquid (^{3}He) retains the normal state below the temperature of quantum degeneration down to very low temperatures.

Homogeneous nucleation in degenerate liquid ^{4}He and ^{3}He takes place at negative pressures and very low temperatures, when the occurrence of quantum nucleation can be expected. The first evaluations of the temperature, T_*, and pressure, p_*, of the transition from thermal activation to nuclei tunneling (quantum-crossover) in HeII were made by Akulichev [101]. Employing the data obtained by Akulichev [101], in the absence of initial cavitation nuclei, we get $T_* \simeq 0.31$ K, the cavitation strength, Δp_*, being equal to 1.46 MPa. Akulichev's results were revised by Maris and Xiong [214, 224, 468], who approximated the spinodal of liquid helium-4 and discovered that at

$T = 0$, $p_{sp} = -0.952$ MPa. This value of $p_{sp}(0)$ was later confirmed in Refs. [241, 469] utilizing the data obtained by Caupin and Balibar [215] and resulting in $p_{sp} = -0.964$ MPa. A value close to this one was also obtained in the latest calculations of Maris and Edwards [228] giving $p_{sp} = -0.962$ MPa. By evaluating the work of formation of a critical nucleus in the framework of the van der Waals capillarity theory [7], Maris and Xiong [214] and Guilleumas et al. [234] obtained the following values for the parameters of crossover in HeII: $T_* = 200$ mK and $p_* = -0.9$ MPa. The results of Refs. [214, 234] were based only on an analysis of the thermodynamic parameters of ^{4}He. The kinetic parameters and the processes in this case were not considered.

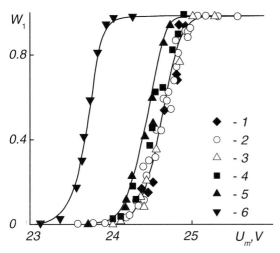

Fig. 6.8 Probability of cavitation in superfluid helium-4 close to the absolute zero of temperature depending on the voltage amplitude on the piezoradiator at different temperatures [232]. 1: $T = 65$ mK; 2: 171; 3: 254; 4: 350; 5: 465; 6: 750.

The first attempt to realize the regime of quantum tunneling of vapor-phase nuclei in experiments on cavitation in superfluid helium-4 was made by Lambare et al. [232]. In studying acoustic cavitation, the authors [232] decreased the temperature down to 65 mK. A measuring cell similar to that used in Refs. [212, 213] was secured on the chamber of a helium mixing refrigerator. For decreasing the thermal diffusion on the radiator, exciting radio-wave pulses were used which were shorter by 30–70 μs than in the experiments discussed in Ref. [213]. As in the paper of Petterson et al. [213], a study was made of the statistical laws of the cavitation process depending on the amplitude of the voltage and the temperature applied to the radiator.

Figure 6.8 presents the probability of emergence of cavitation events, W_1, as a function of the voltage on a piezoradiator for several temperatures in the range (65–750 mK). The voltage $U_{*1/2}$, at which the probability of cavitation $W_1 = 1/2$, was defined as the cavitation threshold. The transfer from the emergence of events of cavitation to the moment of its hundred-per-cent realization takes place in the range of voltages that do not exceed 4% of the threshold value. As it is evident from Fig. 6.8, at temperatures below 600 mK the voltage $U_{*1/2}$ is temperature independent. This fact is illustrated more clearly by Fig. 6.9, which presents temperature dependences of the voltage $U_{*1/2}$ for three different values of the static pressure. All the curves demonstrate the presence of a plateau below ~ 600 mK. At higher temperatures, the voltage $U_{*1/2}$ decreases with increasing temperature $T < 600$ mK, which is regarded by the authors of Ref. [232] as a transition from the thermo-activation mechanism of nucleation to quantum tunneling of nuclei.

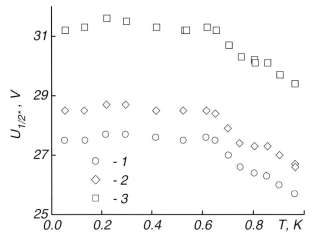

Fig. 6.9 Amplitude of cavitation stress in HeII as a function of temperature at different static pressures [215]. 1: $p = 130$ kPa, 2: 42.6; 3: 3.7.

At quantum cavitation, the nucleation rate is determined by Eq. (6.10). Using Eq. (6.10) in Eq. (3.124) and taking $J_0 = C_0 B$, the extremum action, $I_{*1/2}$, can be determined corresponding to the value of $W_1 = 1/2$. We obtain

$$I_{*1/2} = \ln\left(\frac{J_0 V \bar{\tau}}{\ln 2}\right). \tag{6.55}$$

According to data of Maris [214], for ^{4}He we have $J_0 - 2 \cdot 10^{37}$ m^{-3}s^{-1}. Under the assumption of a weak influence of nonlinear effects on the shape of an acoustic wave, we have

$$V = \lambda^3 \left(\frac{3}{2\pi A}\right)^{3/2}, \quad \bar{\tau} = T_\sim \left(\frac{1}{2\pi A}\right)^{1/2}, \tag{6.56}$$

where λ and $T_\sim$ are the length and the period of an acoustic wave, respectively, and

$$A = \left(\frac{I_{*1/2}}{\hbar}\right)\left(\frac{d \ln I}{d \ln p}\right). \tag{6.57}$$

The derivative in Eq. (6.57) has been calculated by Maris [214]. Using the obtained values of A in Eqs. (6.55)–(6.57), we get $I_{*1/2} = 32\hbar$ and $V\bar{\tau} = 3.1 \cdot 10^{-12}$ m^{3}s. Then from Eq. (6.4) it follows that the limiting stretch corresponding to the transition from the thermo-fluctuation regime to quantum tunneling is $p_* = -0.927$ MPa. This value of $-p_*$ is only 0.035 MPa smaller than the pressure on the spinodal of helium-4 at $T = 0$.

For stretched ^{4}He, theoretical estimations [214, 233] of the value of T_* prove to be approximately three times smaller than the temperature below which the cavitation strength in an experiment [232] becomes independent of T (Fig. 6.9). According to Refs. [215, 240], this contradiction is apparent and can be explained as follows: in a first approximation, an acoustic wave is adiabatic, and therefore the temperature in the cavitation zone changes in the same way as the pressure. At temperatures $T < 0.7$ K, the isentropic process in superfluid helium is characterized by a linear dependence of the sound velocity on temperature [172]. The sound velocity measured at $p_n = -0.923$ MPa (75 m·s^{-1}) is approximately three times lower than that measured at $p = 0$ (238 m·s^{-1}). Thus, the local temperature of the liquid in the piezoradiator focus at the moment preceding the initiation of cavitation has to be three times lower than the temperature of the liquid in the cell, i.e., to be approximately equal to 200 mK, predicted by theory [214, 234]. The data corrected in such a way are presented in Fig. 6.10.

Burmistrov and Dubovskii [467] do not agree with such an interpretation of experimental data performed by Lambare et al. [232], Caupin and Balibar [215]. In order to achieve that temperature variations in a sound wave follow pressure variations, it is necessary to make nonlinear effects negligible, and the average length of phonon–phonon scattering has to be much smaller than the sound wavelength ($l_{ph} \ll \lambda$). Experiments [215, 232] were conducted at rates $\omega \simeq 1$ MHz. The characteristic time of sub-barrier motion in this case is $\tau_0 = 10^{-10}$ s. Thus, $\omega\tau_0 < 1$, and nonlinear effects may really be neglected. The length of phonon–phonon scattering, l_{ph}, in HeII at 0.7 K is 1.3 mm. It increases with decreasing temperature in proportion to T^{-7} and reaches 15 mm at $T = 0.5$ K. At $\lambda \sim 0.1$ mm, it means that the local temperature in a sound wave does not follow the pressure variations, and the temperature in the piezoradiator focus in this case may not differ considerably from the temperature of the liquid in a cell.

Experiments [215, 232] were conducted close to the spinodal, where the sound velocity is small. In this case, the kinetic energy of a growing bub-

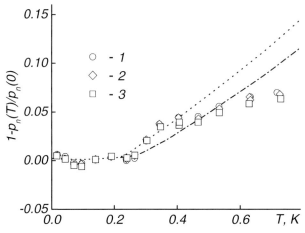

Fig. 6.10 Comparison of experimental data [215, 232] on cavitation in superfluid helium-4 close to the absolute zero of temperature with quantum nucleation theories of Maris [214] (*dashed line*) and the Barcelona group [233] (*dotted line*) after temperature correction in the cavitation zone.

ble decreases considerably, which—according to data of Refs. [430, 433]—leads to an increase in the temperature of the crossover transition, T_*. Thus, $T_* = 0.6$ K may quite be the genuine temperature of the quantum-crossover [467]. Attempts to achieve the regime of quantum tunneling of nuclei of the vapor phase in normal liquid helium-3 were made in Refs. [229–231]. According to Maris [214], the temperature of the crossover transition from thermo-activation nucleation to quantum tunneling in normal liquid helium-3 is 120 mK, and the pressure on the spinodal at $T = 0$ is equal to -0.31 MPa. The data on limiting stretches of ^{3}He obtained in Refs. [229–231] are close to the spinodal ones ($T < 100$ mK) and do not reveal any temperature-independent section on the curve of limiting stretches (see Figs. 3.31 and 3.32). Caupin, Balibar and Maris [235] connect this behavior with the peculiar properties of a Fermi liquid in the region of negative pressures.

In the investigated temperature range (40–1000 mK), ^{3}He is a normal viscous Fermi liquid. In Ref. [214] in order to evaluate T_*, the authors ignored the viscosity of helium-3. According to Eqs. (6.1) and (6.13), the internal friction decreases the probability of quantum nucleation. The dissipation contribution connected with the internal friction is determined by the ratio between the length of the free path of a quasiparticle, l_F, and the characteristic nucleus radius, R_c. In a Fermi liquid, $l_F \sim T^2$ [470]. At $T = 1$ K, the value of $l_F \simeq 0.05$ nm, i.e., $l_F \ll R_c \simeq 0.1$ nm. With decreasing temperature, the value of l_F increases and reaches 5 nm at $T = 0.1$ K. Here, $l_F \gg R_c$. Thus, in the range (1–0.1 K) nucleation proceeds in the hydrodynamic regime (see

Eq. (6.17)), and the limiting temperature, T_l, in the vicinity of the spinodal of ^{3}He is equal to $\simeq 0.18$ K. This is a higher value of T_* than that obtained by Maris [214]. At $T < T_l$, the dissipation contribution to the effective action is determined by the limit (specified by Eq. (6.16)) and the temperature of the quantum crossover can be written as [467]

$$T^* = \frac{16}{27} \frac{\hbar \sigma R_c^2}{\Delta I_{\text{diss}*}}. \tag{6.58}$$

The calculations performed in Ref. [467] have given a value of $T_* \simeq 2$ mK for normal helium-3. This value is much lower than the temperatures at which measurements of the cavitation strength of ^{3}He were made [229–231]. Thus, as distinct from superfluid ^{4}He, in normal ^{3}He dissipative processes connected with the viscosity result in the main contribution to the effective action. In this case, the temperature of the thermo-quantum crossover, T_*, proves to be much lower than that given by the nondissipative model [471].

6.6
Some Other Problems of Phase Metastability

According to the arguments outlined in Ref. [23], a supercooled simple fluid keeps its thermodynamic stability at temperatures as low as desired. If no other kinds of instability appear in this metastable system and the kinetic difficulties connected with the processes of nucleation and vitrification are overcome, the liquid can be cooled to the temperature of its degeneration, T_d, whereupon solidification is no longer obligatory. In liquids composed of molecules which have no spin, Bose condensates are formed at a temperature lower than the temperature of degeneration, which is manifested in the superfluidity of a quantum liquid. For the ideal Bose gas, the temperature of Bose condensation is

$$T_{\lambda_0} = 3.31 \frac{\hbar^2}{k_B} \frac{\rho'^{2/3}}{m g^{2/3}}, \tag{6.59}$$

where m is the atomic (molecular) mass, ρ' is the number density, and g is the statistical weight. In the case of ^{4}He, the temperature is $T_{\lambda_0} \simeq 3.1$ K, whereas—according to experiment—the temperature is $T_\lambda = 2.172$ K. The difference between T_{λ_0} and T_λ is evidently connected first and foremost with the inadequacy of the model of the ideal degenerate gas applied to liquid helium.

As it follows from Eq. (6.59), the smaller the molecular mass of a substance, the higher the temperature of Bose condensation. Ginzburg and Sobyanin [472] draw the attention to the fact that hydrogen, the molecules of which are Bose particles with a mass less than the mass of helium atoms, has a rather high temperature of Bose condensation. From Eq. (6.59), for molecular hydrogen, $T_{\lambda_0} \simeq 6.6$ K is found. This value is approximately half as high as the temperature of fusion of hydrogen, $T_m = 13.8$ K. It is expected [438] that the account of molecular interactions in hydrogen decreases the value of T_{λ_0} by 2–3 K.

After the publication of Ref. [472], several methods were proposed regarding how the temperature of the λ-transfer in molecular hydrogen can be obtained. In addition, first attempts of their experimental realization were made [473, 474]. All methods offered are connected in one way or another with the creation of phase metastability. For the attainment of T_λ, Ginzburg and Sobyanin recommended to use the decrease of the temperature of fusion of hydrogen at its extension into the range of negative pressures. However, as mentioned in Ref. [101], the quasistatistical extension of liquid hydrogen with the following cooling results in its disruption or in crystallization at the temperature of degeneration. At the same time it is possible, according to Ref. [101], to overcome the temperature barrier of 6 K in the dynamic regime when the frequencies of nucleation are not lower than 10^{15} m^{-3}s^{-1}.

The fusion temperature of small samples depends on their size [117]. Bretz and Thomson [473] examined crystallization and fusion of molecular hydrogen in mesopores of Vycor glass whose channel diameter was $\simeq 6$ nm. The minimum value of the temperature of crystallization was 9.5 K, which was 4.3 K lower than the fusion temperature of a massive sample. Another way to achieve superfluidity in molecular hydrogen connected with supercooling of small drops was studied by Maris et al. [474]. The measuring camera was a glass cylinder with cupric lids. The temperature of the top lid was $\simeq 15$ K, of the bottom lid $\simeq 7$ K. The camera was filled with gaseous helium. The temperature gradient led to the appearance of a stationary distribution of the density of helium throughout the height of the camera. Drops of liquid hydrogen from 50 μm to 1 mm in diameter, which floated freely at a certain height of the camera, were injected into the camera. It was possible to vary the position and consequently the temperature of the drops by pressure change. Maris et al. observed drops with the help of a long-focus microscope. A drop descended to a lower level after crystallization. In the experiment, Maris et al. determined the lifetime of the drop and calculated the mean lifetime, $\bar{\tau}$, and the frequency of nucleation, J, for the ensemble of crystallized drops. The maximum value of the achieved supercooling of drops of liquid hydrogen was $\Delta T_n = T_m - T_n \simeq 3.2$ K.

The results of the experiments are shown in Fig. 6.11. The temperature dependence of J calculated according to the theory of homogeneous nucleation is also presented there. In application to crystallization of a supercooled liquid

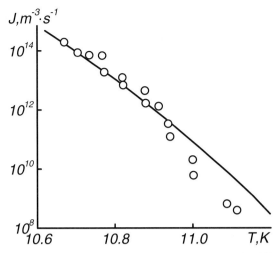

Fig. 6.11 Frequency of nucleation in supercooled liquid hydrogen [474]. *Line*: calculation according to the theory of homogeneous nucleation.

the kinetic factor in Eq. (2.112) is determined by the following expression [117]:

$$B = \frac{k_B T}{\hbar} \exp\left(-\frac{E_a}{k_B T}\right), \quad (6.60)$$

where E_a is the energy of activation of the process of molecule transfer from the liquid to the crystal phase, which in Ref. [474] was set equal to the energy of activation of self-diffusion in liquids. The work of formation of a crystal nucleus is

$$W_* = \frac{16\pi}{3} \frac{\sigma_{ls}^3}{(\mu_l - \mu_s)^2 \rho_s^2}. \quad (6.61)$$

Here, the index l refers to the liquid and s to the crystal. The chemical potential difference of a liquid and a crystal taken at the temperature and the pressure of the metastable phase is approximated by the following expression:

$$\mu_l - \mu_s \simeq \frac{(T_m - T)l_{ls}}{T_m} + \frac{1}{2}\frac{(T_m - T)^2(c_s - c_l)}{T_m}, \quad (6.62)$$

where l_{ls} is the heat of fusion and c is the heat capacity. The interfacial tension, being equal to $\sigma_{ls} = 0.874$ mN·m^{-1}, has been determined from the condition of the best agreement between theory and experiment at high frequencies of nucleation (see Fig. 6.11).

For hydrogen, the curve of the temperature dependence of the frequency of nucleation is dome shaped (Fig. 6.12). If the interfacial tension is temperature independent, then $J_{max} \simeq 10^{22}$ m^{-3}s^{-1}, and $T_{max} \simeq 7$ K. With allowance for the dependence of σ_{ls} on T in the framework of the Woodruff model [475], the maximum shifts in direction of higher temperatures and $J_{max} \simeq 10^{18}$ m^{-3}s^{-1}. Drops of a size 1 μm in diameter at $T = T_{max}$ remain liquid on average during ~ 2 s. Even if drops of molecular hydrogen are supercooled during T_{max}, the question about their vitrification remains vague. Moreover, at rather low temperatures, thermo-fluctuational nucleation is to give way to quantum tunneling. In Fig. 6.12, the lines A' and B' specify the temperature dependence of the frequency of nucleation in the region of change of nucleation regimes for the cases of $\sigma_{ls} = $ const and $\sigma_{ls}(T)$ [474], respectively. In this case the temperature of the quantum crossover is equal to 2–3 K.

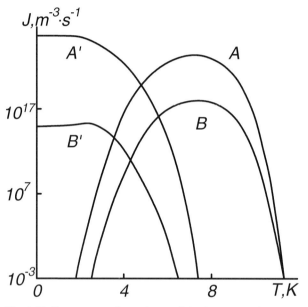

Fig. 6.12 Temperature dependence of the frequency of nucleation in supercooled liquid hydrogen at the thermo activated (A, B) and the quantum (A', B') mechanism of formation of the crystal phase. Curves A, A' are obtained at the approximation σ_{ls} =const.; B, B' at the approximation $\sigma_{ls} = \sigma_{ls}(T)$ [474].

Other systems, where we can expect the manifestation of quantum tunneling of nuclei, are supercooled liquid ^{3}He, ^{4}He, and solid metastable solutions of ^{3}He–^{4}He. Crystallization of supercooled superfluid helium-4 was investigated by the methods of hydrostatic compression [476, 477], and also by the compression in acoustic [478] and electric [479, 480] fields. In all the above-mentioned experiments, crystallization took place at supercompressions $\Delta p_n = p_n - p_m = (0.35 - 10)$ kPa, where p_m is the pressure on the fusion line. The maximum depth of invasion into the range of supercooled states of ^{4}He was achieved in experiments with focused acoustic fields [481–483]. As in the investigation of cavitation [211, 213], acoustic vibrations in liquid helium ($f = 1$ MHz) were created and focused in a volume with a linear dimension $\simeq 0.36$ mm with the help of a hemispherical piezoprojector. The duration of acoustic impulses was 3–6 µs. The amplitude of the focus pressure reached ± 2 MPa. In order to determine the pressure of crystallization of supercompressed superfluid helium, a glass plate was used which was placed in such a way that the acoustic vibrations were focused in the transition helium/glass layer. In the experiment, the intensity of the reflected and the transmitted light was measured from an Ar^+ laser whose beam was directed to the acoustic focus at the helium/glass boundary. Employing transmitted light, the moment of liquid crystallization was registered, and by measurements of the intensity of the reflected light the index of environment refraction was determined. The density of the liquid phase at the moment preceding crystallization was calculated by the index of refraction with the help of the Clausius–Mossotti equation. A static pressure equal to $p_m = 2.5324$ MPa was created in the measuring cell. The pressure of crystallization, p_n, was calculated according to the data on the density from the equation of state of ^{4}He [481].

The experiments [481–483] were carried out in the range of temperatures from 30 mK to 1.5 K. The nucleation of the crystal phase of helium has a stochastic nature. The possibility of crystallization as a function of density was similar to that presented in Fig. 6.8. The density of the liquid phase corresponding to the probability of crystallization, $W_1 = 0.5$, was recognized as the limiting value of the density, ρ_n. The temperature dependence of the limiting density, ρ_n, is given in Fig. 6.13. At $T \leq 300$ mK, the value of ρ_n does not depend on temperature. According to Refs. [481, 482], this result is due to quantum tunneling of nuclei of the crystal phase of ^{4}He. In the limit of low temperatures, $\rho_n(0) = (0.17552 \pm 0.00010)$ g/cm^3 holds. The pressure $p_n = p_m + (0.43 \pm 0.02)$ MPa corresponds to this density. Balibar [240] estimated the value of the activation barrier according to the experimental data on the probability of crystallization of ^{4}He and obtained the following value $G_* = W_*/(k_B T) \simeq 10$. For this value of the Gibbs number and for the frequencies of nucleation realized in the experiments, the theory of homogeneous nucleation gives the value of the supercompression $\Delta p_n \simeq 1.6$ MPa. In this case the radius of a critical nucleus is $R_* \simeq 0.8$ nm and the temperature of the quantum crossover is $T_* \simeq 0.8$ K. If the supercompression $\Delta p_n = 0.43$ MPa registered in the experiment would be in agreement with homogeneous nucleation then $R_* \simeq 10$ nm and $W_*/(k_B T) \simeq 10^4$. Thus the formation of the crystal phase of helium in the experiments [481–483] most likely took place at heterogeneous centers where liquid helium and the glass plate are in contact. However, this does not exclude the possibility of quantum nucleation at $T < 300$ mK. The phenomenon of phase stratification takes place not only in liquid but also in solid solutions of ^{3}He–^{4}He [172]. Stratification is observed

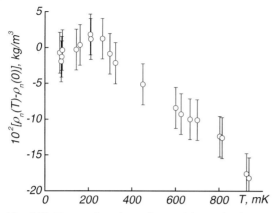

Fig. 6.13 Temperature dependence of the nucleation threshold density, $\rho_c(T)$ [481].

at temperatures lower than 0.4 K. The temperature of stratification depends weakly on the value of the applied pressure.

The initial stage of the kinetics of phase decay of weak solid solutions of ^{4}He in ^{3}He and ^{3}He and ^{4}He was investigated in Refs. [484–486]. A sample of a solid solution of ^{3}He–^{4}He was grown from a gas mixture in a disk-shaped metal cell with an inner diameter of 9 mm and a height of 1.5 mm. After the preparation, the sample was disconnected with the filling system by the block of the supplying capillary. Supersaturation was created by supercooling a crystal over the line of phase stratification to temperatures t_n (200–300 mK). During the process of phase stratification, the pressure in the cell was measured as a function of time [484]. The size of forming drops was registered by nuclear magnetic resonance [486]. In Refs. [484, 485], the concentration of ^{4}He in ^{3}He varied in the interval of 2.2–3.34%. The pressure in the cell was 3.15–3.86 MPa. The dependence $p = p(\tau)$ was an exponential one, i.e., $p(\tau) \sim \exp(-\tau/\tau_0)$. At high supercoolings, the value of τ_0 did not depend on the temperature and the concentration of the solution.

The concentration of nuclei of a new phase at the final stage of nucleation is determined by the following expression [487]:

$$N_m \simeq c_0^{7/4} \beta^{3/8} \exp\left[-\frac{3}{8}\beta^3 \left(\ln \frac{c_0}{c_n}\right)^{-2}\right], \tag{6.63}$$

where c_0 is the initial concentration of the solution and c_n is the equilibrium value of concentration corresponding to the temperature T_n,

$$\beta = \frac{8}{3}\pi \frac{a^2 \sigma}{T_n}, \tag{6.64}$$

σ is the interfacial tension and a is the interatomic distance.

The characteristic time of nucleation is

$$\tau_N = \left(\frac{4c_0}{N_m}\right)^{1/3} \frac{a^2 c}{D_g \beta'}, \tag{6.65}$$

where D_g is the diffusion coefficient. The characteristic time of duration of the stage of diffusional growth of nuclei is

$$\tau_D \simeq \frac{a^2 N_m^{-2/3} c_0^{-1/3}}{D_g}. \tag{6.66}$$

The authors of Refs. [484, 485] estimated the interfacial tension from Eqs. (6.63) and (6.65) according to the experimental data on c_0, c_n, N, τ. According to all investigated values of concentration, the average value of σ was $1.8 \cdot 10^{-2}$ mN·m^{-1}. The estimation of σ from Eqs. (6.63) and (6.66) yielded $1.3 \cdot 10^{-2}$ mN·m^{-1}. The authors of Ref. [484] considered the continuity of the value of σ obtained by two methods of calculation as an evidence of the homogeneous mechanism of nucleation. Similar work was performed for weak solutions of ^{3}He in ^{4}He ($\sim$ 1%) [486]. Using two methods for estimating the interfacial tension, we obtained $\sigma \simeq 1.27 \cdot 10^{-2}$ mN·m^{-1}. When the concentrations of ^{3}He were equal to 2% at the last stages of the phase transition, the characteristic size of the evolved grains of a new phase was growing in time according to the Lifshitz-Slezov law [368].

Superfluidity is typical for liquid ^{3}He, too. At temperatures lower than 3 mK, atoms of ^{3}He, which are fermions, form Cooper pairs (molecules) [488]. The strong anisotropy of Cooper molecules leads to the circumstance that superfluid ^{3}He has several phases, which combine in themselves the properties of an ordered magnetic phase, a liquid crystal, and superfluid ^{4}He. In the absence of a magnetic field there exists two stable superfluid phases, ^{3}He-A and ^{3}He-B (Fig. 6.14). At pressure $p_c \simeq 2.12$ MPa and temperature $T_c \simeq 2.55$ mK, they maintain equilibrium with each other and with the normal phase. If the transition of the normal into the superfluid phase is a phase transition of second order, the transition between the superfluid phases ^{3}He-A and ^{3}He-B at temperatures lower than T_c is of first order.

Considerable supercoolings [489, 490] are characteristic of the superfluid A-phase. According to Ref. [490], at a temperature of 0.15 T_c (0.39 mK), the lifetime of the metastable A-phase is approximately 30 min, which is shorter than that expected from the theory of thermo-fluctuational nucleation [491]. A small difference in the free energies of bulk phases and a relatively large value of the interfacial energy lead to anomalously large values of radii of critical nuclei in the supercooled A-phase. For $T = 0.15 T_c$ and $p = 3.4$ MPa, the radius of a critical nucleus is $R_* \simeq 0.55$ μm, and the number of atoms contained in it is 10^{10}. A spontaneous center of crystallization in supercooled liquid hydrogen contains only a few hundreds of molecules.

6 Nucleation Kinetics Near the Absolute Zero of Temperature

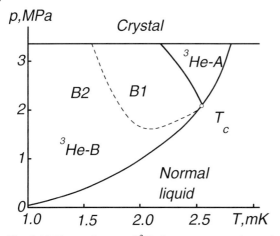

Fig. 6.14 Phase diagram of ^{3}He for temperatures lower than 3 mK.

Leggett [491] assumes that the reason for the "premature" disruption of the metastable state of the superfluid A-phase is the initiating influence of background radiation. When a particle of high energy passes through the metastable A-phase, the normal phase of helium-3 appears on the thermal spikes generated by δ-electrons (see Fig. 6.15). After the cooling of a thermal spike, the B-phase appears in it, which eliminates phase metastability. There are other points of view on the initiation of the decay of the supercooled state of ^{3}He-A [492,493].

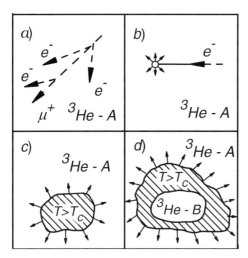

Fig. 6.15 Process of formation of the B-phase in the supercooled A-phase under the influence of cosmic radiation (muons).

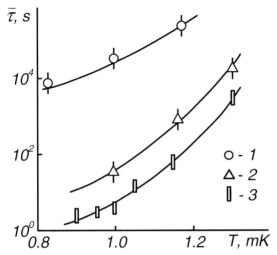

Fig. 6.16 Mean lifetime of the metastable A-phase as a temperature function in a magnetic field with an induction of 28.2 mT [490]. 1: natural conditions; 2: thermal neutron radiation; 3: γ-radiation field.

The influence of γ-radiation and of neutrons on the stability of the metastable A-phase was studied in Ref. [490]. At radiation of the supercooled A-phase by γ-quanta from the source ^{60}Co, the mean lifetime was diminished by a factor of approximately 1650 in comparison with natural conditions (Fig. 6.16). The increase of the γ-background intensity did not change the nature of the temperature dependence, $\bar{\tau}$. At radiation by neutrons and by γ-quanta, the lifetimes differed approximately by the factor of seven. The results of measuring of $\bar{\tau}$ in the γ-radiation field were approximated by the following formula [490]:

$$\bar{\tau} = A \exp\left[\alpha \left(\frac{R_*}{R_0}\right)^\beta\right], \tag{6.67}$$

where $A = 2.11 \cdot 10^{-4}$ s; $\alpha = 5.25$; $\beta = 1.5$; $R_0 = 0.45$ μm is the radius of a critical nucleus at $T = 0$. According to the Leggett model [490], the exponent in Eq. (6.67) should be equal to 3–5.

The addition of the magnetic field, H, significantly extends possible kinds of metastable states in superfluid ^{3}He [494,495]. By overheating the B-phase, the metastable planar P-phase of ^{3}He can be obtained [494]. It is supposed that the overheated B-phase can keep equilibrium with the metastable P-phase. According to Ref. [494], at $T > T_0$ the transition between the metastable B- and P-phases is a transition of second order, and at $T < T_0$ it is a transfer of first order. The point (T_0, H_0) is a tricritical point for these phases. According to the experimental data of Ref. [496], $T_0 \simeq 0.8\, T_c$ and $H_0 \simeq 2$ kGs.

For the same reasons as in HeII, quantum vortices appear in superfluid ^{3}He. Two types of vortices have been observed experimentally in the B-phase. They are different in structure and separated by a first-order phase transition line (dashed line, Fig. 6.14) [496]. The theory predicts the existence of a large number of vortices of other types. In particular, quantum vortices with numerous circulation quanta can exist near the line of phase equilibrium in the B-phase [495]. Such multiquantum vortices have a large core, which consists of the metastable A-phase and is surrounded by the B-phase. This result is contrary to the case of nucleation where the stable phase is surrounded by the metastable phase.

The investigations of phase transitions in ^{3}He are just beginning. Superfluid ^{3}He is a unique model object not only for the studies of phase metastability but also for solving the problems of cosmology. The propagation of sound through superfluid helium is characterized by the same laws as the propagation of light in a gravitational field. The mathematics of cosmic strings is analogous to the mathematics of defects in liquid helium. According to several scientists [497], the superfluid state of ^{3}He-A precisely replicates all space-time properties and the four-dimensional texture of the universe. The disruption of the metastable phases ^{3}He-A in ^{3}He-B can thus explain the skewness of the universe [493].

7
Explosive Boiling-Up of Cryogenic Liquids

7.1
Superheating in Outflow Processes

The outflow of a compressed liquid with an initial temperature higher than the temperature of normal boiling through short channels is characterized by a high degree of metastability of the flow. At low superheatings, vapor bubbles forming on existing and easily activated boiling centers have no time to provide a noticeable vapor content in a channel. The contribution of spontaneous evaporation centers at $T < 0.9T_c$ is not significant. An increase of temperature leads to the boiling-up of a liquid at the exit of the channel and its closing [498].

Aiming at establishing the mechanism of evaporation in boiling-up flows of cryogenic liquids, in Refs. [499–501] the flow-rate characteristics are investigated for cylindrical channels of length $l = (0.48–4.05)$ mm and diameter $d = 0.32–0.5$ mm during the outflow of nitrogen, oxygen, methane, and argon. The main data array has been obtained on a channel with $l = 0.48$ mm and $d = 0.40$ mm ($l/d = 1.2$). The substance under investigation filled the inner space of the bellows ($V \simeq 0.5$ dm^3) [502]. The pressure was created by compressed helium. At specified values of temperature, T_0, and pressure, p_0, in the chamber, the amount of liquid escaping through the channel into the atmosphere was determined by the value of the compression by the bellows. The outflow time was predetermined by an electronic relay and registered by a frequency meter chronometer. The error in the determination of the specific flow rate did not exceed 5%.

A study was performed regarding the dependence of the flow rate on the initial pressure (by isotherms) or the temperature (by isobars). Measurements were carried out in the temperature range from $T_0 = 0.6T_c$ to the saturation temperature and pressures $p_0 = (0.3–1.5)p_c$. The character of the temperature (pressure) dependence of the specific volume, v, and mass, j, flow rates depends on the channel size as well as on the initial values of temperature, T_0, and pressure, p_0. For channels with $l/d \ll 3$ and pressures in the cham-

Explosive Boiling of Superheated Cryogenic Liquids. Vladimir G. Baidakov
Copyright © 2007 WILEY-VCH Verlag GmbH & Co. KGaA, Weinheim
ISBN: 978-3-527-40575-6

ber $p_0 \ll 0.5p_0$, the temperature dependences of v and j are close to straight lines (Fig. 7.1). A linear dependence is also found for the square of the specific flow rate in dependence on the pressure difference on the nozzle (Fig. 7.2). The flow-rate characteristics ($p_0 \gg 0.5p_c$) have typical breaks indicative of the change in the outflow regime. The temperature at which a bend of experimental curves can be observed depends only slightly on pressure, increasing with it. A transfer from short ($l/d \ll 3$) to long channels results in the degeneration of bends on the flow-rate characteristics [499].

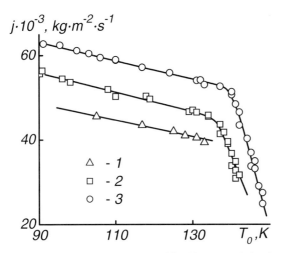

Fig. 7.1 Specific mass flow rate of liquid oxygen during its outflow through a channel with a length of $l = 0.48$ mm and a diameter of $d = 0.4$ mm for different initial pressures in the chamber. 1: $p_0 = 2$ MPa, 2: 3.0, 3: 3.8.

At $T_0 < 0.9T_c$, the specific flow rates are calculated by

$$j = \rho'v = \mu[2\rho'(p_0 - p)]^{1/2}, \qquad (7.1)$$

where $\mu = (0.68\text{--}0.69)$ is the flow-rate coefficient of a nonboiling liquid, are close to those observed in experiment. This result is an indication of an insignificant evaporation in the channel (Fig. 7.3). At the same time, a systematic underestimation of the flow rates registered in experiment is found compared with those values calculated by Eq. (7.1). The latter feature is connected with the liquid evaporation from the free surface of a jet.

A numerical simulation of the process of outflow of nonboiling cryogenic liquids through short channels is performed in Ref. [499]. It was assumed that the vapor forming during evaporation from the jet surface has a velocity equal to the velocity of sound at the exit of the channel and is equal to zero when the temperature of the jet surface becomes equal to the temperature of normal boiling of a liquid. The liquid was considered incompressible, there was no slipping between the liquid and the channel wall. The distributions of pressure, temperature, and liquid velocity in the channel were found from the system of Navier–Stokes, continuity and heat transfer equations

$$(\vec{v}\nabla)\vec{v} = -\frac{1}{\rho'}\nabla p + \frac{\eta}{\rho'}\Delta v, \tag{7.2}$$

$$\Delta \vec{v} = 0, \tag{7.3}$$

$$\vec{v}\nabla T = D_T \Delta T. \tag{7.4}$$

The temperature of the liquid in the channel is close to its value in the chamber, and greater is the value of the excess with respect to atmospheric pressure in the channel, the stronger the liquid is superheated. The results of calculating the specific bulk flow rate of liquid argon by the hydraulic formula (7.1), when the pressure in the channel is determined from Eqs. (7.2)–(7.4), are presented in Fig. 7.3 by dashed-dotted lines.

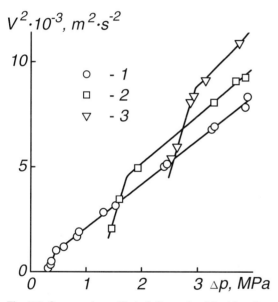

Fig. 7.2 Square of specific bulk flow rate of liquid methane as a function of the pressure difference on the channel for different isotherms. 1: $T_0 = 130$ K; 2: 160; 3: 174. For the values of the channel parameters, see Fig. 7.1.

If $T_0 > 0.9T_c$, the liquid at the exit of the channel is transferred into the zone of intensive fluctuation nucleation. Let us examine the dynamics of spontaneous boiling-up of a liquid passing through a short channel by introducing a number of approximations that simplify the solution of the problem. We shall assume that vapor bubbles formed by homogeneous nucleation are evenly distributed over the considered channel section and have a velocity equal to the velocity of the liquid. The flow is one-dimensional, the current is stationary and isentropic. The internal friction of the liquid and the friction of the flow against the channel walls are absent. The thermodynamic parameters of the vapor in a bubble coincide with their equilibrium values at a given static pressure. The process of outflow with nucleation is described in the approximations mentioned by the Euler equations of continuity, heat transfer and the equation that determines the vapor content in the channel

$$\rho v \frac{dv}{dz} = -\frac{dp}{dz}, \tag{7.5}$$

$$\frac{d}{dz}j = \frac{d}{dz}(\rho v) = \frac{d}{dz}\left[\frac{\rho' v}{1+(\rho'/\rho''-1)x}\right] = 0, \tag{7.6}$$

$$(h''-h')\frac{dx}{dz} + (1-x)c_p'\frac{dT}{dz}$$
$$+ \left[x\left(\frac{dh''}{dp}\right)_s - \frac{1}{\rho} + (1-x)\frac{1}{\rho'}(1-\alpha_p'T)\right]\frac{dp}{dz} = 0, \tag{7.7}$$

$$\frac{dx}{dz} = \frac{d}{dz}\int_{x_0}^{z} m(z',z)[1-x(z')]\frac{J(z')}{\rho'v(z')}dz'. \tag{7.8}$$

Here, ρ is the density of the vapor–liquid mixture, x is the mass vapor content, h is the enthalpy, α_p is the coefficient of thermal expansion, $m(z',z)$ is the mass of a bubble formed at the section z after having reached the section z', and z_s is the section at which a liquid reaches the saturation state ($J(z_s) = 0$).

The system of equations, Eqs. (7.5)–(7.8), allows one a numerical solution exclusively [503]. Owing to a very strong dependence of J on $\Delta p = p_s(T) - p$, there is a sufficiently narrow region in the flow in which the intensity of nucleation of the vapor phase takes such values that the heat flow into growing boiling centers stops any further increase of supersaturation. Let at the section of the flow z, which is achieved at time τ, the maximum supersaturation be $\Delta p = \Delta p_*$. The rate of change of supersaturation Δp in this case is equal to zero, and for the nucleation rate one can write

$$J \simeq J_* \exp\left[-\frac{1}{2}G_p\Delta\ddot{p}_*(\tau-\tau_*)^2\right], \tag{7.9}$$

where $G_p = (\partial G/\partial \Delta p)_*$, $\Delta\ddot{p}_* = (d^2\Delta p/d\tau^2)_*$, J_* is the nucleation rate at $p = p_*$. The half width of the function (7.9) is

$$\Delta\tau_{1/2} \simeq 0.83 \left(\frac{2}{G_p\Delta\ddot{p}_*}\right)^{1/2}. \tag{7.10}$$

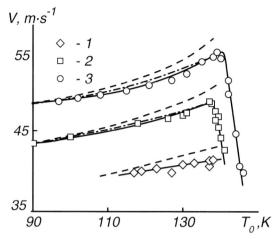

Fig. 7.3 Specific volume flow rate of liquid argon as a function of temperature at different initial pressures. 1: $p_0 = 2.18$ MPa; 2: 2.84; 3: 3.60. *Dashed lines* show results of calculation by Eq. (7.1), *dashed-dotted lines*, by Eqs. (7.1)–(7.4). For the channel parameters, see Fig. 7.1.

Assuming that downstream of z_* nucleation is absent, and upstream nucleation is determined only by nuclei arising on the interval $\tau_* - \Delta\tau_{1/2}$, Eq. (7.8) with allowance for Eq. (7.9) may be integrated, and the system of differential equations, Eqs. (7.5)–(7.8), is reduced to a system of algebraic equations for an isolated flow section with intensive fluctuational nucleation. The vapor content at the section z_* is determined by the value of J_*. The increase of x is accompanied by a pressure increase in the channel, which results in a decreasing nucleation rate. The competition of these factors leads to the realization of the critical regime of outflow of a boiling-up liquid and bends on the flow-rate characteristics (Figs. 7.1–7.3).

Due to the small size of the channels, the experimental determination of the pressure and other parameters of a flow in the channels is extremely difficult. The critical outflow regime of a boiling-up liquid, when there is practically a homogeneous metastable liquid in the channel, makes it possible to evaluate the minimum pressure in the channel from Eq. (7.1) by experimental data for the flow rate. The results of such a calculation in reduced thermodynamic coordinates are given in Fig. 7.4. In data processing, a correction was made for the adiabatic cooling of the liquid. As it is evident from Fig. 7.4, the minimum pressure in the channel lies in the range of state variables limited by the lines of attainable superheating ($J = 10^7$ m^{-3}s^{-1}) and the spinodal, which points to the spontaneous boiling-up of a liquid. Figure 7.4 also gives the results of calculating the pressure in the channel in the absence of boiling-up (Eqs. (7.2)–(7.4)) and during homogeneous nucleation (Eqs. (7.5)–(7.8)). The correlation

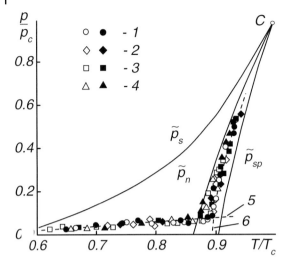

Fig. 7.4 Reduced pressure in the channel during the outflow of argon (1), nitrogen (2), oxygen (3), and methane (4). (5): results of solving the system of Eqs. (7.2)–(7.4); (6): Eqs. (7.5)–(7.8). *Light dots*: $\tilde{p}_0 = 0.58$; *dark dots*: $\tilde{p}_0 = 0.74$. p_s: saturation line; p_n: line of attainable superheating ($J = 10^7$ m^{-3}s^{-1}); p_{sp}: spinodal.

between the minimum pressure in short channels and the pressure along the line of attainable superheating was previously observed during the outflow of high-temperature liquids [498].

7.2
Vapor Explosion at the Interface of Two Different Liquids

The contact of two liquids, when the temperature of one of them exceeds the saturation temperature of the other, may result in an explosive evaporation (vapor explosion). The vapor explosion is a process of vapor generation characterized by a smaller time scale than the time of acoustic relaxation and accompanied by the initiation of shock waves. The pressure difference at the front of a shock wave in the liquid phase reaches thousands of atmospheres or more [504]. Vapor explosions were registered when cryogenic liquids were spread over a water surface [505, 506] and were widely discussed in special literature in connection with the problem of safe water transportation of liquefied natural gas [507–509]. Similar safety problems are encountered in nuclear-power engineering [510], and in metallurgy and paper industries [511, 512].

The probability and the power of vapor explosions increase with increasing volumes of interacting liquids. And although the absence of the similarity between a small-scale (interaction of a small amount of one liquid with a large mass of another) and a large-scale (interaction of large amounts of liquids) vapor explosion has been established, all the experiments have revealed the existence of temperature limits below and above which the phenomenon of explosive evaporation is absent. Usually, models discussed in literature do not predict such limitations on explosive evaporation [513].

The place of contact of two immiscible liquids possesses a lower population of completed or easily activated boiling centers than a liquid/solid interface. If contacting liquids wet each other well and are chemically inactive, a low-boiling liquid can be superheated to the temperature of spontaneous boiling-up. In the absence of the vapor phase at the place of contact of the liquids, a thermal boundary layer is formed, the thickness, δ, of which increases with the square root of time. Nucleation is possible when the value of δ is sufficient for the formation of a critical nucleus.

The temperature in the boundary layer T_i may be evaluated by solving the heat-transfer equation for two half-infinite volumes of a cold and hot liquid. Neglecting the effects of liquid expansion, convection and considering the thermophysical properties of the liquids as unchangeable, we have [514]

$$T_i = T_2 - \frac{T_2 - T_1}{1 - \kappa_0}, \tag{7.11}$$

where

$$\kappa_0 = \left[\frac{(\Lambda_2' c_{p2}' \rho_2')}{(\Lambda_1' c_{p1}' \rho_1')}\right]^{1/2}, \tag{7.12}$$

and the indices 1 and 2 refer, respectively, to the cold and the hot liquid.

Let us examine phenomena proceeding at the interface of two immiscible liquids at their free contact. The necessary conditions of realization of spontaneous nucleation in a boundary layer are reduced to the requirements

$$T_n \leq T_i < T_{sp}, \qquad \delta > 2R_*. \tag{7.13}$$

The homogeneous nucleation rate has to be sufficient in order to guaranty that at least one critical nucleus is formed in the boundary layer during its heating to the temperature T_i. Homogeneous nucleation will also be determining in the presence of completed boiling centers at the place of contact of the liquids if the volume fraction of vapor forming on them is much smaller than the volume of the thermal boundary layer [9–11]. For a drop of propane ($d \leq$ 2–3 mm), the characteristic time of formation of a boundary layer, $\delta \simeq 10$ nm thick with a temperature T_n, is $\tau_* \simeq 10^{-9}$ s. The diameter of a critical bubble relatively weakly depends on superheating, and for the effective nucleation rate in this case we have $J = (\tau_* V_\delta)^{-1} = 10^{24}$ m^{-3}s^{-1}, where V_δ is the volume of the superheated liquid layer.

The conditions of realization of explosive evaporation accounting for the interaction of small amounts of cryogenic and low-boiling liquids with heat-transfer agents are investigated in Refs. [509, 515–517]. Experiments were conducted on liquid hydrocarbons of the methane series (propane, isobutane, n-pentane) and nitrogen [515, 516]. The main data array was obtained on drops from 0.4 to 3 mm. The initial temperature of drops was equal to the temperature of the normal boiling of a cold liquid (nitrogen, propane, isobutane) or varied (experiments with n-pentane). Drops fell from a height $H \simeq (20\text{--}150)$ mm or were brought into free contact ($H \leq 10$ mm) with the surface of a hot liquid (water, glycerin). The temperature of the latter varied from room temperature to that close to the temperature of normal boiling. The character of interaction of drops with the surface of a hot liquid depends on their size, initial temperatures, and the height of the fall.

Statistical analysis was used for the results of observations. For drops with 0.3–1 mm in diameter, three characteristic types of interaction have been revealed: evaporation in regimes of convective heat transfer, film and mixed boiling (A); explosive boiling (B); film boiling of spheroids (C). At fixed values of T_1 and T_2, from 50 to 200 acts of interaction of drops with a hot liquid were registered, and the number of outcomes of the enumerated types was determined. Each of the interaction types was realized in certain temperature limits (Fig. 7.5).

Explosively evaporating drops were observed in the temperature range $0.76 T_{1c} < T_2 < 1.1 T_{1c}$, where T_{1c} is the temperature at the critical point of the cold liquid. An explosion in this context means an abrupt evaporation of a drop accompanied by a characteristic sound effect of the click type. Right after the contact with the glycerin surface, a drop jumped up or was thrown off and evaporated instantaneously. At the place of the explosion, there remained a trace in the form of a rapidly scattering vapor cloud. In a number of cases, before the explosion drops were captured by a hot liquid ($d \leq 0.5$ mm) or lived ($d \geq 0.5$ mm) for some time at the glycerin surface until their diameter became less than $\sim (0.2\text{--}0.5)$ mm, whereupon they boiled up explosively. The rate of realization of explosions of one type or another depended on the height of the drop fall. Explosive evaporation could be initiated by the breaking of a drop, when as a result of an impact upon the surface or intensive bubble boiling a drop broke into several small parts, which then evaporated explosively.

The results of calculating the temperature of the boundary layer, T_i, for propane drops are shown in Fig. 7.5 in the form of an additional axis at the abscissa. The values of T_i^*, at which the probability of explosive boiling-up of drops of a cold liquid reaches its maximum, are given in Table 7.1. These data are compared with the results of experiments determining the temperature of attainable superheating of the investigated liquids ($J = 10^7$ m^{-3}s^{-1}). The proximity of the values of T_i^* and T_n points to the determining role of

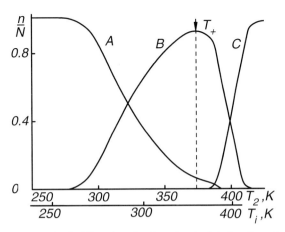

Fig. 7.5 Probability of realization of an isolated regime of evaporation of propane drops at the surface of glycerin. The initial temperature of the drops is 230 K.

spontaneous nucleation in a vapor explosion, at least, regarding the behavior of small drops. With respect to the liquids presented in Table 7.1 an increase of the nucleation rate by five orders corresponds to an increase of the temperature of superheating by approximately 1 K.

Tab. 7.1 Temperatures of the interface of liquids at the moment of explosive boiling-up of a drop and the attainable superheating of the cold liquid.

Liquid	C_3H_8	iC_4H_{10}	nC_5H_{12}	CHF_2Cl
T_i (K)	330 [516]	357 [516]	418 [516]	333 [517]
T_n (K)	327 [148]	361.2 [148]	419.3 [256]	330 [126]

According to Eq. (7.11), with a decrease in the initial temperature of the drop explosive boiling is realized at higher hot-liquid temperatures. In Fig. 7.6, the hatching shows the region of explosive boiling-up for drops of n-pentane. The initial temperature varied in these experiments from 273 to 308 K. The dashed line shows the values of T_i^* corresponding to the temperature $T_2 = T_+$. Ibidem, correlations can be observed between T_1 and T_2 when T_i is equal to the temperature of the attainable superheating, T_n, and the spinodal, T_{sp}. Agreement in the slopes of these lines supports the interpretation in terms of the spontaneous mechanism of vapor generation and the legitimacy of the model employed. To a large extent, the development of explosive evaporation will be determined by the dynamics of growth of vapor bubbles. At the initial stages of growth of microbubbles, the effect of inertial forces (Rayleigh phase) prevails, at the final stages, the heat transfer (thermal phase) dominates. For realization of explosive evaporation it is necessary that the inertial mechanism, at which the vapor pressure in a bubble is higher than the pressure in the surrounding medium, should exist until the bubble reaches relatively large radii. Owing to this mechanism, the breaking of a cold liquid takes place and the initial shock wave is generated. When small drops, which form in the process of breaking, contact a hot liquid, they evaporate explosively in the inertial regime of growth of a vapor phase. Shock waves destroy the vapor film between the cold and the hot liquid, which, coming in direct thermal contact, ensure explosive boiling-up.

An increase in the external pressure decreases the duration of the Rayleigh phase, which results in the suppression of explosive boiling-up. Experimental corroboration of this conclusion is obtained in Refs. [517, 518], where at pressures above 0.3 MPa the cessation of vapor explosions can be observed in systems freon-22–petroleum oil and hydrocarbons of the methane series-petroleum oil. No explosive processes were registered at a contact of nitrogen

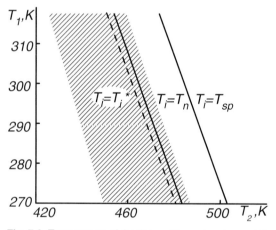

Fig. 7.6 Temperature of the thermal boundary layer at the contact of n-pentane drops with glycerin depending on their initial temperatures. $T_n = 419.3$ K ($J = 10^7$ m^{-3}s^{-1}); $T_{sp} = 435$ K; $T_i^* = 418$ K.

drops with water, the temperature of which varied from 290 to 373 K [516], and spreads over the water surface ($T_2 \simeq 293$ K) of comparatively large masses of nitrogen, methane, and ethane [509]. According to the model of spontaneous nucleation, the temperature at the nitrogen/water interface is $T_i \simeq (280-350)$ K, which considerably exceeds not only the temperature on the spinodal, but also its critical temperature ($T_{1c} = 126.2$ K). At the same time, explosions in nitrogen–water and ethane–water systems took place if shock waves of amplitude up to 5.8 MPa were artificially generated in them. A shock wave caused the collapse of a vapor film with the formation of cold-liquid jets hitting against the surface of a hot liquid and penetrating into its volume [519]. The mechanism of jet formation is connected both with the origination of the

Rayleigh–Taylor instability in the process of bubble collapse and with the development of the Landau instability at the stage of its growth [518, 519]. Theoretical simulation of the destabilizing action of a pressure pulse on a vapor film in the process of vapor explosion is given in Refs. [520, 521].

In a large-scale vapor explosion, spontaneous boiling-up of isolated drops manifests itself only as an initiating event, as a result of which there form divergent weak shock waves leading to explosions of neighboring drops. Breaking of cold-liquid drops to sizes ~ 100 μm at the front of a shock wave and an abrupt (10^3–10^4 times) increase in the contact area of the liquids result in the formation of a detonation wave with a structure similar to that of the detonation wave in chemically reacting media [522, 523]. Behind the front of such a self-sustaining wave, a fine fragmentation of a cold liquid can be observed connected with its fast superheating and subsequent expansion of the explosion products into the surrounding space. The velocity of a blast wave is close to the sound velocity in a mixture of reagents.

Another kind of vapor explosion connected with liquid superheating may be realized under depressurization of high-pressure vessels that contain liquids superheated to temperatures exceeding the temperature of their normal boiling. An abrupt pressure drop results in the superheating of the liquid phase, explosive boiling-up with generation of shock waves. This is most often observed in systems of storage of liquefied gases [524]. Cryogenic liquids are usually stored and transported at pressures close to atmospheric in the two-phase state. An equilibrium vapor content is maintained owing to the action of a large number of completed boiling centers on the container walls. On account of good wettability of solid materials with cryogenic liquids, many of the boiling centers decrease their activity in time or stop their action altogether. In this case, local heat inflows and pressure pulsations may lead to local superheatings of a liquid, to its explosive boiling-up with the development of considerable shock loads in cryogenic systems, which disturb the normal functioning of cryogenic systems, and in a number of cases to their destruction [525].

List of Symbols

A	– surface area of the nucleus, parameter of thermodynamic similarity
B	– kinetic pre-factor
C	– velocity of sound, proportionality factor in the distribution function of nuclei
c	– concentration, mole fraction, critical index
c_p, c_v	– specific heat at constant pressure and constant volume
D	– diffusion coefficient
D_T	– thermal diffusivity
E	– total energy, general notation of thermodynamic potential
e_0	– negative eigenvalue of stability operator
F	– Helmholtz free energy
f	– Helmholtz free energy density
G_*	– Gibbs' number
Gi	– Ginzburg number
$\hbar$	– Planck constant; $\hbar = h/2\pi$
I	– action
J	– nucleation rate
j	– mass flow rate
k_B	– Boltzmann constant
m	– atomic (molecular) mass
N	– number of molecules in a nucleus, total number of liquid boiling-up events
n	– number of liquid boiling-up acts in a chosen interval of time (temperature, concentration)
$P(x_o, \tau)$	– distribution function of nuclei in variables x_0
$P(x_o)$	– stationary distribution function of nuclei in variables x_0
$P_{eq}(x_o)$	– equilibrium distribution function of nuclei in variables x_0
p	– pressure
$\widetilde{p}$	– reduced pressure
$\triangle p$	– liquid stretch at isothermal conditions
q	– wave number

List of Symbols

R	– nucleus radius
$\vec{r}$	– radius vector
S	– entropy
s	– specific entropy, index specifying the binodal
T	– temperature
$\tilde{T}$	– reduced temperature
$\dot{T}$	– rate of change of temperature
T_n	– temperature of spontaneous nucleation
T_{sp}	– spinodal temperature
$\triangle T$	– superheating of the liquid at isobaric conditions
δT	– supercooling at isochoric conditions or at constant concentration
t	– reduced temperature
U	– internal energy
u	– specific internal energy
V	– volume
v	– specific volume, specific volume flow rate, velocity
W	– work of nucleus formation
z	– factor that corrects the normalization of the distribution function of nuclei
β_s, β_T	– adiabatic and isothermal compressibility
Γ	– kinetic factor
ζ	– random force
η, η_v	– shear and bulk viscosity
κ	– influence coefficient
Λ	– thermal conductivity, quantum similarity parameter
λ	– wave length, density of probability of formation of a critical nucleus in a metastable phase
λ_o	– increment of increase of an unstable variable at nucleation
μ	– chemical potential, critical index
ξ	– correlation length
ρ	– number of molecules in a unit volume, mass density
σ	– surface tension
τ	– time
$\bar{\tau}$	– mean lifetime of a metastable phase
Φ	– Gibbs potential
φ	– order parameter
$*$	– critical-nucleus index

References

1 J. W. Gibbs, *The Collected Works*, vol. 2, *Thermodynamics* (Longmans and Green, New York, London, Toronto, 1928).

2 M. Volmer and A. Weber, Z. Phys. Chem. **119**, 277 (1926).

3 L. Farkas, Z. Phys. Chem. **125**, 236 (1927).

4 R. Becker and W. Döring, Ann. Phys. **24**, 719 (1935).

5 Ya. B. Zeldovich, Zh. Eksp. Teor. Fiz. **12**, 525 (1942).

6 Ya. I. Frenkel, *Kinetic Theory of Liquids* (Oxford University Press, Oxford, 1946).

7 J. D. van der Waals and Ph. Kohnstamm, *Lehrbuch der Thermodynamik* (Johann-Ambrosius-Barth Verlag, Leipzig und Amsterdam, 1908).

8 I. M. Lifshitz and Yu. Kagan, Zh. Eksp. Teor. Fiz. [Sov. Phys.- JETP **35**, 206 (1972)] **62**, 385 (1972).

9 V. P. Skripov, *Metastable Liquids* (Wiley, New York, 1974).

10 V. P. Skripov, E. N. Sinitsyn, P. A. Pavlov, G. V. Ermakov, G. N. Muratov, N. V. Bulanov, and V. G. Baidakov, *Thermophysical Properties of Liquids in the Metastable (Superheated) State* (Gordon and Breach Science Publishers, New York, London, Paris, Montreux, Tokyo, Melbourne, 1988).

11 P. A. Pavlov, *Dinamika Vskipaniya Sil'no Peregretykh Zhidkostei* [Dynamics of Boiling-up of Highly Superheated Liquids] (Publishing House of the Ural Branch of the Russian Academy of Sciences, Sverdlovsk, 1988 (in Russian)).

12 V. G. Baidakov, *Peregrev Kriogennykh Zhidkostei* [Superheating of Cryogenic Liquids] (Publishing House of the Ural Branch of the Russian Academy of Sciences, Ekaterinburg, 1994 (in Russian)).

13 V. P. Skripov and M. Z. Faizullin, *Crystal-Liquid-Gas Phase Transitions and Thermodynamic Similarity* (Wiley-VCH, Berlin, Weinheim, 2006).

14 P. G. Debenedetti, *Metastable Liquids: Concepts and Principles* (Princeton University Press, Princeton, New Jersey, 1996).

15 D. Kashchiev, *Nucleation: Basic Theory with Applications* (Butterworth-Heinemann, Oxford, 2000).

16 J. W. P. Schmelzer (Editor), *Nucleation Theory and Applications* (Wiley-VCH, Weinheim, 2005).

17 I. Yu. Ushenin and V. G. Baidakov, Fluid Mech. Sov. Res. **21**, 91 (1992).

18 V. K. Semenchenko, *Izbrannye Glavy Theoreticheskoi Fiziki* [Selected Chapters in Theoretical Physics] (2nd revised and enlarged edition, Prosveshchenie, Moscow, 1966 (in Russian)).

19 G. A. Korn and T. M. Korn, *Mathematical Handbook* (McGraw-Hill, New York, San Francisco, Toronto, London, Sydney, 1968).

20 L. D. Landau and E. M. Lifshitz, *Statistical Physics*, vol. 1 (3rd edition, Pergamon Press, Oxford, 1980).

21 R. Balescu, *Equilibrium and Nonequilibrium Statistical Mechanics* (Wiley, New York, London, Sydney, Toronto, 1975).

22 V. G. Baidakov and V. P. Skripov, Heat Transf.- Sov. Res. **11**, 1 (1979).

23 V. P. Skripov and V. G. Baidakov, Teplofiz. Vys. Temp. **10**, 1226 (1974).

24 O. Penrose and J. L. Lebowitz, J. Stat. Phys. **3**, 211 (1971).

Explosive Boiling of Superheated Cryogenic Liquids. Vladimir G. Baidakov
Copyright © 2007 WILEY-VCH Verlag GmbH & Co. KGaA, Weinheim
ISBN: 978-3-527-40575-6

25 J. L. Lebowitz and O. Penrose, J. Math. Phys. **7**, 98 (1966).

26 M. Cassandro and E. A. Oliveri, J. Stat. Phys. **17**, 229 (1977).

27 K. Millard and L. Lund, J. Stat. Phys. **8**, 225 (1973).

28 H.-T. Yau, J. Stat. Phys. **74**, 705 (1994).

29 R. Brout, *Phase Transitions* (University of Brussels, New York, Amsterdam, 1965).

30 G. E. Uhlenbeck and G. W. Ford, *Lectures in Statistical Mechanics* (American Mathematical Society, Providence, Rhode Island, 1963).

31 A. F. Andreev, Zh. Eksp. Teor. Fiz. **45**, 2064 (1963).

32 J. S. Langer, Ann. Phys. **41**, 108 (1967).

33 V. Ya. Krivnov, B. N. Provotorov, and V. L. Eidus, Teor. Mat. Fiz. **26**, 352 (1976).

34 W. Klein, Phys. Rev. B.: Condens. Matter **21**, 5254 (1980).

35 S. N. Isakov, Commun. Math. Phys. **95**, 427 (1984).

36 V. P. Skripov, In: *Teplofizika Metastabil'nykh Sistem* [Thermal Physics of Metastable Systems] (Publishers of the Ural Branch of the Academy of Sciences of the USSR, Sverdlovsk, vol. 3, 1989).

37 V. G. Baidakov, V. P. Skripov, and A. M. Kaverin, Zh. Eksp. Teor. Fiz. [Sov. Phys.- JETP **40**, 335 (1974)] **67**, 676 (1974).

38 V. G. Baidakov and T. A. Gurina, J. Chem. Thermodyn. **17**, 131 (1985).

39 V. G. Baidakov and T. A. Gurina, J. Chem. Thermodyn. **21**, 1009 (1989).

40 V. G. Baidakov and T. A. Gurina, Physica B **160**, 221 (1989).

41 V. G. Baidakov, Teplofiz. Vys. Temp. **32**, 681 (1994).

42 V. G. Baidakov and V. P. Skripov, Zh. Fiz. Khim. **50**, 1309 (1976).

43 V. G. Baidakov and V. P. Skripov, Zh. Eksp. Teor. Fiz. [Sov. Phys.- JETP] **75**, 1008 (1978).

44 V. G. Baidakov, A. M. Kaverin, and V. P. Skripov, J. Chem. Thermodyn. **14**, 1003 (1982).

45 V. G. Baidakov, A. M. Kaverin, and V. P. Skripov, Physica B **128**, 207 (1985).

46 V. G. Baidakov and A. M. Kaverin, J. Chem. Thermodyn. **21**, 1159 (1989).

47 V. G. Baidakov and A. M. Kaverin, Teplofiz. Vys. Temp. **32**, 837 (1994).

48 V. G. Baidakov and V. G. Kuvayev, Heat Transf.- Sov. Res. **12**, 93 (1980).

49 V. G. Baidakov, Teplofiz. Vys. Temp. **25**, 256 (1987).

50 D. Dahl and M. R. Moldover, Phys. Rev. Lett. **27**, 1421 (1971).

51 V. G. Baidakov and A. M. Rubshtein, Phys. Lett. A **131**, 454 (1988).

52 V. G. Baidakov, *Teplofizicheskie Svoi'stva Peregretykh Kriogennykh i Nizko Kipyashchikh Zhidkostei'. Obzory po Teplofizicheskim Svoi'stvam Veshchestv* [Thermophysical Properties of Superheated Cryogenic and Low-Boiling Liquids. Reviews of Thermophysical Substance Properties], N3 (65), TFZ, Moscow, 1987.

53 V. G. Baidakov, Sov. Techn. Rev.- Therm. Phys. Rev. **5**, 1 (1994).

54 V. G. Baidakov and V. P. Skripov, *Termodinamicheskie Svoistva Zhidkovo Argona v Metastabil'nom (Peregretom) Sostoyanii* [Thermodynamic Properties of Liquid Argon in Metastable (Superheated) State], Preprint TF-001/7801, Ural Science Center of the Academy of Sciences of USSR, Sverdlovsk, 1978.

55 A. J. M. Yang, P. D. Fleming, and J. H. Gibbs, J. Chem. Phys. **67**, 74 (1977).

56 J. S. Rowlinson and B. Widom, *Molecular Theory of Capillarity* (Clarendon Press, Oxford, 1982).

57 V. G. Baidakov, *Mezhfaznaya Granitsa Prostykh Klassicheskikh i Kvantovykh Zhidkostei* [The Interface of Simple Classical and Quantum Liquids] (Nauka, Ekaterinburg, 1994).

58 J. W. Cahn and J. E. Hilliard, J. Chem. Phys. **31**, 688 (1959).

59 L. D. Landau and E. M. Lifshitz, *Kvantovaya Mekhanika. Nerelyativistskaya Teoriya* [Quantum Mechanics. Nonrelativistic Theory] (Fizmatizd, Moscow, 1963).

60 J. S Langer, Lect. Notes Phys. **132**, 12 (1980).

61 E. M. Lifshitz and L. P. Pitaevskii, *Fizicheskaja Kinetika* [Physical Kinetics] (Nauka, Moscow, 1979).

62 A. Z. Patashinskii and V. L. Pokrovskii, *Fluktuazionnaya Teoriya Fazovykh Perekhodov* [Fluctuation Theory of Phase Transitions] (Nauka, Moscow, 1982).

63 S. Ma, *Modern Theory of Critical Phenomena* (Advanced Book Program Reading, W. A. Benjamin, Massachusetts, 1976).

64 V. G. Baidakov, Fluid Mech.- Sov. Res. **16**, 1 (1987).

65 A. Z. Patashinskii and B. I. Shumilo, *Teoriya Zarodysheobrazovaniya pri Fazovom Perekhode Pervogo Roda* [Theory of Nucleation at a First-Order Phase Transition] Preprint IYaF 78–101, Novosibirsk, 1979.

66 E. H. Stanley, *Introduction to Phase Transitions and Critical Phenomena* (Clarendon Press, Oxford, 1971).

67 V. G. Baidakov, Heat Transf.- Sov. Res. **11**, 6 (1979).

68 P. Glansdorff and I. Prigogine, *Thermodynamic Theory of Structure, Stability and Fluctuations* (Wiley-Interscience, London, New York, Sydney, Toronto, 1970).

69 Ya. B. Zeldovich and O. M. Todes, Zh. Eksp. Teor. Fiz. **10**, 1441 (1940).

70 Ya. B. Zeldovich, Zh. Eksp. Teor. Fiz. **80**, 2111 (1981).

71 A. M. Berezhkovskii and V. Yu. Zitserman, Physica A **166**, 585 (1990).

72 B. V. Gnedenko, *Kurs Teorii Veroyatnostei* [A Course of the Theory of Probability] (Nauka, Moscow, 1969).

73 H. Risken, *The Fokker–Planck Equation* (Springer, Berlin, 1984).

74 P. Hänggi, P. Talkner, and M. Borkovec, Rev. Mod. Phys. **62**, 251 (1990).

75 E. A. Brener, V. N. Marchenko, and S. V. Meshkov, Zh. Eksp. Teor. Fiz. **85**, 2107 (1983).

76 A. I. Rusanov, *Fazovye Ravnovesiya i Poverkhnostnye Yavleniya* [Phase Equilibria and Surface Phenomena] (Khimya, Leningrad, 1967).

77 V. A. Shneidman, Zh. Eksp. Teor. Fiz. **91**, 520 (1988).

78 F. M. Kuni and A. A. Melikhov, Teor. Mat. Fiz. **81**, 247 (1989).

79 B. V. Deryagin, A. V. Prokhorov, and N. N. Tunitsky, Zh. Eksp. Teor. Fiz. [Sov. Phys.- JETP **46**, 962 (1977)] **73**, 1831 (1977).

80 H. Trinkaus, Phys. Rev. B.: Condens. Matter **27**, 7372 (1983).

81 G. Shi and J. H. Seinfeld, J. Chem. Phys. **93**, 9033 (1990).

82 V. Yu. Zitserman and L. M. Bereshchkovskii, Zh. Fiz. Khim. **64**, 1795 (1990).

83 H. Kramers, Physica **7**, 284 (1940).

84 V. G. Baidakov and S. P. Protsenko, Teplofiz. Vys. Temp. **41**, 231 (2003).

85 V. G. Baidakov and S. P. Protsenko, Dokl. Akad. Nauk **394**, 752 (2004).

86 V. G. Baidakov, In: *Metastabil'nye Sostoyaniya i Fazovye Perekhody* [Metastable States and Phase Transitions] (Ural Branch of the Russian Academy of Sciences, Ekaterinburg, vol. 1, 61, 1997).

87 V. G. Baidakov, Teplofiz. Vys. Temp. [High Temp. Thermal Physics **36**, 143 (1998)] **36**, 147 (1998).

88 V. G. Baidakov, Dokl. Akad. Nauk **394**, 179 (2004).

89 V. P. Skripov and A. V. Skripov, Usp. Fiz. Nauk **128**, 193 (1977).

90 L. M. Berezhkovskii and V. Yu. Zitserman, J. Chem. Phys. **102**, 3331 (1995).

91 F. C. Collins, Z. Electrochem. **59**, 404 (1955).

92 H. L. Frish, J. Chem. Phys. **27**, 90 (1957).

93 R. P. Andres and M. Boudart, J. Chem. Phys. **42**, 2057 (1965).

94 D. Kashchiev, Surface Sci. **14**, 209 (1969).

95 I. Kanne-Dannetschek and D. Stauffer, J. Aerosol. Sci. **12**, 105 (1981).

96 A. I. Rusanov and F. M. Kuni, Dokl. Akad. Nauk SSSR **185**, 386 (1969).

97 B. V. Deryagin, Zh. Eksp. Teor. Fiz. [Sov. Phys.- JETP **38**, 1129 (1974)] **65**, 2261 (1973).

98 J. D. Gunton and M. Droz, Lect. Notes Phys. **183**, 1 (1983).

99 A. M. Berezhkovskii, A. N. Drozdov, and V. Yu. Zitserman, Zh. Fiz. Khim. **62**, 2599 (1988).

100 V. G. Baidakov, Fiz. Nizk. Temp. **10**, 683 (1984).

101 V. A. Akulichev, *Kavitatsiya v Kriogennykh i Kipyashchikh Zhidkostyakh* [Cavitation in Cryogenic and Boiling Liquids] (Nauka, Moscow, 1978).

102 S. Ono and S. Kondo, *Molecular Theory of Surface Tension in Liquids* (Springer, Berlin, Göttingen, Heidelberg, 1960).

103 V. G. Baidakov, Fluid Mech.- Sov. Res. **13**, 40 (1984).

104 W. Döring, Z. Phys. Chem. **36**, 376 (1937).

105 M. Volmer, *Kinetik der Phasenbildung* (Edwards Brothers, Ann Arbor, 1945).

106 I. N. Stranski and R. Kaischew, Z. Phys. Chem. **38**, 451 (1938).

107 R. Kaischew and I. N. Stranski, Z. Phys. Chem. **26**, 317 (1934).

108 Yu. M. Kagan, Zh. Fiz. Khim. **34**, 92 (1960).

109 A. A. Melikhov, F. M. Kuni, and P. A. Kon'kov, Vestnik Leningrad State University, Fiz. Khim. **3**, 8 (1990).

110 V. I. Roldigin, Teor. Eksp. Khim. **20**, 595 (1984).

111 H. Eyring, J. Chem. Phys. **3**, 107 (1935).

112 A. V. Prokhorov, Dokl. Akad. Nauk SSSR **239**, 1323 (1978).

113 V. G. Baidakov, *Dostizhimyi' Peregrev Kriogennykh i Nizkokipyashchikh Zhidkostei'. Obzory po Teplofizicheskim Svoi'stvam Veshchestv* [Attainable Superheating of Cryogenic and Low-Boiling Liquids. Reviews of Thermophysical Substance Properties], N3 (53), TFZ, Moscow, 1965.

114 B. V. Gnedenko, Yu. K. Belyaev, and A. D. Solov'ev, *Matematicheskie Metody v Teorii Nadeshchnosti* [Mathematical Methods of Reliability Theory] (Nauka, Moscow, 1965).

115 B. I. Kidyarov, *Kinetika Obrazovaniya Kristallov Zhidkoi' Fazy* [Kinetics of Formation of Liquid-Phase Crystals] (Nauka, Novosibirsk, 1979).

116 V. P. Skripov, V. P. Koverda, and G. T. Butorin, Kristallografiya **15**, 1219 (1970).

117 V. P. Skripov and V. P. Koverda, *Spontannaya Kristallizatsiya Pereokhlashchdennykh Zidkostei* [Spontaneous Crystallization of Supercooled Liquids] (Nauka, Moscow, 1982).

118 A. N. Kolmogorov, Izv. Akad. Nauk SSSR, Ser. Mat. **3**, 355 (1937).

119 K. L. Wismer, J. Phys. Chem. **26**, 301 (1922).

120 F. B. Kenrick, C. S. Gilbert, and K. L. Wismer, J. Phys. Chem. **28**, 1297 (1924).

121 V. P. Skripov, V. G. Baidakov, S. P. Protsenko, and V. V. Maltzev, Teplofiz. Vys. Temp. **11**, 682 (1973).

122 V. P. Skripov, V. G., Baidakov, and A. M. Kaverin, Physica A **95**, 169 (1979).

123 A. M. Kaverin and V. G. Baidakov, In: *Teplofizicheskie Svoi'stva Zhidkostei' i Vzryvnoe Vskipanie* [Thermophysical Properties of Liquids and Explosive Boiling-Up] (Ural Science Center of the Academy of Sciences of USSR, Sverdlovsk, 3, 1976)

124 K. Nishigaki and Y. Saji, Jpn. J. Appl. Phys. **20**, 849 (1981).

125 M. Blander, Adv. Colloid Interface Sci. **10**, 1 (1979).

126 Y. Mori, K. Hijikata, and T. Nagatani, Int. J. Heat Mass Transf. **19**, 1153 (1976).

127 W. Porteous and M. Blander, AIChE J. **21**, 560 (1975).

128 V. G. Baidakov, V. P. Skripov, and A. M. Kaverin, Zh. Eksp. Teor. Fiz. **65**, 1126 (1973).

129 V. G. Baidakov and V. P. Skripov, In: *Teplofizicheskie Issledovaniya Zhidkostei'* [Thermophysical Investigations of Liquids] (Ural Science Center of the Academy of Sciences of USSR, Sverdlovsk, 6, 1975).

130 L. C. Brodie, D. N. Sinha, C. E. Sanford, and L. S. Semura, Rev. Sci. Instrum. **52**, 1697 (1981).

131 L. C. Brodie, D. N. Sinha, L. S. Semura, and C. E. Sanford, J. Appl. Phys. **48**, 2882 (1977).

132 D. N. Sinha, L. S. Semura, and L. C. Brodie, Phys. Rev. A.: Gen. Phys. **26**, 1048 (1982).

133 V. G. Baidakov, Teplofiz. Vys. Temp. **23**, 133 (1985).

134 K. Nishigaki and Y. Saji, Cryogenics. **23**, 473 (1983).

135 K. Nishigaki and Y. Saji, J. Phys. Soc. Jpn. **52**, 2293 (1983).

136 H. Wakeshima and K. Takata, J. Appl. Phys. **29**, 1126 (1958).

137 G. R. Moore, AlChE J. **5**, 458 (1959).

138 V. P. Skripov and G. V. Ermakov, Zh. Fiz. Khim. **38**, 396 (1964).

139 E. N. Sinitsyn and V. P. Skripov, Prib. Tekh. Eksp. **4**, 178 (1966).

140 V. G. Baidakov, A. M. Kaverin, A. M., Rubshtein, and V. P. Skripov, Pis'ma Zh. Tekh. Fiz. **3**, 1150 (1977).

141 V. G. Baidakov, A. M. Kaverin, and V. P. Skripov, Kolloid. Zh. **42** 314 (1980).

142 V. G. Baidakov and A. M. Kaverin, Teplofiz. Vys. Temp. **19**, 321 (1981).

143 N. G. Bruijn, *Asymptotic Methods in Analysis* (North-Holland, Amsterdam, 1958).

144 A. Yu. Dianov, S. A. Mal'tsev, V. G. Baidakov, and V. P. Skripov, Heat Transf.-Sov. Res. **12**, 75 (1980).

145 A. M. Kaverin, V. G. Baidakov, and V. P. Skripov, Inzh.-Fiz. Zh. **38**, 680 (1980).

146 V. G. Baidakov and V. P. Skripov, Zh. Fiz. Khim. **56**, 818 (1982).

147 D. N. Sinha, L. C. Brodie, and J. S. Semura, Phys. Rev. B.: Condens. Matter **36**, 4082 (1987).

148 V. G. Baidakov, A. M. Kaverin, and V. P. Skripov, Zh. Fiz. Khim. **60**, 444 (1986).

149 D. Lezek, L. C. Brodie, J. S. Semura, and E. Bodegon, Phys. Rev. B.: Condens. Matter **37**, 150 (1988).

150 N. M. Semenova and G. V. Ermakov, J. Low Temp. Phys. **74**, 119 (1989).

151 V. E. Vinogradov and E. N. Sinitsyn, Zh. Fiz. Khim. **51**, 2704 (1977).

152 V. G. Baidakov, A. M. Kaverin, and I. I. Sulla, Teplofiz. Vys. Temp. **27**, 410 (1989).

153 V. G. Baidakov, Kolloid. Zh. **44**, 409 (1982).

154 V. G. Baidakov, Ukrain. Fiz. Zh. **27**, 1332 (1982).

155 V. G. Baidakov, G. N. Muratov, and K. V. Khvostov, Zh. Fiz. Khim. **55**, 2941 (1981).

156 V. G. Baidakov, K. V. Khvostov, and G. N. Muratov, Zh. Fiz. Khim. **56**, 814 (1982).

157 V. G. Baidakov and I. I. Sulla, Ukrain. Fiz. Zh. **32**, 885 (1987).

158 V. G. Baidakov and I. I. Sulla, Zh. Fiz. Khim. **59**, 955 (1985).

159 V. G. Baidakov, Fiz. Nizk. Temp. **10**, 677 (1984).

160 V. G. Baidakov, K. V. Khvostov, and V. P. Skripov, Fiz. Nizk. Temp. **7**, 957 (1981).

161 V. G. Baidakov and K. V. Khvostov, Fiz. Nizk. Temp. **8**, 476 (1982).

162 J. L. Brown, D. A. Glaser, and M. L. Perl, Phys. Rev. **102**, 586 (1956).

163 M. H. Medvedev, *Stintillyatsionnye Detektory* [Scintillation Detectors] (Atomizdat, Moscow, 1977).

164 L. P. Filippov, *Podobie Svoistv Veshchestv* [Similarity of Substance Properties] (Izd. Moscow State University, Moscow, 1978).

165 V. G. Baidakov, A. M. Kaverin, and V. P. Skripov, Zh. Fiz. Khim. **54**, 2119 (1980).

166 R. D. Finch and M. L. Chu, Phys. Rev. **161**, 202 (1967).

167 I. A. Gachicheladze, K. O. Keshishev, and A. I. Shalnikov, Pis'ma Zh. Eksp. Teor. Fiz. **12**, 231 (1970).

168 V. N. Lebedenko and B. U. Rodionov, Pis'ma Zh. Eksp. Teor. Fiz. **16**, 583 (1972).

169 A. G. Khrapak and I. T. Yakubov, Usp. Fiz. Nauk **129**, 45 (1979).

170 H. T. Davis and R. G. Brown, Adv. Chem. Phys. **31**, 329 (1975).

171 T. Miyakawa and D. L. Dexter, Phys. Rev. **184**, 166 (1969).

172 B. N. Esel'son, V. N. Grigor'ev, V. G. Ivanzov et al., *Rastvory Kvantovykh Zhidkostei' ^{3}He–^{4}He* [Solutions of Quantum Liquids ^{3}He–^{4}He] (Nauka, Moscow, 1973).

173 Yu. A. Aleksandrov, G. S. Voronov, V. M. Gorbunkov, N. B. Delone, and Yu. I. Nechaev, *Puzyr'kovye Kamery* [Bubble Chambers] (Gosatomizdat, Moscow, 1963).

174 R. H. Hildebrand and D. E. Nagle, Phys. Rev. **92**, 517 (1953).

175 J. Hord, R. B. Jakobs, C. C. Robinson, and L. L. Sparks, Trans. ASME J. Eng. Power **86**, 485 (1964).

176 N. M. Semenova and G. V. Ermakov, Teplofiz. Vys. Temp. **24**, 870 (1986).

177 L. J. Rubarcyk and J. T. Tough, J. Low Temp. Phys. **43**, 197 (1981).

178 K. Nishigaki and Y. Saji, Phys. Rev. B.: Condens. Matter **33**, 1657 (1986).

179 M. Iino, M. Suzuki, and A. J. Ikushima, J. Low Temp. Phys. **63**, 495 (1986).

180 V. G. Baidakov, V. P. Skripov, A. M. Kaverin, and K. V. Khvostov, Dokl. Akad. Nauk SSSR **260**, 858 (1981).

181 V. G. Baidakov and V. P. Skripov, Zh. Fiz. Khim. **56**, 1234 (1982).

182 V. G. Baidakov and G. Sh. Boltachev, J. Chem. Phys. **121**, 8594 (2004).

183 V. G. Baidakov, G. Sh. Boltachev, and G. G. Chernykh, Phys. Rev. E **70**, 011603 (2004).

184 V. G. Baidakov, G. Sh. Boltachev, and J. W. P. Schmelzer, J. Colloid Interface Sci. **231**, 312 (2000).

185 J. W. P. Schmelzer and V. G. Baidakov, J. Phys. Chem. B **105**, 11595 (2001).

186 V. G. Baidakov and G. Sh. Boltachev, Zh. Fiz. Khim. **69**, 515 (1995).

187 V. G. Baidakov and G. Sh. Boltachev, Dokl. Akad. Nauk **363**, 753 (1998).

188 V. G. Baidakov and G. Sh. Boltachev, Phys. Rev. E **59**, 469 (1999).

189 J. A. Wingrave, R. S. Schechter, and W. H. Wade, In: *The Modern Theory of Capillarity* (Khimia Publishers, Leningrad, 1980).

190 L. R. Fisher and J. N. Israelachvili, J. Chem. Phys. **81**, 530 (1984).

191 L. R. Fisher and J. N. Israelachvili, Nature **277**, 548 (1979).

192 M. Kornfeld, *Uprugost' i Prochnost' Zhidkostei* [Elasticity and Strength of Liquids] (Gostekhteorizdat, Moscow, Leningrad, 1951).

193 A. D. Misener and F. T. Hedgcock, Nature **171**, 835 (1953).

194 A. D. Misener and G. Hebert, Nature **177**, 946 (1956).

195 J. W. Beams, Phys. Rev. **104**, 880 (1956).

196 J. W. Beams, Phys. Fluids. **2**, 1 (1959).

197 K. L. McCloud and C. F. Mate, Bull. Am. Phys. Soc. **12**, 96 (1967).

198 M. F. Wilson, D. O. Edwards, and J. T. Tough, Bull. Am. Phys. Soc. **12**, 96 (1967).

199 J. A. Nissen, E. Bodegem, L. C. Brodie, and J. S. Semura, Adv. Cryogenics Eng. **33**, 999 (1988).

200 P. L. Marston, J. Low Temp. Phys. **25**, 407 (1976).

201 V. E. Vinogradov and P. A. Pavlov, In: *Metastabil'nye Sostoyaniya i Fazovye Perekhody* [Metastable States and Phase Transitions] (Ural Branch of the Russian Academy of Sciences, Ekaterinburg, N 3, 14. 1999).

202 R. E. Apfel, J. Acoust. Soc. Am. **48**, 1179 (1970).

203 R. D. Finch, R. Kagiwado, M. Barmatz, and I. Rudnick, Phys. Rev. **134**, 1425 (1964).

204 R. D. Finch and T. G. J. Wang, J. Acoust. Soc. Am. **39**, 511 (1966).

205 R. D. Finch, T. G. J. Wang, R. Kagiwado, and M. Barmatz, J. Acoust. Soc. Am. **40**, 211 (1966).

206 A. Mosse, M. L. Chu, and R. D. Finch, J. Acoust. Soc. Am. **47**, 1258 (1970).

207 P. D. Jarman and K. J. Taylor, J. Low Temp. Phys. **2**, 389 (1970).

208 A. Mosse and R. D. Finch, J. Acoust. Soc. Am. **49**, 156 (1971).

209 E. A. Neppiras and R. D. Finch, J. Acoust. Soc. Am. **52**, 335 (1972).

210 E. Bodegem, J. A. Nissen, J. S. Semura, and L. C. Brodie, Cryogenics **29**, 207 (1980).

211 J. A. Nissen, E. Bodegem, L. C. Brodie, and J. S. Semura, Phys. Rev. B.: Condens. Matter **40**, 6617 (1989).

212 Q. Xiong and H. J. Maris, J. Low Temp. Phys. **82**, 105 (1991).

213 M. S. Petterson, S. Balibar, and H. J. Maris, Phys. Rev. B.: Condens. Matter **49**, 12062 (1994).

214 H. J. Maris, J. Low Temp. Phys. **98**, 403 (1995).

215 F. Caupin and S. Balibar, Phys. Rev. B **64**, 064507 (2002).

216 J. C. Fisher, J. Appl. Phys. **19**, 1062 (1948).

217 X. Chavanne, S. Balibar, and F. Caupin, J. Low Temp. Phys. **126**, 615 (2002).

218 R. B. Dean, J. Appl. Phys. **15**, 446 (1944).

219 R. D. Finch and M. L. Chu, Phys. Rev. **161**, 202 (1967).

220 P. M. McConnel, M. L. Chu, and R. D. Finch, Phys. Rev. A.: Gen. Phys. **1**, 411 (1970).

221 H. C. Dhingra and R. D. Finch, J. Acoust. Soc. Am. **59**, 19 (1976).

222 H. J. Maris, J. Low Temp. Phys. **94**, 125 (1994).

223 F. Dalfovo, Phys. Rev. B.: Condens. Matter **46**, 5482 (1992).

224 H. J. Maris and Q. Xiong, Phys. Rev. Lett. **63**, 1078 (1989).

225 K. W. Schwarz and C. W. Smith, Phys. Lett. A **82**, 251 (1981).

226 V. P. Skripov, Zh. Fiz. Khim. **68**, 1382 (1994).

227 G. H. Bauer, D. M. Ceperley, and N. Goldenfeld, Phys. Rev. **61**, 9055 (2000).

228 H. J. Maris and D. O. Edwards, J. Low Temp. Phys. **129**, 1 (2002).

229 S. Balibar, F. Caupin, P. Roche, and H. J. Maris, J. Low Temp. Phys. **113**, 459 (1998).

230 F. Caupin, P. Roche, S. Marchand, and S. Balibar, J. Low Temp. Phys. **113**, 473 (1998).

231 F. Caupin and S. Balibar, Physica B **284–288**, 212 (2000).

232 H. Lambare, P. Roche, S. Balibar, H. J. Maris, O. A. Andreeva, C. Guthmann, K. O. Keshishev, and E. Rolley, Eur. Phys. J. B **2**, 381 (1998).

233 M. Guilleumas, M. Pi, M. Barranco et al., Phys. Rev. B.: Condens. Matter **47**, 9116 (1993).

234 M. Guilleumas, M. Barranco, D. Jezek, R. Lombard, and M. Pi, Phys. Rev. B **54**, 16135 (1996).

235 F. Caupin, S. Balibar, and H. J. Maris, J. Low Temp. Phys. **126**, 91 (2002).

236 M. A. Solis and J. Navarro, J. Phys. Rev. B **45**, 13080 (1992).

237 J. Casulleras and J. Boronat, Phys. Rev. Lett. **84**, 3121 (2000).

238 P. R. Roach et al., J. Low Temp. Phys. **52**, 433 (1983).

239 F. Caupin, S. Balibar, and H. J. Maris, J. Low Temp. Phys. **126**, 73 (2002).

240 S. Balibar, J. Low Temp. Phys. **129**, 363 (2002).

241 C. Su, C. E. Cramer, and H. J. Maris, J. Low Temp. Phys. **113**, 479 (1998).

242 J. Classen, C.-K. Su, M. Mohazzab, and H. J. Maris, Phys. Rev. B **57**, 3000 (1998).

243 D. Konstantinov and H. J. Maris, Phys. Rev. Lett. **90**, 025302 (2003).

244 H. J. Maris, A. Ghosh, D. Konstantinov, and M. Hirsch, J. Low Temp. Phys. **134**, 227 (2004).

245 V. G. Baidakov, Dokl. Akad. Nauk **394**, 179 (2004).

246 V. G. Baidakov, S. P. Protsenko, G. G. Chernykh, and G. Sh. Boltachev, Phys. Rev. E **65**, 041601 (2002).

247 V. G. Baidakov and G. Sh. Boltachev, Teplofiz. Vys. Temp. **43**, 420 (2005).

248 A. S. Besov, B. K. Kedrinskii, and E. I. Pal'chikov, Pis'ma Zh. Tekh. Fiz. **15**, 23 (1989).

249 V. E. Vinogradov, In: *Neravnovesnye Fazovye Perekhody i Teplofizicheskie Svoistva Veshchestv* [Nonequilibrium Phase Transitions and Thermophysical Properties of Substances] (Ural Branch of the Russian Academy of Sciences, Ekaterinburg, 101, 1996).

250 V. E. Vinogradov and P. A. Pavlov, In: *Metastabil'nye Sostoyaniya i Fazovye Perekhody* [Metastable States and Phase Transitions] (Ural Branch of the Russian Academy of Sciences, Ekaterinburg, N 2, 60. 1998).

251 V. P. Skripov, P. A. Pavlov, V. G. Baidakov et al., *Kondensirovannye Fazy pri Otritsatel'nykh Davleniyakh* [Condensed Phases at Negative Pressures] Scientific Report of the Institute of Thermal Physics of the Ural Branch of the Russian Academy of Sciences, Ekaterinburg, 2004.

252 P. A. Pavlov, J. Eng. Thermophys. **12**, 25 (2003).

253 D. Glaser, Phys. Rev. **87**, 665 (1952).

254 D. Glaser, Phys. Rev. **91**, 496 (1953).

255 V. P. Skripov and E. N. Sinitsyn, Usp. Fiz. Nauk [Sov. Phys. Usp. **7**, 887 (1965)] **84**, 727 (1964).

256 E. N. Sinitsyn and V. P. Skripov, Zh. Fiz. Khim. **43**, 875 (1969).

257 V. G. Baidakov, A. M. Kaverin, and V. P. Skripov, *Acoustic Cavitation and Ultrasound Velocity in Superheated Liquid Xenon*, Proc. 10th All Union Acoustic Conference, Moscow, 1977.

258 V. G. Baidakov, A. M. Kaverin, and V. P. Skripov, Akust. Zh. **27**, 697 (1981).

259 D. N. Sinha, L. C. Brodie, J. S. Semura, and D. Lezak, Cryogenics **22**, 271 (1982).

260 D. Lezak, L. C. Brodie, and J. S. Semura, Cryogenics **23**, 659 (1983).

261 D. Lezak, L. C. Brodi, and J. S. Semura, Cryogenics **24**, 211 (1984).

262 D. A. Labuntsov, Izv. Akad. Nauk SSSR. Energet. Transp. N 1, 58 (1963).

263 R. Riethmüller, Nucl. Eng. Design **43**, 295 (1977).

264 A. M. Kaverin, V. G. Baidakov, V. P. Skripov, and A. N. Kat'yanov, Zh. Tekhn. Fiz. **55**, 1220 (1985).

265 V. G. Baidakov and A. M. Kaverin, Teplofiz. Vys. Temp. **28**, 90 (1990).

266 B. Widom and A. S. Clarke, Physica **168**, 149 (1990).

267 G. Navascues and P. Tarazona, J. Chem. Phys. **75**, 2441 (1981).

268 R. D. Gretz, Surface Sci. **5**, 239 (1966).

269 V. A. Grigor'ev, Yu. M. Pavlov, and E. V. Ametistov, *Kipenie Kriogennykh Zhidkostei* [Boiling of Cryogenic Liquids] (Energiya, Moscow, 1977).

270 P. Tarazona and G. Navascues, J. Chem. Phys. **75**, 3114 (1981).

271 E. N. Sinitsyn, Teplofiz. Vys. Temp. **22**, 400 (1984).

272 E. N. Sinitsyn, In: *Teplofizika Metastabil'nykh Sistem v Svyazi s Yavleniyami Kipeniya i Kristallizatsii* [Thermal Physics of Metastable Systems in Connection with Boiling and Crystallization Phenomena] (Ural Science Center of the Academy of Sciences of the USSR, Sverdlovsk, 39, 1987).

273 H. Reiss, J. Chem. Phys. **18**, 840 (1950).

274 N. N. Danilov, E. N. Sinitsyn, and V. P. Skripov, In: *Teplofizika Metastabil'nykh Sistem* [Thermal Physics of Metastable Systems] (Ural Science Center of the Academy of Sciences of the USSR, Sverdlovsk, 28, 1977).

275 C. A. Ward, A. Balakrishnan, and F. C. Hooper, J. Basic Eng. **92**, 27 (1970).

276 A. H. Falls, L. E. Skriven, and H. T. Davis, J. Chem. Phys. **78**, 7300 (1983).

277 W. Döring and K. Neumann, Z. Phys. Chem. **186**, 193 (1940).

278 E. I. Nesis and Ya. I. Frenkel, Zh. Tekh. Fiz. **22**, 1500 (1952).

279 D. Stauffer, J. Aerosol Sci. **7**, 319 (1976).

280 A. A. Melikhov, V. B. Kurasov, Yu. Sh. Dshchikaev, and F. M. Kuni, Zh. Tekh. Fiz. **61**, 27 (1991).

281 D. T. Wu, J. Chem. Phys. **99**, 1990 (1993).

282 B. V. Deryagin and A. V. Prokhorov, Kolloid. Zh. **44**, 847 (1982).

283 F. M. Kuni, V. M. Ogenko, L. M. Ganyuk, and L. G. Grechko, Kolloid. Zh. **55**, 22 (1993).

284 F. M. Kuni, V. M. Ogenko, L. N. Ganyuk, and L. G. Grechko, Kolloid. Zh. **55**, 28 (1993).

285 Yu. V. Melikhov, Yu. V. Trofimov, and F. M. Kuni, Kolloid. Zh. **56**, 201 (1994).

286 V. G. Baidakov, In: *Metastabil'nye Sostoyaniya i Fazovye Perekhody* [Metastable States and Phase Transitions] (Ural Branch of the Russian Academy of Sciences, Ekaterinburg, N 2, 12, 1998).

287 V. G. Baidakov, J. Chem. Phys. **110**, 3955 (1999).

288 V. G. Baidakov, Teplof. Vys. Temp. [High Temp. **37**, 565 (1999)] **37**, 595 (1999).

289 T. A. Renner, G. H. Kucera, and M. Blander, J. Colloid Interface Sci. **52**, 319 (1975).

290 I. I. Sulla, A. M. Kaverin, and V. G. Baidakov, *Kinetics of Nucleation in Solutions of Liquified Gases*, Proc. 7th Int. Conf. Surface and Colloid Sci., Compieque, France, 1991.

291 V. G. Baidakov, A. M. Kaverin, and G. Sh. Boltachev, J. Chem. Phys. **106**, 5648 (1997).

292 A. M. Kaverin, V. G. Baidakov, and V. I. Andbaeva, *Spontaneous boiling-up and surface tension of solutions of cryogenic liquids*, Proc. 3rd Russian Workshop "Metastable States and Fluctuation Phenomena". Ekaterinburg, Institute of Thermal Physics of the Ural Branch of the Russian Academy of Sciences, 2005.

293 V. G. Baidakov and A. M. Kaverin, Teplof. Vys. Temp. [High Temp. **38**, 852 (2000)] **38**, 886 (2000).

294 V. G. Baidakov, A. M. Kaverin, and G. Sh. Boltachev, J. Phys. Chem. B **106**, 167 (2002).

295 V. G. Baidakov and A. M. Kaverin, Explosive boiling-up of superheated solutions of cryogenic liquids, Proc. National Conf. on Heat Power Engineering, Kasan, Russia, 2006.

296 E. A. Hemmingsen, Science **167**, 1493 (1970).

297 W. A. Gerth and E. A. Hemmingsen, Z. Naturforsch. A **31**, 1711 (1976).

298 Y. Finkelstein and A. Tamir, AIChE J. **31**, 1409 (1985).

299 P. G. Bowers, C. Hofstetter, H. L. Ngo, and R. T. Toomey, J. Colloid Interface Sci. **215**, 441 (1999).

300 M. B. Rubin and R. M. Noyes, J. Phys. Chem. **91**, 4193 (1987).

301 P. G. Bowers, C. Hofstetter, C. R. Letter, and R. T. Toomey, J. Phys. Chem. **99**, 9632 (1995).

302 P. G. Bowers, K. Bar-Eli, and R. M. Noyes, J. Chem. Soc., Faraday Trans. **92**, 2843 (1996).

303 H. Kwak and R. L. Panton, J. Chem. Phys. **78**, 5795 (1983).

304 G. Sh. Boltachev and V. G. Baidakov, Zh. Fiz. Khim. [Russ. J. Phys. Chem. **75**, 1455 (2001)] **75**, 1597 (2001).

305 J. W. P. Schmelzer and V. G. Baidakov, J. Chem. Phys. **119**, 10759 (2003).

306 T. W. Forest and C. A. Ward, J. Chem. Phys. **69**, 2221 (1978).

307 C. A. Ward, A. Balakrishnan, and F. C. Hooper, J. Basic Eng. **92 D**, 695 (1970).

308 V. P. Skripov and P. A. Pavlov, Zh. Fiz. Khim. **59**, 2451 (1985).

309 P. A. Pavlov and P. V. Skripov, Teplofiz. Vys. Temp. **23**, 70 (1985).

310 R. N. Herring and P. L. Barrick, Int. Adv. Cryogenic Eng. **10**, 151 (1965).

311 F. F. Kharakhorin, Zh. Tekh. Fiz. **10**, 1533 (1940).

312 M. G. Gonikberg and V. G. Fastovskii, Zh. Fiz. Khim. **14**, 257 (1940).

313 I. I. Sulla and V. G. Baidakov, Vysokochist. Veshchest. N **2**, 51 (1988).

314 V. G. Baidakov and I. I. Sulla, Int. J. Thermophys. **16**, 909 (1995).

315 I. I. Sulla and V. G. Baidakov, In: *Teplovye Protsessy i Metastabilnye Sostoyaniya* [Thermal Processes and Metastable States] (Ural Branch of the Academy of Sciences of the USSR, Sverdlovsk, 51, 1990).

316 V. G. Baidakov and A. M. Kaverin, Zh. Fiz. Khim. **78**, 1150 (2004).

317 A. M. Kaverin, V. N. Andbaeva, and V. G. Baidakov, Zh. Fiz. Khim. **80**, 665 (2006).

318 V. G. Baidakov, I. I. , Sulla, and A. M. Kaverin, *Nucleation Kinetics in Superheated Solutions of Cryogenic Liquids*, Proc. 2nd Liquid Matter Conf., Florence, Italy, 1993.

319 A. A. Vasserman and A. Ya. Keizerova, Teplofiz. Vys. Temp. **16**, 1185 (1978).

320 V. V. Sichov, A. A. Vasserman, A. D. Koslov, G. A., Spiridonov, and V. A. Tsimarnii, *Thermodynamical Properties of Nitrogen* (Izdat. Standartov, Moskva, 1977 (in Russian)).

321 V. V. Sichov, A. A. Vasserman, A. D. Koslov, G. A. Spiridonov, and V. A. Tsimarnii, *Thermodynamical Properties of Oxygen* (Izdat. Standartov, Moskva, 1981).

322 J. A. Schouten, A. Deerenberg, and N. J. Trappeniers, Physica A **81**, 151 (1975).

323 S. F. Barreiros, J. S. G. Calado, P. Clancy et al., J. Phys. Chem. **86**, 1722 (1982).

324 R. N. Herring and P. L. Barrick, Int. Adv. Cryogenic Eng. **10**, 151 (1965).

325 E. W. Lemmon, R. T. Jacobsen, S. G. Penoncello, and D. G. Friend, J. Phys. Chem. Ref. Data **29**, 331 (2000).

326 Ch. Tegeler, R. Span, and W. Wagner, J. Phys. Chem. Ref. Data **28**, 779 (1999).

327 R. Span, E. W. Lemmon, R. T. Jacobsen, and W. Wagner, Int. J. Phys. Thermophys. **14**, 1121 (1998).

328 R. Schmidt and W. Wagner, Fluid Phase Equilib. **19**, 175 (1985).

329 G. Sh. Boltachev and V. G. Baidakov, Zh. Fiz. Khim. [Russ. J. Phys. Chem. **80**, 501 (2006)] **80**, 594 (2006).

330 G. M. Wilson, P. M. Silverberg, and M. G. Zellner, Adv. Cryo. Eng. **10**, 192 (1965).

331 G. B. Narinskii, Russ. J. Phys. Chem. **40**, 1093 (1966).

332 G. Sh. Boltachev, V. G. Baidakov, and J. W. P. Schmelzer, J. Colloid Interface Sci. **264**, 228 (2003).

333 G. Sh. Boltachev and V. G. Baidakov, Kolloid. Zh. **62**, 5 (2000).

334 V. G. Baidakov and G. Sh. Boltachev, Zh. Fiz. Khim. **71**, 1965 (1997).

335 I. I. Sulla and V. G. Baidakov, Zh. Fiz. Khim. **68**, 63 (1994).

336 I. I. Sulla and V. G. Baidakov, Zh. Fiz. Khim. **68**, 67 (1994).

337 K. C. Nadler, J. A. Zollweg, W. B. Street, and I. A. McLure, J. Colloid Interface Sci. **122**, 530 (1988).

338 C. D. Holcomb and J. A. Zollweg, J. Phys. Chem. **97**, 4797 (1993).

339 V. G. Baidakov and G. Sh. Boltachev, Zh. Fiz. Khim. [Russ. J. Phys. Chem. **75**, 21 (2001)] **75**, 27 (2001).

340 V. G. Baidakov and G. Sh. Boltachev, Zh. Fiz. Khim. [Russ. J. Phys. Chem. **75**, 27 (2001)] **75**, 33 (2001).

341 J. W. P. Schmelzer, V. G. Baidakov, and G. Sh. Boltachev, J. Chem. Phys. **119**, 6166 (2003).

342 V. G. Baidakov, A. M. Kaverin, E. A. Turchaninova, and V. N. Andbaeva, Experimental investigations of spontaneous boiling-up of superheated solutions of cryogenic liquids, Proc. 4th Russ. National Conf. on Heat Transfer. Moskau, Russia, 2006.

343 V. G. Baidakov, A. M. Kaverin, and G. Sh. Boltachev, *The Boiling-Up Kinetics of a Gas-Filled Liquid*, Proc. 5th Minsk Int. Forum Heat- and Mass Exchange, Minsk, Belorussia, 2004.

344 A. M. Kaverin, V. N. Andbaeva, and V. G. Baidakov, *The Surface Tension of Solutions of Liquefied Gases*, Proc. 11th Russ. Conf. Thermophys. Proper., St.-Petersburg, Russia, 2005.

345 M. A. Anisimov, *Kriticheskie Yavleniya v Zhidkostyakh i Zhidkikh Kristallakh* [Critical Phenomena in Liquids and Liquid Crystals] (Nauka, Moscow, 1987 (in Russian)).

346 A. Z. Patashinskii and B. I. Shumilo, Fiz. Tver. Tela **22**, 1126 (1980).

347 Yu. E. Kuzovlev, T. K. Soboleva, and A. E. Filippov, Zh. Eksp. Teor. Fiz. **103**, 1742 (1993).

348 Yu. E. Kuzovlev, T. K. Soboleva, and A. E. Filippov, Pis'ma Zh. Eksp. Teor. Fiz. **58**, 353 (1993).

349 A. Z. Patashinskii and B. I. Shumilo, Zh. Eksp. Teor. Fiz. **77**, 1417 (1979).

350 J. S. Langer and L. A. Turski, Phys. Rev. A.: Gen. Phys. **8**, 3230 (1973).

351 J. S. Langer and A. J. Schwartz, Phys. Rev. A.: Gen. Phys. **21**, 948 (1980).

352 N. J. Gunther, D. A. Nicole, and D. J. Wallace, J. Phys. A.: Math. Gen. **13**, 1755 (1980).

353 K. Kawasaki, J. Stat. Phys. **12**, 365 (1975).

354 K. Binder and D. Stauffer, Adv. Phys. **25**, 343 (1976).

355 K. W. Sarkies and N. E. Frenkel, Phys. Rev. A.: Gen. Phys. **11**, 1724 (1975).

356 S. Fisk and B. Widom, J. Chem. Phys. **50**, 3219 (1969).

357 K. K. Mon and D. Jasnov, Phys. Rev. Lett. **59**, 2983 (1987).

358 M. Fisher, In: *Ustoichivost' i Fazovye Perekhody* [Stability and Phase Transitions] (Izd. Inostrannoi Literatury, Moscow, 245, 1973).

359 R. McGraw and H. Reiss, J. Stat. Phys. **20**, 385 (1979).

360 R. McGraw, J. Chem. Phys. **91**, 5655 (1989).

361 J. L. Lebowitz and E. Helfand, J. Chem. Phys. **43**, 774 (1965).

362 R. B. Heady and J. W. Cahn, J. Chem. Phys. **58**, 896 (1973).

363 D. W. Heermann, J. Stat. Phys. **29**, 631 (1982).

364 H. Furukawa, Phys. Rev. A.: Gen. Phys. **28**, 1729 (1983).

365 V. G. Baidakov, A. M. Rubstein, T. A. Evdokimova, and V. P. Skripov, Phys. Lett. A **88**, 196 (1982).

366 I. Edrei and M. Gitterman, J. Phys. A.: Math. Gen. **19**, 3279 (1986).

367 M. Gitterman, I. Edrei, and Y. Rabin, Lect. Notes in Physics **216**, 295 (1985).

368 I. M. Lifshitz and V. V. Slezov, Zh. Eksp. Teor. Fiz. **35**, 479 (1958).

369 R. G. Havland, N.-C. Wong, and C. M. Knobler, J. Chem. Phys. **73**, 522 (1980).

370 J. V. Sengers, Ber. Bunsenges. Phys. Chem. **76**, 234 (1972).

371 J. S. Huang, W. I. Goldburg, and M. R. Moldover, Phys. Rev. Lett. **34**, 639 (1975).

372 V. G. Baidakov, A. M. Rubstein, and V. P. Skripov, *Nucleation and the Caloric Equation of State of a Superheated Liquid in the Vicinity of a Liquid-Vapor Critical Point*. Proc. All-Union Conference on "Modern Problems of Statistical Physics", Lvov, 1987.

373 B. E. Sundquist and R. A. Oriani, J. Chem. Phys. **36**, 2604 (1962).

374 J. S. Huang, S. Vernon, and N.-C. Wong, Phys. Rev. Lett. **33**, 140 (1974).

375 E. D. Siebert and C. M. Knobler, Phys. Rev. Lett. **52**, 1133 (1984).

376 A. J. Schwartz, S. Krishnamurthy, and W. I. Goldburg, Phys. Rev. A.: Gen. Phys. **21**, 1331 (1980).

377 S. Krishnamurthy and W. I. Goldburg, Phys. Rev. A.: Gen. Phys. **22**, 2147 (1980).

378 R. Strey, J. Wagner, and D. Woermann, Ber. Bunsenges. Phys. Chem. **86**, 306 (1980).

379 P. Alpern, Th. Benda, and P. Leiderer, Phys. Rev. Lett. **49**, 1267 (1982).

380 J. Bodensohn, S. Klesy, and P. Leiderer, Europhys. Lett. **8**, 59 (1989).

381 J. K. Hoffer, L. I. Campbell, and R. I. Bartlett, Phys. Rev. Lett. **45**, 912 (1980).

382 J. K. Hoffer and D. N. Sinha, Phys. Rev. A.: Gen. Phys. **33**, 1918 (1986).

383 A. M. Rubstein, V. G. Baidakov, and V. A. Kondyurin, In: *Fazovye Perekhody v Metastabil'nykh Sistemakh* [Phase Transitions in Metastable Systems] (Ural Science Center of the Academy of Sciences of the USSR, Sverdlovsk, 76, 1983).

384 V. G. Baidakov, A. M. Rubshtein, and V. R. Pomortsev, Fluid Mech. Res. **21**, 89 (1992).

385 O. M. Lapteva, G. I. Pozharskaya, and Yu. D. Kolpakov, In: *Termodinamika i Kinetika Fazovykh Perekhodov* [Thermodynamics and Kinetics of Phase Transitions] (Ural Branch of the Russian Academy of Sciences, Ekaterinburg, 127, 1992).

386 A. Z. Patashinskii and B. I. Shumilo, *Metastabil'naya Sistema Vblizi Oblasti Neustoichivosti* [Metastable System in the Vicinity of an Instability Region] (Preprint IYaF 79–157, Novosibirsk, 1979).

387 K. Binder, Phys. Rev. A.: Gen. Phys. **29**, 341 (1984).

388 D. W. Heermann and W. Klein, Phys. Rev. Lett. **50**, 1062 (1983).

389 D. W. Heermann, W. Klein, and D. Stauffer, Phys. Rev. Lett. **49**, 1262 (1982).

390 C. Unger and W. Klein, Phys. Rev. B.: Condens. Matter **29**, 2698 (1984).

391 D. W. Heermann W. Klein, Phys. Rev. B.: Condens. Matter **27**, 1732 (1983).

392 J. W. Cahn, Acta Metall. **9**, 795 (1961).

393 J. W. Cahn, Trans. Metall. Soc. AIME. **242**, 166 (1968)

394 J. W. Cahn, J. Chem. Phys. **42**, 93 (1965).

395 J. S. Langer, Ann. Phys. **65**, 53 (1971).

396 J. S. Langer, M. Bar-on, and H. D. Miller, Phys. Rev. A.: Gen. Phys. **11**, 1417 (1975).

397 T. Ohta, Progr. Theor. Phys. Suppl. **79**, 141 (1984).

398 H. Furukawa, Progr. Theor. Phys. **73**, 585 (1985).

399 H. Furukawa, Phys. Rev. A.: Gen. Phys. **31**, 1103 (1985).

400 H. E. Cook, Acta Metall. **18**, 297 (1970).

401 A. Z. Patashinskii and I. S. Yakub, Zh. Eksp. Teor. Fiz. **73**, 1954 (1977).

402 M. Grant, M. S. Miguel, J. Vinals, and J. D. Gunton, Phys. Rev. B.: Condens. Matter **31**, 3027 (1985).

403 J. Marro, A. B. Bortz, M. H. Kalos, and J. L. Lebowitz, Phys. Rev. B.: Condens. Matter **12**, 2000 (1979).

404 K. Kawasaki, Progr. Theor. Phys. **57**, 826 (1977).

405 K. Kawasaki and T. Ohta, Progr. Theor. Phys. **67**, 147 (1982).

406 K. Binder, J. Chem. Phys. **79**, 6387 (1983).

407 K. Binder, In: *Stochastic Nonlinear Systems in Physics, Chemistry, and Biology* (Springer, Berlin, Heidelberg, New York, 62, 1981).

408 E. D. Siggia, Phys. Rev. A.: Gen. Phys. **20**, 595 (1979).

409 H. Furukawa, Physica **123 A**, 497 (1984).

410 H. Furukawa, Adv. Phys. **34**, 703 (1985).

411 F. F. Abraham, Phys. Reports **53**, 93 (1979).

412 W. I. Goldburg, C.-H. Shaw, J. S. Huang, and M. S. Pilant, J. Chem. Phys. **68**, 484 (1978).

413 W. I. Goldburg, A. J. Schwartz, and M. W. Kim, Progr. Theor. Phys. Suppl. **64**, 477 (1978).

414 A. Stein, S. I. Davidson, J. C. Allegra, and G. F. Allen, J. Chem. Phys. **56**, 6164 (1972).

415 D. N. Sinha and J. K. Hoffer, Rev. Sci. Instrum. **55**, 875 (1984).

416 J. S. Huang, W. I. Goldburg, and A. W. Bjerkaas, Phys. Rev. Lett. **32**, 921 (1974).

417 N.-C. Wong and C. M. Knobler, J. Chem. Phys. **69**, 725 (1978).

418 Y. C. Chou and W. I. Goldburg, Phys. Rev. A.: Gen. Phys. **20**, 2105 (1979).

419 P. S. Hohenberg and D. R. Nelson, Phys. Rev. B.: Condens. Matter **20**, 2665 (1979).

420 G. Dee, J. D. Gunton, and K. Kawasaki, Progr. Theor. Phys. **65**, 365 (1981).

421 M. S. Miguel, J. D. Gunton, G. Dee, and P. S. Sahni, Phys. Rev. B.: Condens. Matter **23**, 2334 (1981).

422 A. A. Tager, In: *Teplofizika Metastabil'nykh Sostoyanii v Svyazi s Yavleniyami Kipeniya i Kristallizatsii* [Thermal Physics of Metastable States in Connection with Boiling and Crystallization Phenomena] (Ural Branch of the Academy of Sciences of the USSR, Sverdlovsk, 12, 1987).

423 T. Hashimoto, Phase Transit. **12**, 47 (1988).

424 E. Gulari, A. F. Collings, R. L. Schmidt, and C. J. Pings, J. Chem. Phys. **56**, 6169 (1972).

425 P. Leiderer, D. R. Watts, and W. W. Webb, Phys. Rev. Lett. **33**, 483 (1974).

426 P. Leiderer, D. R. Nelson, D. R. Watts, and W. W. Webb, Phys. Rev. Lett. **34**, 1080 (1975).

427 Th. Benda, P. Alpern, and P. Leiderer, Phys. Rev. B.: Condens. Matter **26**, 1450 (1982).

428 Y. C. Chou and W. I. Goldburg, Phys. Rev. A.: Gen. Phys. **23**, 858 (1980).

429 N.-C. Wong and C. M. Knobler, Phys. Rev. Lett. **43**, 1733 (1979).

430 S. N. Burmistrov and L. B. Dubovskii, Zh. Eksp. Teor. Fiz. **93**, 733 (1979).

431 A. O. Caldeira and A. J. Legget, Phys. Rev. Lett. **46**, 211 (1981).

432 A. I. Larkin and Yu. N. Ovchinnikov, Zh. Eksp. Teor. Fiz. **86**, 719 (1984).

433 S. E. Korshunov, Fiz. Nizk. Temp. [Sov. J. Low Temp. Phys. **14**, 316 (1988)] **14**, 575 (1988).

434 A. I. Larkin and Yu. N. Ovchinnikov, Pis'ma Zh. Eksp. Teor. Fiz. [JETP Lett. **37**, 382 (1983)] **37**, 322 (1983).

435 M. Uwaha, J. Low Temp. Phys. **52**, 15 (1983).

436 I. M. Lifshitz, V. N. Polesskii, and V. A. Khokhlov, Zh. Eksp. Teor. Fiz. [Sov. Phys.- JETP **47**, 137 (1978)] **74**, 268 (1978).

437 D. Bailin and A. Love, J. Phys. A.: Math. Gen. **13**, 271 (1980).

438 H. Maris, G. M. Seidel, and T. E. Huler, J. Low Temp. Phys. **51**, 471 (1983).

439 A. F. Andreev, *Quantum Theory of Solids* (Mir, Moscow, 1982).

440 Yu. Kagan and A. J. Leggett, *Quantum Tunneling in Condensed Medium* (North-Holland, Amsterdam, 1992).

441 A. I. Larkin and Yu. N. Ovchinnikov, Zh. Eksp. Teor. Fiz. **85**, 1510 (1983).

442 G. G. Ihas, O. Avenel, R. Aarts et al., Phys. Rev. Lett. **69**, 327 (1992).

443 A. F. Andreev, Zh. Eksp. Teor. Fiz. **50**, 1415 (1966).

444 W. F. Saam, Phys. Rev. A.: Gen. Phys. **5**, 335 (1972).

445 G. E. Watson, J. O. Reppy, and R. C. Richardson, Phys. Rev. **188**, 388 (1969).

446 J. Landau, J. T. Tough, N. R. Brubaker, and D. O. Edwards, Phys. Rev. Lett. **23**, 283 (1969).

447 N. R. Brubaker and M. R. Moldover, *Nucleation of Phase Separation in* ^{3}He–^{4}He (Proc. 13th Int. Conf. Low Temp. Phys., New York, London, 1972).

448 V. A. Mikheev, E. Ya. Rudavskii, V. K. Chagavets, and G. A. Sheshin, Physica B **169**, 511 (1991).

449 V. A. Maidanov, V. A. Mikheev, N. P. Mikhin, N. F. Omelaenko, E. Ya. Rudavskii, V. K. Chagovets, and G. A. Sheshin, Fiz. Nizk. Temp. [Sov. J. Low Temp. Phys. **18**, 663 (1992)] **18**, 943 (1992).

450 V. A. Mikheev, E. Ya. Rudavskii, V. K. Chagovets, and G. A. Sheshin, Fiz. Nizk. Temp. **20**, 621 (1994).

451 T. Satoh, M. Morishita, M. Ogata et al., Physica **169 B**, 513 (1991).

452 T. Satoh, M. Morishita, M. Ogata, and S. Katoh, Phys. Rev. Lett. **69**, 335 (1993).

453 T. Satoh, M. Morishita, S. Katoh et al., Physica **197 B**, 397 (1994).

454 E. M. Lifshitz and L. P. Pitaevskii, *Statistical Physics, Part II* (Pergamon, Oxford, 1980).

455 V. G. Baidakov, Fiz. Nizk. Temp. [Sov. J. Low Temp. Phys. **20**, 971 (1994)] **20**, 1239 (1994).

456 S. N. Burmistrov, L. B. Dubovskii, and V. L. Tsymbalenko, J. Low Temp. Phys. **90**, 363 (1993).

457 S. N. Burmistrov and L. B. Dubovskii, Zh. Eksp. Teor. Fiz. [Sov. Phys.- JETP **73**, 1020 (1991)] **100**, 1844 (1991).

458 J. R. Clow and J. D. Reppy, Phys. Rev. Lett. **19**, 291 (1967).

459 E. Varoquaux, M. W. Meisel, and O. Avenel, Phys. Rev. Lett. **57**, 2291 (1986).

460 R. P. Feynman, *Statistical Mechanics: A Set of Lectures* (W. A. Benjamin, Massachusetts, 1972).

461 S. V. Iordanskii, Zh. Eksp. Teor. Fiz. **48**, 708 (1965).

462 J. S. Langer and M. E. Fisher, Phys. Rev. Lett. **19**, 560 (1967).

463 J. S. Davis, J. Steinhauer, Yu. Mukharsky et al., Phys. Rev. Lett. **69**, 323 (1992).

464 P. C. Hendry, N. S. Lawson, P. V. McClintock et al., Phys. Rev. Lett. **60**, 604 (1988).

465 C. M. Muirhead, W. F. Vinen, and R. J. Donnelly, Philos. Trans. Roy. Soc. London **311 A**, 433 (1984).

466 E. Varoquaux, C. G. Ihas, O. Avenel, and R. Aarts, Phys. Rev. Lett. **70**, 2114 (1993).

467 S. N. Burmistrov, L. B. Dubovskii, and T. Satoh, In: *Nucleation Theory and Applications* (Joint Institute for Nuclear Research Publishing Department, Dubna, Russia, 273, 2002).

468 Q. Xiong and H. J. Maris, J. Low Temp. Phys. **77**, 251 (1989).

469 J. Boronat, J. Casulleras, and J. Navarro, Phys. Rev. **1350**, 3427 (1994).

470 I. M. Khalatnikov, J. Low Temp. Phys. **52**, 433 (1983).

471 S. N. Burmistrov and L. B. Dubovskii, Zh. Eksp. Teor. Fiz. **118**, 885 (2000).

472 V. L. Ginzburg and A. A. Sobyanin, Pis'ma Zh. Eksp. Teor. Fiz. **15**, 345 (1972).

473 M. Bretz and A. L. Thomson, Phys. Rev. B.: Condens. Matter **24**, 467 (1981).

474 H. J. Maris, G. M. Seidel, and F. L. B. Williams, Phys. Rev. B.: Condens. Matter **36**, 6799 (1987).

475 D. P. Woodruff, *The Liquid–Solid Interface* (Cambridge University Press, London, 1972).

476 J. P. Ruutu, P. J. Hakonen, J. S. Penttila, A. V. Babkin, J. P. Saramaki, and E. B. Sonin, Phys. Rev. Lett. **77**, 2514 (1996).

477 V. L. Tsymbalenko, J. Low Temp. Phys. **88**, 55 (1992).

478 X. Chavanne, S. Balibar, and F. Caupin, Phys. Rev. Lett. **86**, 5506 (2001).

479 V. L. Tsymbalenko, J. Low Temp. Phys. **121**, 53 (2000).

480 Y. Sasaki and T. Mizusaki, J. Low Temp. Phys. **110**, 491 (1998).

481 X. Chavanne, S. Balibar, and F. Caupin, J. Low Temp. Phys. **125**, 155 (2001).

482 S. Balibar, X. Chavanne, and F. Caupin, Physica B **329–333**, 380 (2003).

483 S. Kimura, F. Ogasawara, R. Nomura, and Y. Okuda, J. Low Temp. Phys. **134**, 145 (2004).

484 A. Ganshin, V. Grigor'ev, V. Maidanov, N. Omelaenko, A. Penzev, E. Rudavskii, and A. Rybalko, J. Low Temp. Phys. **116**, 349 (1999).

485 A. Penzev, A. Ganshin, V. Grigor'ev, V. Maidanov, E. Rudavskii, A. Rybalko, V. Slezov, and Ye. Syrnikov, J. Low Temp. Phys. **126**, 151 (2002).

486 M. Poole and B. Cowan, J. Low Temp. Phys. **134**, 211 (2004).

487 N. P. Mikhin, V. N. Grigor'ev, V. A. Maidanov, A. A. Penzev, E. Ya. Rudavskii, A. S. Rybalko, V. V. Slezov, and Ye. V. Syrnikov, J. Low Temp. Phys. **134**, 205 (2004).

488 V. P. Mineev, Usp. Fiz. Nauk **139**, 303 (1983).

489 P. J. Hakonen, M. Krusius, M. M. Salomaa, and J. T. Simola, Phys. Rev. Lett. **54**, 245 (1985).

490 P. Schiffer, M. T. O'Keefe, M. D. Hildreth et al., Phys. Rev. Lett. **69**, 120 (1992).

491 A. J. Leggett, Phys. Rev. Lett. **53**, 1096 (1984).

492 S. Balibar, T. Mizusaki, and Y. Sasaki, J. Low Temp. Phys. **120**, 293 (2000).

493 Yu. M. Bunkov and O. D. Timofeevskaya, J. Low Temp. Phys. **110**, 45 (1998).

494 G. E. Volovik, Pis'ma Zh. Eksp. Teor. Fiz. **52**, 972 (1990).

495 G. E. Volovik and T. Sh. Misirpashaev, Pis'ma Zh. Eksp. Teor. Fiz. **51**, 475 (1990).

496 J. M. Kyynarainen, J. P. Pekola, A. J. Manninen, and K. Torizuka, Phys. Rev. Lett. **64**, 1027 (1990).

497 A. Linde, *Particle Physics and Inflationary Cosmology* (Harwood, Switzerland, 1990).

498 V. P. Skripov and N. A. Shuravenko, Teplofiz. Vys. Temp. **16**, 563 (1978).

499 V. G. Baidakov, S. A. Maltzev, G. I. Pozharskaya, and V. P. Skripov, Teplofiz. Vys. Temp. **21**, 959 (1983).

500 V. G. Baidakov, A. M. Kaverin, S. A. Maltzev, and V. P. Skripov, *Superheat and Explosive Boiling-Up of Cryogenic Liquids* (Proc. 3rd All Union Conf. Cryo. Engin., Balashikha, 1982).

501 S. A. Maltzev, V. G. Baidakov, and V. P. Skripov, Teplofiz. Vys. Temp. **24**, 820 (1986).

502 S. A. Maltzev and V. G. Baidakov, In: *Fazovye Prevrashcheniya v Metastabil'nykh Sistemakh* [Phase Transitions in Metastable Systems] (Ural Science Center of the Academy of Sciences of the USSR, Sverdlovsk, 80, 1983).

503 V. G. Baidakov, *Metastabil'nye Kriogennye i Nizkokipyashchie Zhidkosti* [Metastable Cryogenic and Low-Boiling Liquids] (Doctoral Dissertation on Physics and Mathematics, Sverdlovsk, 1985).

504 E. V. Stepanov, *Fizicheskie Aspekty Yavleniya Parovogo Vzryva* [Physical Aspects of the Phenomenon of Vapor Explosion] (Preprint IAE-5450/3, Moscow, 1991).

505 D. L. Katz and C. M. Sliepcevich, Hydr. Proc. **11**, 240 (1971).

506 G. Fröhlich, Fiz. Nizk. Temp. **16**, 562 (1990).

507 E. Nakanishi and R. C. Reid, Chem. Eng. Progr. **67**, 36 (1971).

508 K. Yang, Nature **243**, 221 (1973).

509 W. M. Porteous and R. C. Reid, Chem. Eng. Progr. **72**, 83 (1976).

510 L. C. Witte and J. E. Cox, Adv. Nucl. Sci. Techn. **7**, 329 (1973).

511 L. C. Witte, J. E. Cox, and J. E. Bouvier, J. Metall. **22**, 39 (1970).

512 J. A. Sallack, Pul. Pap. Magaz. Canada **56**, 114 (1955).

513 P. Groenveld, J. Heat Transf. **94**, 122 (1972).

514 G. Carslow and D. Eger, *Teploprovodnost Tverdykh Tel* [Thermal Conductivity of Solids] (Nauka, Moscow, 1964).

515 V. G. Baidakov and S. A. Maltzev, In: *Teplomassoobmen v Dvukhfaznykh Protokakh* [Heat Exchange in Two-Phase Flows] (ITMO, Minsk, 31, 1988)

516 V. G. Baidakov and S. A. Maltzev, In: *Teplovye Prozessy i Metastabil'nye Sostoyaniya* [Thermal Processes and Metastable States] (Ural Branch of the Academy of Sciences of USSR, Sverdlovsk, 28, 1990).

517 R. E. Henry and H. K. Fauske, J. Heat Transf. **101**, 107 (1979).

518 D. Frost and B. Sturtevant, J. Heat Transf. **108**, 418 (1986).

519 D. J. Buchanan, J. Phys. D.: Appl. Phys. **7**, 1441 (1974).

520 M. L. Corradini, Nucl. Sci. Eng. **84**, 196 (1983).

521 A. H. Rausch and A. D. Levine, Cryogenics **14**, 139 (1974).

522 S. J. Boad, R. W. Hall, and R. S. Hall, Nature **254**, 319 (1975).

523 G. R. Fowles, Science **204**, 168 (1979).

524 J. L. Manas, Fire Int. **8**, 27 (1984).

525 M. A. K. Lodhi and R. W. Mires, Int. J. Hydrogen Energy **14**, 35 (1989).

Index

a
adiabatic spinodal 14, 15
argon 14, 29, 80, 81, 87, 91, 92, 95, 96, 105, 106, 114, 124, 137–139, 141–144, 146, 147, 178, 179, 182, 190, 191, 196–198, 202, 203, 205, 206, 214, 215, 309, 311, 313, 314
attainable superheating 61–158, 172–173, 213–217

b
binary solution 15, 165–172, 195–204, 213, 219, 230, 234, 238, 240, 242, 243
binodal curve 185, 205
boundary of essential instability 3, 13, 16, 22, 96, 134, 220, 253, 284
butane 93, 172, 173

c
cavitation strength 116, 123–138, 144, 294, 297, 299
correlation length 30, 64, 219, 221, 255, 262, 263
correlation radius 16, 22, 26, 64, 220, 221, 226, 228, 240, 255, 269
critical bubble 61–65, 67, 68, 72, 78, 80, 100, 103, 116–118, 120–122, 137, 142–144, 157, 160, 161, 164, 195–204, 206–211, 217, 245, 246
critical cluster 64, 273
critical point 3, 14, 16, 18, 22, 75, 79, 80, 95, 104, 106, 107, 109, 112, 116, 143, 148, 159, 161, 184, 191, 192, 195, 204, 209, 219–222, 226, 230–237, 240–243, 246–248, 251–253, 255, 259, 265, 266, 316
critical region 32, 43, 47, 53, 54, 58, 74, 76, 240
cryogenic liquids 19, 75, 77, 81, 85, 86, 88, 94, 95, 99, 107, 117, 123–138, 151, 152, 157, 158, 174–176, 186, 204, 213–217, 309–319

d
distribution function 31–34, 37, 39–41, 45, 46, 49–52, 57–59, 66, 73–76, 100, 119, 129, 169, 230, 257, 286
dynamic methods 80, 81, 88–91, 138

e
electron bubble 3
equation of state 2, 8, 20, 30, 62, 190–195, 201, 233, 303
equilibrium 1–4, 7–9, 12–14, 17–19, 22, 25–27, 29–33, 36, 37, 40, 41, 45–48, 50, 51, 53–55, 57, 58, 61–68, 74, 76, 100, 108–111, 116, 117, 119, 121, 131, 152, 157, 159–161, 163, 166, 174, 176, 183, 184, 193–195, 204, 207–211, 214, 221–225, 228, 229, 232, 234–236, 243, 255, 257–259, 264, 279–281, 283, 285, 294, 304, 305, 307, 308, 312, 319, 321
ethane 85, 92, 102, 105, 172, 318
explosive boiling 88, 140, 309–319

f
Fokker–Planck equation 31, 32, 36, 37, 44, 48, 75, 257, 286

g
gas-saturated solutions 3, 159, 184, 186
Gibbs number 35, 68, 103, 142–144, 183, 203, 204, 209, 227, 232, 233, 251, 290, 292, 303
Gibbs potential 7, 53, 61–63, 159, 161
Gibbs theory of capillarity 64, 117

h
helium 81, 82, 86, 88, 89, 91, 96, 107–109, 111–116, 124–127, 129, 130, 132–134, 136–138, 150, 151, 175, 186–195, 199, 200, 203–205, 207, 208, 215, 216, 246, 247, 267–270, 276, 278, 284, 288, 289, 294, 295, 298–300, 303, 306, 308
Helmholtz free energy 17, 19–22, 25, 27, 31, 32, 65, 162, 195, 210, 219, 222, 224, 253
heterogeneous nucleation 59, 81, 130, 151–158, 178, 242, 281
heterophase fluctuations 8, 18, 22–27, 30–35, 58, 107, 220, 225, 232, 233, 235, 263, 273, 289

Explosive Boiling of Superheated Cryogenic Liquids. Vladimir G. Baidakov
Copyright © 2007 WILEY-VCH Verlag GmbH & Co. KGaA, Weinheim
ISBN: 978-3-527-40575-6

h

homogeneous nucleation 2, 3, 47, 51, 58, 59, 75, 77, 80, 81, 91, 100, 102, 103, 109, 110, 112, 113, 115, 117, 118, 120, 124, 125, 127, 130, 133, 137, 142, 147, 149, 151, 152, 155, 173, 177, 178, 180, 183–185, 187, 188, 204, 205, 207–209, 216, 234, 240, 247, 250, 251, 282, 300, 301, 303, 311, 313, 315
homophase fluctuations 8, 13
hydrogen 3, 82, 83, 97, 107–113, 115, 278, 299–302, 305

i

induced nucleation 4
isotopes 89, 107, 110, 111, 115, 138, 267–270, 278

k

kinetic equations 40, 48, 49, 230
kinetic theory of nucleation 165
krypton 14, 29, 91, 92, 95, 97, 103, 105, 114, 156, 178, 179, 182, 191, 196–198, 202, 203, 205, 206

l

limiting stretching 106
limiting superheating 85, 89, 95, 113, 141, 185, 188, 234
limiting supersaturation 184, 250, 278–284, 292–294
line of attainable superheating 91, 107, 109, 111, 181, 208, 314
liquefied gases 2–4, 81, 88, 95, 100, 104, 105, 159–217, 245, 319
low-boiling liquids 81, 84, 91, 316

m

mean lifetime 86–88, 113, 115, 125, 146–149, 153–156, 176–180, 186–189, 300
mechanical spinodal 14, 15, 29
metastable states 2, 5, 8, 13, 15–17, 26, 30, 48, 77–79, 86, 96, 114, 124, 145, 149, 164, 183, 190, 192, 201, 209, 227, 229, 235, 236, 240, 243, 244, 252, 266, 273, 279, 284, 306, 307
methane 83, 91, 92, 98, 99, 102, 105, 120–122, 185, 309, 311, 314, 316, 318
method of continuous heating 82, 91, 247
method of lifetime measurements 87, 214

n

neon 92, 107, 109, 111, 112, 115, 117, 125
nitrogen 84, 87, 91–93, 95, 98, 99, 105, 106, 118–122, 124, 139, 152–155, 157, 175, 179–183, 185, 186, 188–191, 194, 195, 213–215, 241, 309, 314, 316, 318
nucleation rate 40, 49, 50, 55, 56, 73, 76, 78–80, 82, 84, 86, 88–90, 92, 93, 95, 96, 99, 100, 103, 106, 109, 110, 113–116, 118, 120, 126, 128, 138, 141, 143, 144, 146–149, 165, 169–172, 174, 177, 179–181, 183, 185, 187–189, 203, 205–207, 213, 215, 217, 228, 230, 231, 233, 236, 237, 246, 248, 250, 251, 257, 291, 292, 296, 312, 313, 315, 317
nucleation theory 35–47, 56–59, 74–78, 100–102, 112, 118, 127, 133, 137, 138, 144, 145, 165–172, 183, 187, 240, 244, 251, 252, 281

o

one-component liquids 61–158, 161, 166, 186, 187, 240, 248, 265
outflow of boiling liquids 309–314
oxygen 91, 92, 98, 105, 106, 120–122, 124, 149, 150, 152–154, 179–183, 186–188, 190–195, 199, 200, 203–205, 207, 208, 214, 215, 309, 310, 314

p

phase diagram 14, 174, 192–195, 214, 215, 227, 243, 244, 279, 306
propane 93, 95, 99, 101, 102, 105, 118, 172, 173, 315–317

q

quantum liquids 107–116, 137, 145, 291
quantum nucleation 4, 273, 275–277, 289–294, 298, 303
quantum tunneling 4, 107, 273–278, 282–284, 290, 293, 295, 297, 298, 302, 303
quantum vortices 131, 132, 278, 284–289, 308

r

relaxation 2–4, 26–30, 43, 47, 54, 55, 58, 66, 67, 69, 70, 72, 108, 167, 186, 219, 221, 224, 230, 235, 253, 256, 257, 259, 263, 265, 267, 268, 291, 292, 314

s

spinodal decomposition 4, 259–272
spontaneous boiling 3, 99, 104, 111, 113, 115, 119, 121, 147, 150, 155, 165, 177, 179, 183, 185, 202, 203, 208, 215, 311, 313, 315, 319
stability 2, 3, 7–23, 26, 27, 29, 33–35, 109, 111–113, 136, 137, 151, 178, 183, 190–195, 202, 208, 210, 222–229, 240, 254, 258, 278, 299, 307
stable state 3, 8, 17, 32, 64, 182, 219, 253, 259, 267
steady-state nucleation 228–232, 289
supercooled liquid 16, 78, 278, 300–302, 305

supercooling   106, 227, 228, 233, 234, 236, 243, 244, 246, 247, 250–252, 278, 300, 304, 305
superfluid helium   3, 107, 114, 124, 125, 129–132, 137, 145, 284–289, 295, 297, 302, 303, 305, 308
superheated liquid   47, 59, 61, 62, 66, 67, 77, 78, 80, 84, 85, 87, 95, 96, 99–102, 109, 116, 119, 121, 143, 145–149, 151–156, 241, 245, 315
superheated liquids   65–74, 78–81
superheating   2, 4, 19, 59, 63, 66, 77, 78, 80–96, 100, 101, 105
supersaturation   3, 22, 29, 36, 47, 64, 72, 77, 79, 123, 155, 165, 170, 182, 183, 202, 211, 223, 227–233, 237, 239, 247–251, 275, 278–284, 289, 293, 294, 304, 312

t
thermodynamic potential   4, 7, 11, 16, 18, 20, 23, 26, 32, 53, 62, 65, 159, 160, 195, 209, 211, 219, 221, 230, 259
tricritical point   4, 159, 240–244, 252, 267, 268, 270, 293, 307

u
unstable states   3, 19, 118, 259, 261, 269

v
van der Waals equation of state   2, 17, 21
van der Waals theory of capillarity   3, 20, 64, 118, 138, 202, 210, 295

x
xenon   91, 92, 95, 97, 99, 103, 105, 114, 117, 118, 147, 149, 178, 236, 240–242, 245–248

Related Titles

Skripov, V. P., Faizullin, M. Z.

Crystal-Liquid-Gas Phase Transitions and Thermodynamic Similarity

183 pages with 85 figures and 30 tables
2006, Hardcover
ISBN-13: 978-3-527-40576-3
ISBN-10: 3-527-40576-3

Schmelzer, J. W. P.

Nucleation Theory and Applications

472 pages with 157 figures
2005, Hardcover
ISBN-13: 978-3-527-40469-8
ISBN-10: 3-527-40469-4

Kostorz, G. (Ed.)

Phase Transformations in Materials

724 pages with 392 figures and 33 tables
2001, Hardcover
ISBN-13: 978-3-527-30256-7
ISBN-10: 3-527-30256-5

Colinet, P., Legros, J. C., Velarde, M. G.

Nonlinear Dynamics of Surface-Tension-Driven Instabilities

With a Foreword by I. Prigogine

527 pages with 189 figures and 14 tables
2001, Hardcover
ISBN-13: 978-3-527-40291-5
ISBN-10: 3-527-40291-8

Jacob, D., Sachs, G., Wagner, S. (Eds.)

Basic Research and Technologies for Two-Stage-to-Orbit Vehicles

Collaborative Research Centres

666 pages with 412 figures and 36 tables
2005, Hardcover
ISBN 3-527-27735-8